Lecture Notes in Physics

353

M. Barthes J. Léon (Eds.)

Nonlinear Coherent Structures

Proceedings of the 6th Interdisciplinary
Workshop on Nonlinear Coherent Structures
in Physics, Mechanics, and Biological Systems
Held at Montpellier, France, June 21–23, 1989

Springer-Verlag
Berlin Heidelberg GmbH

Editors

M. Barthes

J. Léon

Université de Montpellier II, Sciences et Techniques du Languedoc
F-34095 Montpellier Cédex 5, France

ISBN 978-3-662-13791-8 ISBN 978-3-540-46937-7 (eBook)
DOI 10.1007/978-3-540-46937-7

Originally published by Springer-Verlag Berlin Heidelberg New York in 1990
Softcover reprint of the hardcover 1st edition 1990

2158/3140-543210 – Printed on acid-free paper

It is a pleasure to aknowledge the **CENTRE NATIONAL DE LA RECHERCHE SCIENTIFIQUE**, the **REGION LANGUEDOC-ROUSSILLON** and the **UNIVERSITE DES SCIENCES ET TECHNIQUES DU LANGUEDOC** which, by providing the workshop with financial support, have contributed to the opportunity of gathering in MONTPELLIER leading figures in both experimental and theoretical physics.

We are grateful to O. ALBERNHE and F. DUCEAU for their active collaboration in the meeting, and to all of our colleagues who helped us in many ways.

Montpellier,
June 1989

M. BARTHES
J. LEON

PREFACE

The 1970 s were golden years for progress in the mathematical theory of solitons. The relationships between the inverse scattering method, the Bäcklund transformation, and an infinite number of conservation laws for a nonlinear wave equation became understood, and a rather large number (i.e. several dozen) of equations were found to fit into this new picture. It is important that many of these equations had arisen in practice, prior to the development of soliton theory. As the mathematical theory of solitons becomes complete, the applications of that theory to real problems in physics, mechanics, and biology become even more important, and these applications are the subject of the present volume.

In such applications it is important, in my view, that the research include three interacting components: experimental studies, numerical studies and theoretical analysis, as is indicated below.

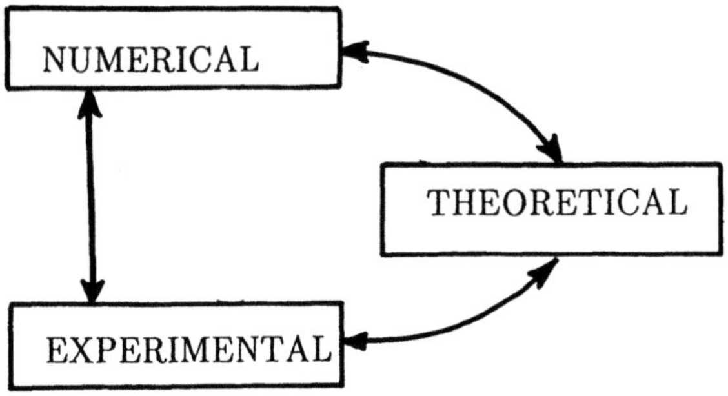

In the "best of all possible worlds" one would find all three components interacting harmoniously within the same research group, but in our (non-Panglossian) reality this is seldom the case. Thus it is important to encourage effective interactions between numerical, experimental, and theoretical scientists who are separated by institutional and geographic barriers. Such interactions are demonstrated by many of the research studies reported here.

But there are problems to consider. If any one of these three research components gets too far ahead of the others, science suffers. Think of it. We all know of studies that have been too theoretical, too numerical, or even too experimental to be really good science. Also it is important that researchers in nonlinear science have the conviction, often contrary to conventional wisdom, that their research is important. In connection with this point, let me cite some examples.

One of the first scientists to work on the Josephson transmission line was Wayne Johnson who completed his doctoral research in the area twenty-one years ago this summer. When he submitted his results for publication in the Journal of Applied Physics, it was rejected because, according to the referee's report, it had "nothing to do with physics". Papers by Brian Henry on "local modes" in small molecules were similarly rejected in the mid 1970 s as being obviously incorrect. Today nonlinear coherent dynamical structures on the Josephson transmission line and small molecules are fully accepted as interesting and important subjects in applied science.

A similar situation is currently being faced by physical scientists who are studying the possibility of nonlinear coherent structures in biological molecules such as protein and DNA. A recent invited commentary on these efforts appearing in a prestigious journal (which I will not name) referred to such physical scientists as "hijackers". We cannot accept this. There are, in fact, two components of biomolecular dynamics related as indicated below.

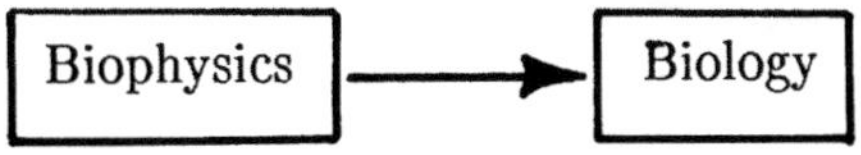

(Note the direction of the arrow.) It is the responsibility of physical scientists, under "biophysics", to discover what dynamic effects are possible in biomolecules. As Michel Peyrard has emphasized, this search should not be limited to merely looking for solitons; it should be a broad effort to understand the true nature of the nonlinear dynamics. After this understanding has been achieved, it is the responsibility of the biological scientist to decide which of these effects actually participate in the functions of living organisms. I am happy to see that several examples of such a responsible approach to the study of biomolecular dynamics are included in this volume.

In his opening survey, Michel Remoissonet points out that more work is needed on techniques to detect solitons (especially nontopological solitons) at the microscopic level. This suggests that the "theoretical" activity shown in my first diagram should include quantum effects.

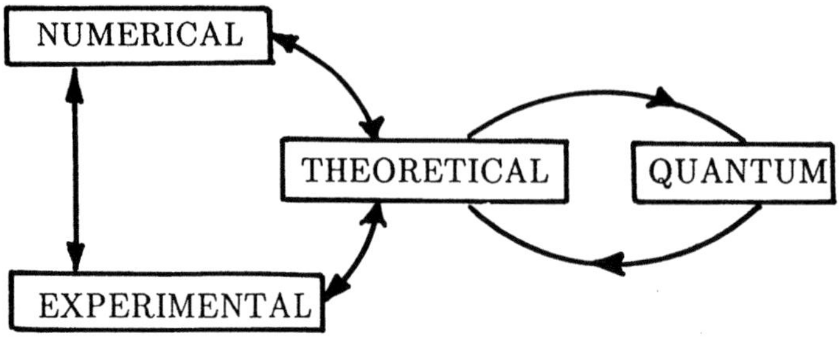

We should not expect much help from the applied mathematicians here. They seem, strangely, unconcerned with such questions. But we can't ignore quantum effects, and the present volume shows that this is not the case.

In summary, this book demonstrates that the study of "nonlinear coherent structures in physics, mechanics, and biological systems" is in a state of healthy growth. Experimental, numerical, and theoretical activities are interacting vigorously in studies of ordered and disordered solid-state systems. The well-established field of fluxon dynamics on the long Josephson junction is still showing new and unexpected phenomena. Finally a corresponding study of biomolecular dynamics seems to be well established. The future of applied soliton research is indeed bright.

Lyngby, Denmark

July 1989

Alwyn Scott

CONTENTS

INTRODUCTION

Solitons and their observable signature in quasi-one-dimensional systems
M. REMOISSENET 3

PART I : HYDROGEN-BONDED CHAINS

Local modes in molecules
A.C. SCOTT 19

Dynamics of nonlinear excitations in DNA
M. PEYRARD, A.R. BISHOP 29

Nonlinear excitations in chains of hydrogen-bonded molecules : incoherent neutron scattering in acetanilide and derivatives
M. BARTHES, R. ALMAIRAC, J.L. SAUVAJOL, J. MORET, R. CURRAT, J. DIANOUX 43

NMR study of nonlinear excitations in prebiological systems
F. MASIN, G. GUSMAN, P. BROEKAERT, C. MAERSCHALK 57

Finite temperature behaviour of the exciton bound states in the compressible molecular chain
D.V. KAPOR, M.J. SKRINJAR, S.D. STOJANOVIC 65

Solitons on a diatomic chain with alternating interactions
F.G. MERTENS, D. HOCHSTRASSER 75

Longitudinal and transverse proton collective dynamics on a two-dimensional multistable substrate
S. PNEVMATIKOS, A.V. SAVIN, A.V. ZOLOTARYUK 83

Toda lattice with transverse degree of freedom
P.L. CHRISTIANSEN, P.S. LOMDAHL, V. MUTO 93

PART II : SOLID-STATE PHYSICS

Charge density waves and superconductivity : Localized bipolarons versus extended Cooper pairs ?
S. AUBRY, G. ABRAMOVICI, D. FEINBERG, P. QUEMERAIS, J. RAIMBAULT 103

Modulated phases in crystals : symmetries of walls and wall lattices ; example of quartz
P. SAINT-GREGOIRE, V. JANOVEC 117

Strain solitary waves in ferroelastic-martensitic transformations; lattice model and continuum approximation
J. POUGET 127

Critical fluctuations in the 2D XY antiferromagnet $BaNi_2(PO_4)_2$: NMR measurements
P. GAVEAU, J.P. BOUCHER, L.P. REGNAULT, Y. HENRY 141

1/S quantum corrections for the magnetic solitons in various Heisenberg models
D.V. KAPOR, M.J. SKRINJAR, S.D. STOJANOVIC 147

PART III : NONLINEAR WAVE PROPAGATION : NUMERICAL AND THEORETICAL STUDIES

Defects in nonlinear waves in convection
A. JOETS, R. RIBOTTA 159

Gap solitons in nonlinear electrical transmission lines
J.M. BILBAULT, M. REMOISSENET 171

Can a nontrivial solitonic mode be observed in a square Josephsen junction ?
E. TURLOT, D. ESTEVE, C. URBINA, M. DEVORET, J.C. FERNANDEZ, R. GRAUER, H. POLITANO, G. REINISCH 181

Radiative damping of driven sine-Gordon breathers
R. DÖTTLING, J. ESSLINGER, W. LAY, A. SEEGER 193

Fluxon trapping by inhomogeneities in long Josephsom junctions
T. BOUNTIS, S. PNEVMATIKOS, S. PROTOGERAKIS, G. SOHOS 203

Dynamical regimes phase-locked to an external microwave field in a long, unbiased Josephson junction
J.C. FERNANDEZ, R. GRAUER, G. REINISCH 213

Existence and stability of driven and damped phase-locked NLS-solitons and SG-breathers
K.H. SPATSCHEK, M. TAKI, T. EICKERMANN 221

Spectral characterization of 2D nonlinear coherent structures
J. LEON, M. BOITI, F. PEMPINELLI 241

The stochastic Φ^4 model
A. SANCHEZ, L. VAZQUEZ 251

Invariants for the generalized Lotka-Volterra equations
L. CAIRO, M.R. FEIX, J. GOEDERT 261

Slow relaxation effects in the evolution of nonlinear quantum systems
V.M. KENKRE 271

INTRODUCTION

SOLITONS AND THEIR OBSERVABLE SIGNATURES IN QUASI-ONE-DIMENSIONAL SYSTEMS

M. Remoissenet

Laboratoire O. R. C. Université de Bourgogne. 21100, Dijon, France.

Abstract. We give an overview of the experimental signatures of nonlinear waves: notably topological and non topological solitons, in specific quasi-one-dimensional devices and condensed matter systems. Non topological solitons can be easily observed and manipulated, on a macroscopic scale, in optical fibers and electrical transmission lines. Topological solitons have been clearly identified as fluxons in Josephson transmission lines and as domain walls in condensed matter systems such as magnetic chains and synthetic polymers. By contrast, at the present time the observable signatures of nonlinear excitations such as pulse or envelope solitons and polarons, which are predicted to occur on a microscopic scale in condensed matter systems, are not conclusive.

1.Introduction

We are living in a nonlinear world where solitary waves and solitons provides a continuous source of fascination and excitement for the physicist. These excitations have been observed in various real systems, but while a considerable number of papers devoted to theoretical and numerical studies, has been published, the number of papers devoted to their experimental identification is relatively very small (Remoissenet, 1989).

Here, we give a survey of specific quasi one dimensional systems where soliton excitations present obervable signatures. In the following, by a soliton we understand a nonlinear wave which has properties close to those of a strict soliton in the mathematical sense. Within the family of solitons, we make a distinction between *topological* and *nontopological* solitons : in the former case, the system is modified by the passage of the excitation, in the second case it returns to its original state. The outline of this paper is as follows. In section 2 and 3 we consider the optical fibers and electrical transmission lines where the exploitation of dispersion and nonlinear effects has led to the demonstration of various properties optical and electrical *non topological* envelope and pulse solitons. The last three sections are concerned with Josephson transmission lines, magnetic chains and polymers where *topological* kink-solitons have been clearly identified experimentally.

2.Optical fibers.

The optical fiber is an interesting (Gloge, 1979) guiding structure(Mollenauer and Stolen, 1982. Doran and Blow , 1983, 1987). The dispersive character of the material of the optical fiber, causes the pulse to spread out an eventually overlap to such an extend that all the information is lost. To overcome this limitation Hasegawa and Tappert (1973a,1973b) proposed to compensate the dispersive effect by the nonlinear change (Kerr effect) of the refractive index of the fiber material. When the frequency shift due to the Kerr effect is balanced with that due to the dispersion, the initial optical pulse may tend to form a nonlinear stable pulse called "optical soliton" which in fact is a quasi-soliton. This prediction and subsequent observation of optical solitons propagation (Mollenauer et al, 1980) in a single fiber has stimulated theoretical and experimental studies on nonlinear guided waves. Namely, with lasers producing high intensity short duration optical pulses, the exploitation of the typical nonlinear effects in optical devices has led to the demonstration of various (quasi) soliton properties such as pulse compression (Mollenauer et al, 1983), the soliton laser (Mollenauer and Stolen, 1984). The consequences for future long distance high bit rate communications systems are the subject of much study (Mollenauer and Smith, 1988).

An optical fiber is a cladded cylindrical waveguide which guides light by total internal reflection : this is achieved by a refractive index variation decreasing in the radial direction from the center to the periphery. One can design the fiber in such a way that it transmits only one mode and attenuates all other modes by absorption or leakage, in this case one has a monomode fiber. Let us consider a monomode fiber where or a sake of simplicity we ignore the transverse (x, y) dependence of the electric field vector E. The one-dimensional wave equation for a linearly polarized optical wave obtained (Jain and Tzoar, 1978. 1987) from the Maxwell equations (Karpman, 1975), is given by

$$\frac{\partial^2 \mathbf{E}}{\partial z^2} - \frac{1}{c^2}\frac{\partial^2 \mathbf{D}_L}{\partial t^2} - \frac{2n_o n_2}{c^2}\frac{\partial^2}{\partial t^2}\mathbf{E}^2\mathbf{E} = 0 \qquad (2.1)$$

Here $\mathbf{D}_L$ is the linear part of the electric displacement, n_o and n_2 are respectively the linear part and nonlinear part (intensity dependent) of the refactive index n, the Fourier transform of which is given by : $n(\omega, |E|^2) = n_o(\omega) + n_2|E|^2$, $c = (\varepsilon_o \mu_o)^{-1/2}$ is the speed of light. We consider a monochromatic field amplitude : $E(z,t)= A(z,t)\, e^{i(k_0 z - \omega_0 t)} + c.c$, with the z axis as the direction of propagation. A is the slowly varying transverse component of the electric field and c.c denotes the complex conjugate, k_0 the wavenumber and ω_0 the frequency.Describing the evolution of A by using a coordinate system which moves at the group velocity $v_g = \partial\omega/\partial k$, where $s = z$, $t' = t - z/v_g$, and making use of the fact that the envelope is slowly varying, from eq

(2.1) one gets (Hasegawa and Tappert, 1973. Tzoar and Jain, 1981, Phys Rev, A, 23,1266) a Nonlinear Schrödinger (NLS) equation

$$iA_s + PA_{tt} + Q\,|A|^2 A = 0 \qquad (2.2)$$

where the nonlinear coefficient $Q = (\partial k/\partial |A|^2)$ and the group velocity dispersion (GVD) coefficient $P = -(1/2)(\partial^2 k/\partial\omega^2)$, are evaluated at $\omega=\omega_0$. Equation (2.2) support N soliton solutions (Zakharov and Shabat, 1972. Yajima and Satsuma, 1974). The relative signs of the dispersive term and nonlinear term are important , for optical fibers Q is always positive, P can be negative or positive (GVD > 0 or GVD < 0) . When PQ>0, eq (2.2) admits an envelope (bright) soliton solution. Hasegawa and Tappert (1973) first modelled the propagation of a guided mode in a perfect nonlinear monomode fiber by the NLS equation. They find that the optical pulse is a *non topological* envelope soliton solution of eq (2.2) and they predicted the stationnary transmission of the pulse at the anomalous dispersion (negative group dispersion : P < 0) regime.Then this prediction was successfully verified by the experiments of Mollenauer et al (1980). Using a monomode fiber they demonstrated a dispersionless transmission of an optical pulse for a distance of about seven hundred meters. Moreover, when PQ>0 a plane wave in unstable for modulation : it is an example of the Benjamin Feir instability (1967) first discussed in the context of water waves. This modulational instability was predicted theoretically (A. Hasegawa and Brinkman,1980. Hasegawa, 1984) and recently observed in optical fibers (Tai and al, 1986).

For P Q < 0. (P< 0) eq (2.2) admits soliton solutions called dark solitons because they are characterized by the absence of light (Tomlison et al, 1989) in a bright background. They were recently observed in optical fibers (Krökel et al, 1988).

Although the quality of optical fibers has dramatically improved over recent years, the fibers do exhibit some attenuation resulting in an attenuation of the pulse transmission which can no longer be neglected for long distance telecommunications. In the linear regime of a monomode optical fiber, the loss is typically in the range 02-1.0 dB/km. There are, however special wavelengths or spectral windows where the attenuation is particularly low . At high powers a nonlinear loss term would result from stimulated processes such as Brillouin scattering (Cotter, 1982a 1982b). Linear absorptive or scattering loss simply adds a term $i\gamma q$ to eq (2.2).

For pulse widths of several picoseconds or more, a description of pulses in optical fibers in term of the NLS equation seems to be reasonable to fit the experimental findings well. However experiments in the high power and the ultrashort pulse regimes, as seen from the studies of pulse compression and frequency shift requires further study to determine the effects of the higher order terms on the NLS model (Kodama et al, 1987. Zhao et al, 1988).

3. Electrical transmission lines.

Nonlinear electrical transmission lines (ETL) or lumped electrical networks (Scott, 1970. Lonngreen and Scott, 1978. Peterson, 1984)) , are efficient to simulate the propagation and properties of nonlinear waves in various physical systems They have received much attention since a pioneering work on a simulation line (Hirota and Suzuki, 1970, 1973) of the integrable Toda lattice. The electrical transmission lines have also direct applications to engineering and modern electronic systems .As we have seen in the preceeding section nonlinear wave motion with dispersion (of the optical index) produce solitons in optical fibers which are continuous systems. Here the dispersion results from the periodic structure of the network . Namely, we consider a simple electrical network where the nonlinearity is caused by the capacitance C(V) of a semiconductor junction , the inductance L is linear (figure 3.1). The resistance R and the conductance G represent the dissipative elements which correspond to the smal but finite losses of the inductor and the capacitor. From the Kirchoff's law we get a set of 2N differential-difference equations which can be combined to give

$$L\frac{d^2Q_n}{dt^2} + R\frac{dQ_n}{dt} + LG\frac{dV_n}{dt} = V_{n+1} + V_{n-1} - 2V_n \qquad (3.1)$$

where V_n (t) is the voltage across the nth capacitance, I_n (t) is the current through the nth inductance and $Q_n(V_n)$ denotes the charge stored in the nth capacitor.

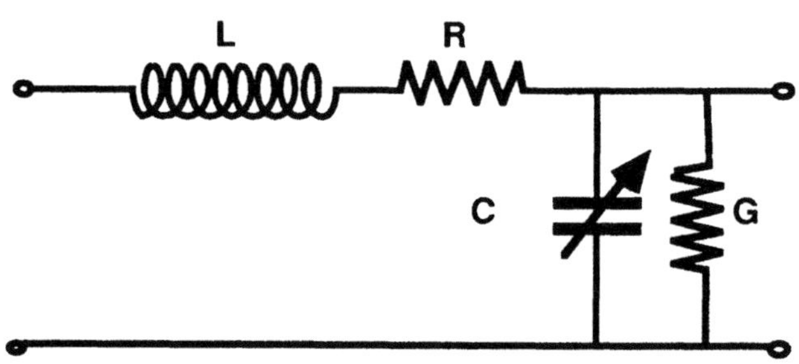

Figure 3.1 . Unit cell of a simple transmission line with dissipative elements.

In a first approximation, if the length of the network is not too important we can neglect the dissipative terms. If the charge voltage-relation is logarithmic equation (2.1) reduces (without dissipation terms) to the Toda lattice equation (Hirota and Susuki, 1970. Watanabe, 1982. Kuusela et al, 1989) . Therefore the experiment of the Toda lattice using a LC transmission line is realizable if the nonlinear charge-voltage relation satisfies exactly a logaritmic function. Usually it is satisfied only in a limited voltage range or it is not satisfied at all. Consequently instead of eq (3.1) one assumes : $Q_n(V_n) = C_o (V_n - a V_n^2 + bV_n^3 + ...)$ and in the following, we consider the case where $bV_n^3 << aV_n^2$. Starting from eq(3.1) , we use the continuum approximation :$V_n (t) \rightarrow V (n, t)$ and expand V in terms of its derivatives, and introduce the expansion and new variables (coordinate system moving with the velocity v_o) :

$$V = \varepsilon V_1 + \varepsilon^2 V_2 + .., \qquad X = \varepsilon^{1/2} (n - v_o t), \qquad T = \varepsilon^{3/2} v_o t /24 \qquad (3.2)$$

where $\varepsilon << 1$ and $v_o = (LC_o)^{-1/2}$. Keeping terms of order ε^3 (Fukushima et al, 1980) and puttting $\alpha = 24a$, we obtain the well known KdV equation which admits soliton solutions

$$V_{1T} + \alpha V_1 V_X + V_{1XXX} = 0 \qquad (3.3)$$

In terrms of the original variables (n, t), the one soliton solution of eq (2.3) becomes

$$V = \varepsilon V_1 (n, t) = \varepsilon \frac{3}{a} (\frac{u}{v_o} - 1) \operatorname{sech}^2 \{ [6(\frac{u}{v_o} - 1)]^{1/2} (n - ut) \} \qquad (3.4)$$

where $u = v_o(1 + \varepsilon v/24)$ .This integrable KDV equation describes the weakly nonlinear and weakly dispersive waves . In the continuous limit it represents the zeroth order nonlinear approximation of the physical system. Many experiments have been performed on KdV solitons on eletrical transmission lines.

The weak losses (see eq 3.1) can be taken into account, leading to an additional dissipative term in the KdV equation which can be treated as a perturbation.

For higher amplitudes pulses one must take into account the higher order nonlinear and dispersion terms in the expansion of the discrete equations (3.1) and one has additional terms in the continuous limit . In this case the propagation of a solitary wave in a LC line was experimentally studied (Watanabe,1982). It was shown that the waveform approaches that of the Toda lattice soliton only when the amplitude is small, but gradually deviates from it when the amplitude is increased.

Untill now, we have considered pulses propagation, but modulated waves can also exist : like for optical fibers, we can obtain a NLS equation which models the propagation of the envelop of voltage V(n,t) [or current I(n,t)] along the ETL (Fukushima et al, 1980 ,1983) with a good agreement with experiments.

Besides the elementary transmission line considered above, various other ETL can be realized by a proper choice of the elements For example, ETLcan be constructed where two or more different wave modes can interact and exchange energy (Yoshinaga et al, 1981). Bi-inductances networks have been constructed to simulate nonlinear diatomic lattices (Kofane et al, 1988).

4. Josephson transmission lines.

In the experimental study of solitons in nonlinear systems the long superconducting Josephson junctions or Josephson transmission lines (JTL), have received an increasing interest over the past years as a model system (Barone and Paterno, 1982. Likharev, 1986. Scott, 1970, Parmentier, 1978. Lomdahl, 1985. Pedersen, 1988) . In the JTL the physical quantity of interest is a quantum of magnetic flux, or a fluxon, which is *a topological* kink-soliton. It is a remarkably robust and stable object , which can be easily manipulated at high speed and stored electronically. Consequently it should be used as a basic bit in information processing systems. The JTL could become a very attractive device in the near future with the new high-temperature superconductors.

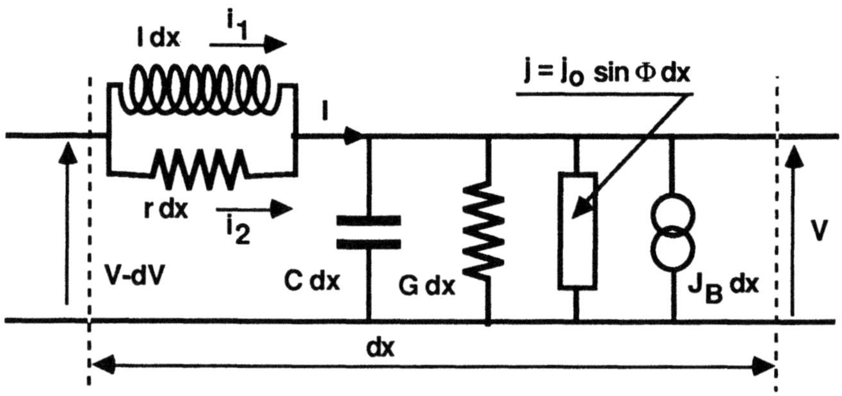

Figure 4.1. Equivalent electrical transmission line of the long Josephson junction with losses.

To the lowest order of approximation a JTL may be modelled (Swihart,1961) by a continuous electrical transmission line structure (Figure4.1) where L is the inductance per unit length, C the capacitance per unit length, $j_o \sin \phi$ the Josephson tunneling current per unit length where ϕ is the phase difference between the macroscopic quantum wave functions, V is the voltage drop across the insulating barrier. Dealing with real transmission lines we must take into account : losses, bias and junction irregularities which influence the motion of real fluxons. On the equivalent transmission line model the dissipative effects can be represented (Figure 4.1) by the following elements : a serie resistance R which represents the scattering of quasiparticles in the surface layers of the two superconductor, a parallel conductance G which is referred to as the quasiparticle tunnelling current (Josephson, 1964). Another source of dissipation originates from the quasi particle pair interference current, more simply it is called the "cosϕ" term originally predicted by Josephson (1962).With the presence of these losses, one also consider a uniformly distributed bias current j_B per unit length.

Applying Kirschoff's laws to the model of figure (3.1) yields a modified SG equation

$$\phi_{XX} - \phi_{TT} - \alpha\phi_T + \beta\,\phi_{XXT} = \sin\phi - \gamma \tag{3.1}$$

Here we have introduced the transformation : $X = X/\lambda_J$, $Y = Y/\lambda_J$, $T = t\,\omega_J$, where :$\lambda_J = (\Phi_o/2\pi\, j_o\, L)^{1/2}$ is the Josephson length and $\omega_J = (2\pi\, j_o/\Phi_o\, C)^{1/2}$ is the Josephson plasma frequency,here Φ_o is the elementary flux quantum . The constants α , β, and γ are defined by : $\alpha = G/\omega C$, $\beta = \omega_J/LR$, $\gamma = j_B/j_o$. When the coefficients α, β and γ are neglected, eq (3.1) reduces to the Sine Gordon equation which has the well known (kink-antikink) soliton solutions propagating with velocity v :

$$\phi_{\pm}(X,T) = 4\,\mathrm{tg}^{-1}\Big[\pm\frac{(X-vT)}{(1-v^2)^{1/2}}\Big] \tag{2.4}$$

Physically the kink (antikink) correspond to a $\pm\,2\pi$ jump in phase difference ϕ accross the insulating barrier separating the two superconductors (Figure 2.2). In other words it is a flux quantum vortex, a current loop which consists of a surface current and a tunneling supercurrent, connecting the two surface layers via the barrier. This current loop supports one quantum Φ_o of magnetic flux and the soliton (antisoliton) is therefore often called a fluxon (antifluxon). The moving soliton described by eq (2.4) is accompanied by a voltage pulse and a current pulse.

Experimentally the fluxon motion, modeled by the modified SG equation, manifests itself in such a way that important information on its dynamics can be extracted from voltage-current characteristics measurements (Parmentier,1978.Pedersen et al, 1984, Davidson et al,1986). Very recently, direct measurements of fluxon propagation (Matsuda et al, 1982. Nitta, 1984) and fluxon-antifluxon collisions (Fujimaki et al, 1987. Matsuda, 1986) were sucessfully performed confirming the dynamical behavior predicted by the modified SG model.

5. Magnetic chains.

Quasi one-dimensional (1D) magnetic compounds where moving domain walls can be viewed as *topological* kink-solitons (Balucani et al,1987. Boucher et al,1987) are now currently used to carry out accurate experiments on nonlinear excitations in condensed matter. The domain walls are highly localized nonlinear objects, they correspond to transition regions between two different but energetically degenerate ground states. In a ferromagnetic chain such walls connect for example neighboring spin up and spin down regions, whereas in an antiferromagnet they form the liaison between the two ground state configurations obtained by interchange of the two sublattices A and B (Figure 5.1). Among the interesting 1D magnets one has the ferromagnetic compound $CsNiF_3$, the antiferromagnetic compound $(CH_3)_4\ NM_nCl_3$ also called TMMC, and $CsCoCl_3$. With spins S = 1 and S = 5/2 $CsNiF_3$ and TMMC are usually treated respectively as classical ferromagnetic and antiferromagnetic spin chains, i.e. quantum effects (Maki, 1981. Balakrisnan et al, 1985) in the spin dynamics can be neglected (Mikeska, 1978, 1980, 1981.) In these systems the nonlinear excitations (domain walls) have relatively large spatial extensions and the dynamics of the discrete spins can be approximated, in the continuum limit, by nonlinear partial differential equations. On the other hand $CsCoCl_3$ is described by an antiferromagnetic quantum spin (S = 1/2) chain (Villain, 1975). Unlike $CsNiF_3$ and TMMC, in the Ising-like spin chain such as $CsCoCl_3$ or $CsCoBr_3$ the widths of the solitons are very narow and the continuum approximation cannot be used (Boucher, 1988).

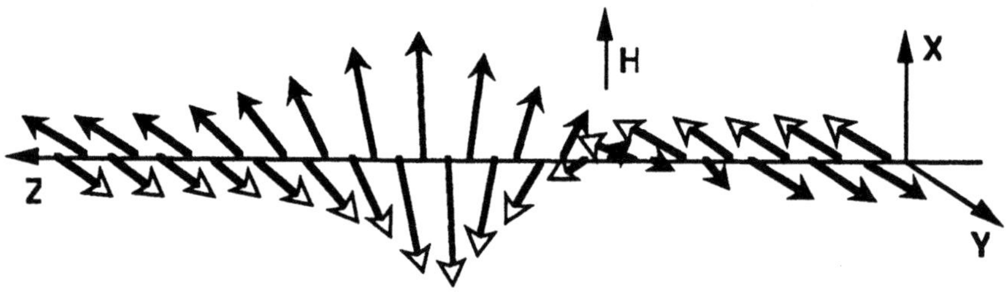

Figure 5.1. Representation of a xy domain wall (SG soliton) in an antiferromagnetic chain.

Adopting the Heisenberg model for $CsNiF_3$ and TMMC in the classical approximation the spins are treated as vectors $\mathbf{S}_n$. The model Hamiltonian is

$$H = -2J \sum_n S_n S_{n+1} + A \sum_n (S^z{}_n)^2 - g\, \mu_B H_x S^x{}_n \tag{5.1}$$

Here, J is the exchange interaction between the spins on adjacent lattice sites n and n+1 . One has : $J > 0$ for $CsNiF_3$ and $J < 0$ for TMMC. The second term represents the influence of crystalline anisotropy : A is the single ion anisotropy constant, the third term represents the Zeeman energy of the spins in an external symetry breaking magnetic field ,$H = H_x$, perpendicular to the chain axis **z** . The quantities g and μ_B are respectively the Landé factor and the Bohr magneton.

From the model Hamiltonian (5.1), in the continuum limit, the spin dynamics of $CsNiF_3$ and TMMC can be described by complicated nonlinear partial differential equations. However, like for the JTL examined in the preceeding section, to lowest nonlinear order of approximation , the nonlinear excitations are frequently taken to be solutions of an integrable prototype equation which is the Sine Gordon equation. As investigated by neutron experiments, the Sine Gordon picture (Mikeska ,1978,1980, 1982) is reliable (Boucher et al,1987) in a restricted in a certain domain of fields and temperature: it is limited to low external magnetic field for $CsNiF_3$ and to low or high external magnetic field for TMMC . For intermediate fields the out of plane motions of the spins play an important role (Wysin et al, 1982, 1984, 1986. Gouvea et al, 1986).

Besides the large amplitude kink-solitons considered above, weak nonlinear excitations, resulting from a field-induced nonlinear coupling of in plane and out of plane spins fluctuations, have been predicted and identified experimentally (Boucher et al, 1989) very recently in TMMC. Nevertheless, the experimental identification of non topological excitations, which correspond to the solutions of coupled NLS equations (Remoissenet , 1989), remains an open problem.

6. Polymers.

Nonlinear self localized excitations are characteristic features of one-dimensional conducting polymers (Heeger et al, 1988), their unique charge-spin relations, light masses and new electronic states lead to interesting experimental consequences in the physical and chemical properties of this growing class of novel materials . Such linear chain polyenes have long been of considerable theoretical and experimental interest, but particularly intense interest has been focused upon the prototype polyene: polyacetylene where the dominant nonlinear excitations are kinks and polarons, and more recently upon polythiophene where the dominant nonlinear excitations are polarons and bipolarons.

Polyacetylene is the simplest case of conjugated polymers and has been studied in especial detail, including possible quasi-soliton (defect) based doping and transport characteristic which are of great technological potential. Polyacetylene $(CH)_x$ consists of weakly coupled chains of CH (Carbon-Hydrogen) units forming a quasi-one-dimensional lattice (Figure 6.1). The two isomers trans-$(CH)_x$ and cis-$(CH)_x$ have respectively two and four CH monomers per unit cell . A crucial difference between trans- and cis-$(CH)_x$ is the presence of a ground state degeneracy in trans-$(CH)_x$. Namely, by symmetry, in trans-$(CH)_x$ one could interchange the double and single bonds (Longuet-Higgins and Salem, 1959, 1960) without changing the energy. Thus, one has a twofold degeneracy with two lowest energy states, A and B, having two different bonding structures. In cis-$(CH)_x$, the ground state is non degenerate. This dimerization in two different patterns in trans-$(CH)_x$ leads to the possibilty of domain walls or kink-solitons (Rice,1979. Su et al, 1979,1980) separating these two regions (Figure 6.1).

When a topological quasi-soliton will move it will convert the A pattern into the B pattern (compare to the kinks of magnetic and antiferromagnetic chains described in section 5).

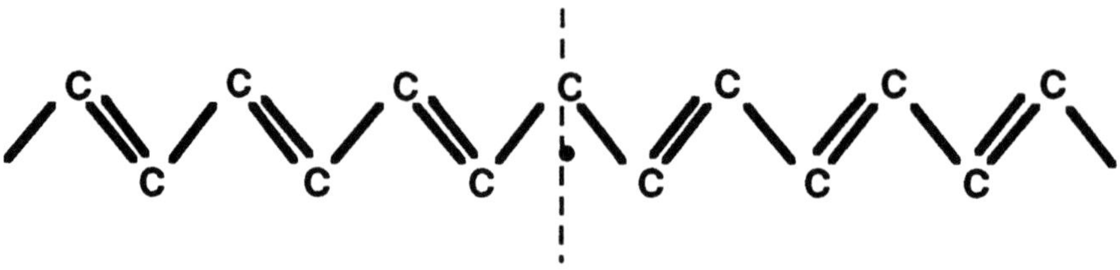

Figure 5.1. Sketch of a defect (domain wall or quasi-soliton) separating the A and B phases or two ground state structures of trans $(CH)_x$. This defect can move along the chain.Here, the signs — and = represent respectively a single long bond and a double short bond (dimerization).

From a discrete model Hamiltoninan, Su, Schrieffer and Heeger (SSH, 1979) have calculated the actual width of this kink, they obtained about 15 lattice spacings. This relatively large value indicates that a continuum model (continuum limit of the SSH model) is also reasonable to describe the excitations (Brazovskii, 1980. Takayama et al, 1980). Associated with the kink or structural distorsion, the electron spectrum is strongly altered in the vicinity of the quasi-soliton leading to reversed spin-charge relations which are a fundamental feature of the soliton model of polyacetylene and are supported by experiments. Namely, a neutral kink, has charge $Q = 0$ and spin $S = \pm 1/2$.The positively or negatively charged solitons have respectively charge $q = \pm e$ and spin $S = 0$.

Recently it was discovered that the kink solutions are not the only nonlinear excitations predicted by the theoretical models of $(CH)_x$. Indeed in both numerical studies of lattice models and analytical studies of continuous models a nonlinear polaron excitation was found (Su et al, 1980b, Campbell and Bishop, 1981. Campbell et al, 1982) . The polaron is an example of self trapped state in condensed matter physics (Fröhlich et al, 1950. Holstein, 1959) . In this case an electron (or a hole) moves through a crystal as a localized wave function rather than an extended Bloch state ; since the electron is localized, it polarizes the medium in its vicinity, thereby, lowering its energy which keeps it localized (Ziman, 1980). In other terms the polaron is a quasi particle arising from phonon dressing of an electron or a hole. In contrast to the kink (anti-kink) soliton the spin charge relation is the conventional one for the polaron in polyacetylene.

The conducting polymers are of a so great technological importance that a considerable number of experimental results have been reported. Then, we just give a short summary of the typical experimental facts [for details the reader is refered to the review of Heeger et al, 1988]. In neutral trans ($CH)_x$ the unpaired spin of the neutral kink soliton makes possible magnetic probes such as nuclear magnetic resonance, electron-nuclear double resonance, dynamic nuclear polarization. These observations support the above theoretical predictions which indicate that the spin is associtated with a neutral quasi-soliton (defect) which is stable and very mobile at highe temperatures. The electrical neutrality of these spin 1/2 kinks was also identified by infrared experiments showing localized infrared active vibrational modes. In contrast to the neutral compounds, charged (doped) trans ($CH)_x$ shows new infrared modes (localized phonons) induced by the quasi-solitons as predicted by the models. They also leads to the appearance of new peaks in the infrared absorption spectrum and optical absorption spectrum. Information of the intrisinc properties of soliton-like excitations was obtained from a serie of photoinduced abdorption experiments, the photoinduced changes confirmed the predictions and established the reversed spin-charge relations of the quasi-soliton model.

While the experimental signatures of the *topological* kink-solitons now seems to be clearly established, there is no sufficient evidence of *non topological* polaron excitations in $(CH)_x$.

References.

Balakrishnan, R. and Bishop, A. R. (1985) Phys Rev Lett, 55, 537.

Balucani, U. Lovesey, S. W. . Rasetti, M. G. and Tognetti, V. Eds. (1987) Magnetic excitations and fluctuations II. Springer Proceedings in Physics 23. Springer Verlag,Berlin.

Barone, A. and Paterno, G. (1982) Wiley, New York.

Benjamin, T.B. and Feir, J.F. (1967) J Fluid Mech, 27,417.

Blow, J. and Doran, N. J.(1987)IEEE Proc, 134, 138.

Blow,K. J. and Doran, N. J. (1985) Opt Commun, 52, 367.

Boucher, J. P. (1989) To be published in Proceedings of Nuclear in magnetism. Munich, August 1988.

Boucher, J. P. Regnault, L. P. and Benner, H. (1987) p 24 In Nonlinearity in condensed matter. Eds A.R. Bishop, D.K. Campbell, P. Kumar and S.E. Trullinger. Springer, Berlin.

Brazovskii,S.A. (1980) Sov Phys JETP, 51 (2), 342.

Campbell, D. K and Bishop, A.R. (1981)Phys Rev B, 24, 4859.

Campbell, D. K and Bishop, A.R. (1982)Nucl. Phys. B 200, 297.

Cotter, D. (1982) Electron Lett., 18, 495.

Cotter, D. (1982) Electron Lett., 18, 504.

Davidson, A. Ducholm, B. and Pedersen, N. F. (1986) J. Appl. Phys., 60, 1447.

Doran, N. J and Blow, K. J. (1983) IEEE J Qant Electron. QE-19, 1883.

Fröhlich, H. Pelzer. H and Zienan, S .(1950) Phil Mag 41, 221.

Fujmaki, A. Nakajima, K. and Sawada, Y. (1987) Phys. Rev. Lett., 59, 2895.

Fukushima, K. (1983) J Phys Soc Japan, 52, 376.

Fukushima, K. Wadati, M. and Narahara, Y. (1980) J Phys Soc Japan, 49, 1593.

Gloge, D. (1979) Rep Prog Phys, 42, 1777.

Hasegawa, A. and Brinkman, W. F.(1980) IEEE J. Quant Electron, QE-16, 694.

Hasegawa,A. and Tappert, F. (1973a) Appl. Phys. Lett, 23, 142.
Hasegawa,A. and Tappert, F. (1973b) Appl. Phys. Lett. 23, 146.
Heeger, A. J. Kivelson, S. Schrieffer, J. R and Su, W. P. Rev Mod Phys, 60, 781, 1988.
Hirota, R. and K. Susuki, K. (1973) Proc IEEE , 61, 1483.
Hirota, R. and Susuki, K . (1970) J Phys Soc Japan, 28, 1366.
Holstein, T. (1959) Ann. Phys., 8, 325.
Jain, M. and Tzoar, N. (1978)J. Appl Phys, 49, 4649.
Jain,M. and Tzoar, N.(1987)IEEE J Qant Elec, QE-23, 510.
Josephson, B. D. (1962) -Phys. Lett. 1, 251.
Josephson, B. D. (1964) Rev Mod Phys, 36, 216.
Karpman, V. I. (1975) Pergamon Press, New York.
Kodama, Y. and Hasegawa, A. (1987) IEEE J Quant Elec. QE - 23, 510.
Kofane, T. Michaux, B. and Remoissenet, M. (1988) J Phys C , 21, 1395.
Krökel,D. Halas, N. J. Gianlani, G and Grischowsky,D. (1988) Phys Rev Lett, 60, 29.
Kuusela, T and Hietarinta, J. (1989) Phys Rev Lett, 62, 700.
Likharev,K .K. (1986) Dynamics of Josephson junctions and circuits. Gordon and Breach, New York.
Lomdahl, P. S. (1985) , J Stat Phys, 39, 5/6, 551.
Longreen K. E. (1978) in Lonngreen, K. and Scott , A. C. (Eds) Solitons in action. Academic Press, New York, pp 127.
Longuet-Higgins, H. C and Salem, L. (1959).Proc Roy Soc, Ser A 251, 172.
Longuet-Higgins, H. C and Salem, L. (1960) Proc Roy Soc, Ser A 255, 435.
Maki, K.(1981) Phys Rev B, 24, 3991.
Matsuda, A. (1986) Phys. Rev. B, 34, 3127.
Matsuda, A. and Uheara, S. (1982) Appl. Phys. Lett., 41, 770.
Mikeska, H. J. (1981) J. Appl. Phys., 52, 1950 .
Mikeska, H. J. (1982) Phys Rev B, 26, 5213.
Mikeska,H.J. (1978) J. Phys. C11, L29 .
Mikeska,H.J. (1980) . J. Phys. C, 13, 2913.
Mollenauer and K. Smith. (1988) Opt. Lett., 13, 675.
Mollenauer,L. F. and Stolen, R. H. (1984) Opt lett. 9, 13.
Mollenauer,L. F. and Stolen, R.H. (1982) Fiberoptic Technology , April, 193.
Mollenauer,L. F. Stolen, R.H. and Gordon, J. P . (1980) Rev. Lett, 45, 1095.
Mollenauer,L. F. Stolen, R.H. and Gordon, J. P. and Tomlinson, W.J.(1983) Optics Letter, 8, 289.
Nitta, J. Matsuda, A and Kawakami, T. (1984) 55, 2758.

Parmentier (1978) in Solitons in action, Eds K. Lonngreen and A. C. Scott, Academic Press, New York, pp 173.
Pedersen, N. F and Welner, D. (1984) Phys. Rev.B, 29, 2551.
Pedersen, N. F. (1989) Proceedings of the ASI Summerschool on superconducting electronics, Ciccio, Italy Plenum .
Peterson, G. E. (1984) AT§TBell Lab Tech J , 63, 901.
Remoissenet, M. (1989) in Proceedings of Workshop "Integrable systems and applications", Oleron, France June 1988. To be published in Lecture Notes in Physics. Springer, Berlin.
Remoissenet, M. (1989) To be published in Proceedings of Winter School on Theoretical Physics" Partially integrable nonlinear equations and their physical applications " March 21-29, 1989, Les Houches France. Kluver Academic, Dordrecht , The Netherlands.
Rice, M. J.(1979) Phys Lett, 71A, 152, 1979.
Satsuma, J. and Yajima, S. (1974) Prog. Theor. Phys. Suppl., 55, 284.
Scott, A. C . Chu, F. Y. F. and Reible, S. A. (1976) J. Appl. Phys. 47, 3272.
Scott, A. C. (1970) Active and nonlinear wave propagation in electronics. Wiley Interscience, New York.
Su, W.P. Schrieffer, J.R. and Heeger, A.J. (1980)Proc Nat Acad Sci, USA, 77, 5626.
Su, W.P. Schrieffer, J.R. and Heeger, A.J.(1979) Phys Rev Lett, 42, 1698.
Su, W.P. Schrieffer, J.R. and Heeger, A.J.(1980) Phys Rev B, 22, 2099, 1980.
Swihart, J. C. (1961). J. Appl. Phys. 32 461.
Tai,K. Hasegawa, A. and Tomita, A. (1986) Phys. Rev. Lett., 56, 135.
Takayama, H. Lin-Liu, Y.R. and Maki, K.(1980) Phys Rev B, 21, 2388, 1980.
Villain, J. (1975) Physica, 79B,1.
Watanabe, S. (1982) J Phys Soc Japan , 51, 1030.
Wysin, G. Bishop, A. R and Kumar, P. (1982) J. Phys. C, 15, L337.
Wysin, G. Bishop, A. R and Kumar, P. (1984) J. Phys. C, 17, 5975.
Wysin, G. Bishop, A. R and Oitmaa, A. R. (1986) J Phys C, 19, 225.
Yoshinaga, T.Sugimoto, N and Kakutani, T. (1981) J. Phys Soc Japan, 50, 2122.
Zakharov, V. E. and Shabat, A. B. (1972) Sov. Phys. JETP, 34,62.
Zhao, W and Bourkoff, E. (1988) IEEE J. Quant Electron., 24, 365-372.
Ziman, J. (1980) Electrons and phonons : the theory of transport phenomena in Solids. Clarendon, Oxford.

PART I

HYDROGEN-BONDED CHAINS

LOCAL MODES IN MOLECULES

Alwyn C. Scott
Laboratory of Applied Mathematical Physics
The Technical University of Denmark
DK-2800 Lyngby, Denmark

1. INTRODUCTION

In recent years the scientific community has become increasingly aware that classical, nonlinear dynamical systems exhibit peculiar and interesting behaviors: chaos, localization of energy (solitons), and "blow-up" in finite time are specific examples.

When dynamical systems are used to model molecular vibrations, the corrections of quantum theory become important. Since quantum theory is linear, the following question arises: Does some aspect of classical, nonlinear behavior survive quantization, and, if so, how?

There are two basic categories of classical nonlinearity in molecular systems. The first I choose to call intrinsic because it arises from nonlinear force constants in molecular bonds. Intrinsic anharmonicity can be modeled by the discrete self-trapping (DST) system [1,2]

$$\left(i\frac{d}{dt} - \Omega_0\right)\bar{A} + \varepsilon M\bar{A} + \gamma\,\mathrm{diag}(|A_1|^2, \cdots, |A_m|^2)\bar{A} = 0 \tag{1.1}$$

where $\bar{A} = \mathrm{col}(A_1,\cdots,A_m)$, the A_j are complex mode amplitudes, and M is a square, $m\times m$, real, symmetric matrix. For $m \leq 2$, the DST equation is exactly integrable [1,3-5]. For $m \geq 3$, quasiperiodic and chaotic trajectories are readily observed [1,6-8].

The second category of classical nonlinearity is called extrinsic because it arises from phonon interactions. It can be modeled by the system

$$\begin{aligned}\left(i\frac{d}{dt} - \Omega_0\right)\bar{A} + \varepsilon M\bar{A} + \chi\,\mathrm{diag}(q_1, \cdots, q_m)\bar{A} &= 0\\ \left(i\frac{d}{dt} - \omega_0\right)q_j &= -\chi|A_j|^2\end{aligned} \tag{1.2a,b}$$

where the $\{q_j\}$, $j = 1,2,\cdots,m$, represent an optical phonon system of frequency ω_0.

If the time evolution of $|A_j|^2$ is slow on the scale of ω_0^{-1}, then the time derivative can be neglected in (1.2b) and (1.2a) reduces to (1.1) with

$$\gamma = \chi^2/\omega_0 \ . \tag{1.3}$$

Thus the classical behaviors of (1.1) and (1.2) are quite similar. The quantum mechanical behaviors, however, are distinctly different.

2. INTRINSIC NONLINEARITY

The quantum theory of the DST equation (1.1) is surprisingly simple and straightforward [9,10]. To see this, note that it has two conserved quantities, the number

$$N = \sum_{j=1}^{m} |A_j|^2 \tag{2.1}$$

and the energy

$$H = \Omega_0 N - \frac{1}{2}\gamma \sum_{j=1}^{m} |A_j|^4 - \sum_{i\neq j} m_{ij} A_i^* A_j \ . \tag{2.2}$$

Under quantization, the complex mode amplitudes (A_j^* and A_j) become boson creation and annihilation operators ($\hat{B}_j^\dagger$ and $\hat{B}_j$). Thus, if $|n_j\rangle$ is an eigenfunction of a particular mode $\hat{B}_j^\dagger|n_j\rangle = \sqrt{n_j + 1}\ |n_j + 1\rangle$, $\hat{B}_j|n_j\rangle = \sqrt{n_j}|n_j - 1\rangle$, and $\hat{B}_j|0\rangle = 0$.

Since the ordering is not specified in Eqs. (2.1) and (2.2) one must use the averages $|A_j|^2 \rightarrow \frac{1}{2}\left(\hat{B}_j^\dagger\hat{B}_j + \hat{B}_j\hat{B}_j^\dagger\right)$ and $|A_j|^4 \rightarrow \frac{1}{6}\left(\hat{B}_j^\dagger\hat{B}_j^\dagger\hat{B}_j\hat{B}_j + \hat{B}_j^\dagger\hat{B}_j\hat{B}_j^\dagger\hat{B}_j + \hat{B}_j^\dagger\hat{B}_j\hat{B}_j\hat{B}_j^\dagger + \hat{B}_j\hat{B}_j^\dagger\hat{B}_j\hat{B}_j^\dagger + \hat{B}_j\hat{B}_j\hat{B}_j^\dagger\hat{B}_j^\dagger + \hat{B}_j\hat{B}_j^\dagger\hat{B}_j^\dagger\hat{B}_j\right)$.

With the boson commutation rule $B_iB_i^\dagger = B_i^\dagger B_i + 1$, (2.1) becomes the number operator

$$\hat{N} = \sum_{j=1}^{m} \left(B_i^\dagger \hat{B}_i\right) \tag{2.3}$$

and Eq. (2.2) becomes the energy operator

$$\hat{H} = \left(\Omega_0 - \frac{1}{2}\gamma\right)\hat{N} - \frac{1}{2}\gamma \sum_{j=1}^{m} \hat{B}_j^\dagger\hat{B}_j\hat{B}_j^\dagger\hat{B}_j - \sum_{i\neq j} m_{ij}\hat{B}_i^\dagger\hat{B}_j \ . \tag{2.4}$$

$\hat{H}$ and $\hat{N}$ operate upon product wavefunctions of the form $|n_1\rangle|n_2\rangle\cdots|n_m\rangle$ which, for typographical convenience, we write as $[n_1,n_2,\cdots,n_m]$.

Thus the quantized discrete self-trapping (QSDT) equation describes an assembly of bosons [11]. It has been discussed in some detail for nearest neighbor interactions in the limit $\gamma \ll \varepsilon$ and $m \rightarrow \infty$ [12-14]. This is the limit in which the classical DST reduces to the nonlinear Schrödinger equation [15].

Stationary states of the QDST equation must be eigenfunctions of both $\hat{N}$ and $\hat{H}$. Consider an n-boson state (i.e. the nth excited level).

With $n < m$, an eigenfunction of $\hat{N}$ with eigenvalue n can be constructed as

$$|\psi_n\rangle = c_1[n\ 0\ 0\ \cdots\ 0] + c_2[0\ n\ 0\ \cdots\ 0] + \cdots + c_m[0\ 0\ 0\ \cdots\ n]$$

$$+ c_{m+1}[(n-1)1\ 0\ \cdots\ 0] + c_p[0\ 0\ \cdots\ 0\ \underbrace{1\ 1\ \cdots\ 1}_{n\ \text{times}}] . \qquad (2.5)$$

The number of terms in (2.5) is equal to the number of ways that n quanta can be placed on m sites. Thus

$$p = \frac{(n + m - 1)!}{(m - 1)!n!} . \qquad (2.6)$$

The wavefunction $|\psi_n\rangle$ in (2.5) is an eigenfunction of $\hat{N}$ for any value of the c_i's. Thus we are free to choose these coefficients so that

$$\hat{H}|\psi_n\rangle = E|\psi_n\rangle . \qquad (2.7)$$

Eq. (2.7) requires that the column vector $\bar{c} = \text{col}(c_1, c_2, \cdots, c_p)$ satisfies the matrix equation

$$[H - IE]\bar{c} = \bar{0} \qquad (2.8)$$

where H is a $p \times p$, symmetric matrix with real elements. The eigenvectors of (2.8) enable us to determine the eigenfunctions in (2.5), and the eigenvalues of (2.8) are the corresponding energies.

This quantized DST equation has been used to calculate the fifteen experimentally observed C - H overtones in the dihalomethanes [16]. With the parameter values (in cm^{-1}) given below,

	$CH_2C/2$	CH_2Br_2	CH_2I_2
ε	29.54	32.80	33.69
γ	127.44	125.45	124.25
Ω_0	3147.51	3152.25	3130.86

the rms error between theory and experiment can be compared with that obtained in reference [16].

	CH_2Cl_2	CH_2Br_2	CH_2I_2
Quantum DST	9.0	6.9	8.0
Reference [16]	12.5	11.2	18.8

These results, obtained in [17], are well within experimental uncertainty. Similar agreement has been obtained in using the quantum DST to describe local modes of C-H oscillation in benzene [18,19].

The local mode behavior of molecular vibrations [20,21] has a characteristic quantum signature. With two degrees of freedom (i.e. with m = 2 for a dihalomethane), the splitting (ΔE) between the two lowest energy levels at each value of n is [17]

$$\frac{\Delta E}{\varepsilon} \leq \frac{2n}{(n-1)!} \left(\frac{\varepsilon}{\gamma}\right)^{n-1} . \tag{2.9}$$

For the first excited level (n = 1) (2.9) gives $\Delta E = 2\varepsilon$. From a linear calculation (assuming $\gamma = 0$) one also finds $\Delta E = 2\varepsilon$. This is an example of the fact that <u>nonlinearity has no effect at the first excited level of the QDST equation</u>. In general, the structures of the two eigenfunctions of lowest energy are

$$|\psi_n\rangle^+ = \frac{1}{\sqrt{2}}([n,0] + [0,n]) + O(\varepsilon)$$

$$|\psi_n\rangle^- = \frac{1}{\sqrt{2}}([n,0] - [0,n]) + O(\varepsilon) \tag{2.10a,b}$$

where $\Delta E = |E^+ - E^-|$. For large n, the two eigenfunctions in (2.10) become approximately degenerate and a local mode [n,0] can be constructed which persists for a time of order $h/\Delta E$.

3. EXTRINSIC NONLINEARITY

Now consider quantization of the extrinsically nonlinear system (1.2). As before A_j and A_j^* become annihilation and creation operators, $\hat{B}_j$ and $\hat{B}_j^\dagger$, for the Ω_0 oscillators. The number operator

$$\hat{N} = \sum_{j=1}^{m} (\hat{B}_j^\dagger \hat{B}_j) \tag{3.1}$$

still commutes with the energy operator. To simplify the problem somewhat, assume the Ω_0 oscillators to be in a linear chain with nearest neighbor interactions of strength ε. Numbering these oscillators by the integers $\ell = 1,2,\cdots,L$, the energy operator is

$$\hat{H} = \hat{H}_0 + \varepsilon\hat{V} \tag{3.2}$$

$$\hat{H}_0 = \sum_{\ell=1}^{L} [\Omega_0 \hat{B}_\ell^\dagger \hat{B}_\ell + \omega_0 \hat{b}_\ell^\dagger \hat{b}_\ell + \chi(\hat{b}_\ell + \hat{b}_\ell^\dagger)\hat{B}_\ell^\dagger \hat{B}_\ell] \tag{3.3}$$

$$\hat{V} = \sum_{\ell=1}^{L} (\hat{B}_{\ell+1}^\dagger \hat{B}_\ell + \hat{B}_{\ell+1}^\dagger \hat{B}_\ell) . \tag{3.4}$$

In (3.3) $\hat{b}_\ell^\dagger$ and $\hat{b}_\ell$ are creation and annihilation operators for the optical phonon system.

An expression for the eigenfunctions of $\hat{H}$ is not known, but several approximate approaches have been used. One approach, proposed by Davydov [22-24], is to write an <u>ansatz</u> eigenfunction as

$$|\psi_D\rangle = \sum_{\ell=1}^{L} |\phi\rangle A_\ell(t) B_\ell^\dagger |0\rangle \tag{3.5}$$

where $|0\rangle$ is the vacuum state of the Ω_0 oscillators and $|\phi\rangle$ is a coherent state wavefunction for the ω_0 oscillators. Thus

$$\hat{b}_\ell |\phi\rangle = q_\ell(t) |\phi\rangle \tag{3.6}$$

and the requirement that $|\psi_D\rangle$ satisfies Schrödinger's equation implies that the $\{A_\ell(t)\}$ and $\{q_\ell(t)\}$ are solutions of (1.2) [25].

A second approach is to note that an exact eigenfunction can be constructed for $\hat{H}_0$ [26]. Thus if $\hat{H}_0|\psi_n\rangle = E(N,n)|\psi_n\rangle$

$$E(N,n) = N\Omega_0 + n\omega_0 - N^2\chi^2/\omega_0 \tag{3.7}$$

$$|\psi(N,n)\rangle = \sqrt{n!}\, \exp(-\tfrac{1}{2}\alpha^2) \sum_{m=0}^{\infty} \frac{(-\alpha)^{(m-n)}}{\sqrt{m!}} L_n^{m-n}(\alpha^2)|N\rangle|m\rangle \ . \tag{3.8}$$

In (3.8), $\alpha \equiv N\chi/\omega_0$, $L_n^{m-n}(\cdot)$ is a Laguerre polynomial, $|N\rangle$ is an eigenfunction of $\hat{B}^\dagger\hat{B}$, and $|m\rangle$ is an eigenfunction of $\hat{b}^\dagger\hat{b}$.

Assuming periodic boundary conditions and N = 1, first order perturbation theory for eigenfunctions of wave number k gives the result

$$E(k) = \Omega_0 - \chi^2/\omega_0 + 2\varepsilon \exp(-\alpha^2) \cos k + O(\varepsilon^2) \ . \tag{3.9}$$

Thus the energy width of a wave packet is

$$\Delta E = 4\varepsilon \exp(-N^2\chi^2/\omega_0^2) \tag{3.10}$$

and the factor $\exp(-N^2\chi^2/\omega_0^2)$ emerges as a <u>bandwidth narrowing factor</u>. In the limit $\omega_0 \to 0$, the k-states become degenerate and an exact, localized wave packet can be constructed. This is the limit in which the approximate wavefunction of Davydov, $|\psi_D\rangle$ in (3.5), becomes exact.

The factor $\exp(-N^2\chi^2/\omega_0^2)$ is also the <u>Franck-Condon factor</u> which is proportional to the low temperature absorption intensity. Thus, in the limit $\omega_0 \to 0$, the absorption intensity goes to zero and the radiative lifetime of $|\psi_D\rangle$ goes to infinity.

These ideas have been applied to the study of local modes of the amide-I band in crystalline acetanilide (ACN) [27]. The basic experimental fact is that an "anomalous" amide-I peak appears at 1650 cm^{-1} which increases in intensity with decreasing temperature and is red shifted from the normal, temperature insensitive peak by about 15 cm^{-1}. In reference [27] this 1650 cm^{-1} peak is assigned to a local mode which is described analytically by the quantized extrinsic nonlinearity introduced above. In addition to the reasons presented in [27] for this assignment, there are two later experimental studies that should be considered.

i) <u>The overtone spectrum</u>. Assignment of the 1650 line to a transition from E(0,0) to E(1,0) in (3.7) implies a low temperature overtone spectrum of the form

$$\nu(N) = AN - BN^2 \tag{3.11}$$

with B = 22.4 cm^{-1}. An overtone spectrum is observed experimentally which is in exact agreement with (3.11) with B = 24.7 cm^{-1} [28].

ii) <u>Temperature dependence</u>. Using (3.7) and (3.8), temperature dependence of the absorption intensity can be calculated by assuming that the phonon modes are thermally populated in the ground state. Assuming that the self-trapping arises from coupling to about ten optical phonon modes clustered near 70 cm^{-1}, one finds good agreement with experimental measurements of intensity vs. temperature as is indicated in Fig. 1.

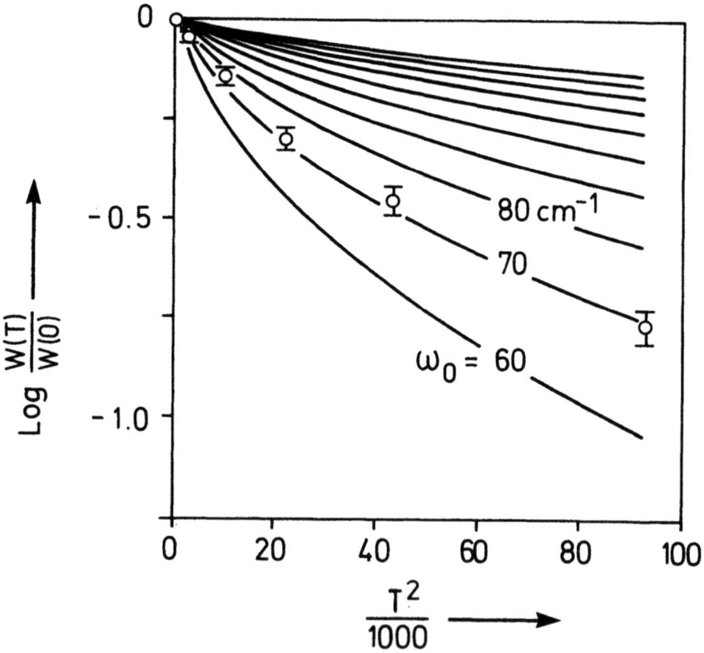

<u>Figure 1</u>. A comparison of measured and calculated values of the relative intensity, W(T)/W(0), of the 1650 cm^{-1} line as a function of Kelvin temperature, T. Solid lines are calculated for self-trapping by ten optical phonons at frequency ω_0. See [26] for details.

Finally it should be noted that three papers which question the assignment of the 1650 cm^{-1} line in ACN to self-trapping by phonons (i.e. extrinsic nonlinearity) are flawed. In particular: i) An assignment of the 1650 cm^{-1} line to a Fermi resonance mechanism [29] is in disagreement with measurements of the line shifts under high pressure [30], ii) Molecular dynamics calculations indicating that the 1650 cm^{-1} line is primarily a hydrogen bending vibration [31] are in disagreement with nitrogen [27] and carbon [30] isotope substitution measurements, and iii) Speculation that a temperature dependent line at 3250 cm^{-1} is evidence against the self-trapping assignment of the 1650 cm^{-1} line [32] misses the point that it (the 3250 line) is the first overtone of the series described in reference [28].

4. LOCAL MODES IN PROTEIN

Over the past decade, the primary aim of my research has been to understand self-trapping as a mechanism for storage and transport of biological energy in protein. As proposed by Davydov [22-24] the self-trapping is effected by acoustic rather than optical phonons. From (3.10) it is clear that the quantum line width and the Franck-Condon factor are much smaller for acoustic mode trapping. Although this insures the stability of the self-trapped state (soliton), it precludes the possibility of optical transition into such a state as has often been emphasized by Davydov [22-24]. Since optical mode self-trapping is observed in ACN [26-28], the following sequence of events seems possible in protein [33]:

i) A photon at about 1650 cm^{-1} excites a local mode which is self-trapped by optical phonons. This state is described by the quantized extrinsic theory sketched in Section 3.

ii) The optical mode self-trapped state relaxes into a state which is self-trapped by acoustic modes. From (3.5) and (3.6) this state is described by using an unquantized, extrinsic system similar to (1.2) to calculate the time evolution of the parameters $\{A_\ell(t)\}$ and $\{q_\ell(t)\}$ in Davydov's approximate wavefunction [34-36].

iii) Using a relationship similar to (1.3), the unquantized, extrinsic DST equation (1.1) can be used to calculate the time evolution of the $\{A_\ell(t)\}$.

REFERENCES

1. J.C. Eilbeck, P.S. Lomdahl and A.C. Scott, Physica D 16 (1985) 318.

2. A.C. Scott, P.S. Lomdahl and J.C. Eilbeck, Chem. Phys. Lett. 113 (1985) 29.

3. V.M. Kenkre and D.K. Campbell, Phys. Rev. B 34 (1986) 4959.

4. L. Cruzeiro-Hansson, P.L. Christiansen and J.N. Elgin, Phys. Rev. B 37 (1988) 7896.

5. A.C. Scott and P.L. Christiansen, "A generalized discrete self-trapping equation," (submitted).

6. J.H. Jensen, P.L. Christiansen, J.N. Elgin, J.D. Gibbon and O. Skovgaard, Phys. Lett. A 110 (1985) 429.

7. S. De Filippo, M. Fusco Girard and M. Salerno, Physica D 26 (1987) 411; ibid. 29 (1988) 421.

8. L. Cruzeiro-Hansson, H. Feddersen, R. Flesch, P.L. Christiansen and M. Salerno, "Classical and quantum analysis of chaos in the discrete self-trapping equation," (submitted).

9. A.C. Scott and J.C. Eilbeck, Phys. Lett. A 119 (1986) 60.

10. S. De Filippo, M. Fusco Girard and M. Salerno, "Avoided crossing and next neighbor spacings for the quantum DST equation," Nonlinearity (to appear).

11. P.A.M. Dirac, The Principles of Quantum Mechanics (4th ed.) Clarendon Press, Oxford (1958) pp. 227-232.

12. E.K. Sklyanin and L.D. Faddeev, Sov. Phys. Dokl. 23 (1987) 902.

13. H.B. Thacker and D. Wilkinson, Phys. Rev. D 11 (1979) 3660.

14. M. Wadati in Dynamical Problems in Soliton Systems (S. Takeno, ed.) Springer, Berlin (1985) pp. 68-80.

15. V.E. Zakharov and A.B. Shabat, Sov. Phys. JETP 34 (1972) 62.

16. O.S. Mortensen, B.R. Henry and M.A. Mohammadi, J. Chem. Phys. 75 (1981) 4800.

17. L. Bernstein, J.C. Eilbeck and A.C. Scott, "The quantum theory of local modes in a coupled system of nonlinear oscillators, " (submitted).

18. A.C. Scott and J.C. Eilbeck, Chem. Phys. Lett. 113 (1985) 29.

19. A.C. Scott, L. Bernstein and J.C. Eilbeck, "Energy levels of the quantized discrete self-trapping equation," J. Biol. Physics (in press).

20. B.R. Henry and W.J. Siebrand, J. Chem. Phys. 49 (1968) 5369.

21. B.R. Henry, J. Chem. Phys. 80 (1976) 2160.

22. A.S. Davydov, Phys. Scr. 20 (1979) 387.

23. A.S. Davydov, Biology and Quantum Mechanics, Pergamon, Oxford (1981).

24. A.S. Davydov, Sov. Phys. Usp. 25 (1982) 899; Usp. Fiz. Nauk 138 (1982) 603.

25. W.C. Kerr and P.S. Lomdahl, Phys. Rev. B 33 (1987) 3629.

26. A.C. Scott, I.J. Bigio and C.T. Johnston, "Polarons in ACN," Phys. Rev. B (in press).

27. G. Careri, U. Buontempo, F. Galuzzi, A.C. Scott, E. Gratton and E. Shyamsunder, Phys. Rev. B 30 (1984) 4689.

28. A.C. Scott, E. Gratton, E. Shyamsunder and G. Careri, Phys. Rev. B 32 (1985) 5551.

29. C.T. Johnston and B.I. Swanson, Chem. Phys. Lett. 114 (1985) 547.

30. C.T. Johnston, private communication.

31. A. Tenenbaum, A. Campa and A. Giansanti, Phys. Lett. A 121 (1987) 126.

32. G.B. Blanchet and C.R. Fincher, Phys. Rev. Lett. 54 (1985) 1310.

33. A.C. Scott, in Synergetics of the Brain, (E. Bazar, F. Flohr, H. Haken and A.J. Mandell, eds.) Springer, Berlin (1983) 345.

34. A.C. Scott, Phys. Rev. A 26 (1982) 578.

35. A.C. Scott, Phys. Scr. 29 (1984) 279.

36. L. MacNeil and A.C. Scott, Phys. Scr. 29 (1984) 284.

Dynamics of Nonlinear Excitations in DNA

Michel PEYRARD
Physique Non Linéaire : Ondes et Structures Cohérentes,
Faculté des Sciences, 6 blvd Gabriel, 21000 Dijon, France

and

A. R. BISHOP
Theoretical Division and Center for Nonlinear Studies,
Los Alamos National Laboratory, Los Alamos, New Mexico, USA

Abstract

We investigate the dynamics of two lattice models for the denaturation of the DNA double helix associated with different degrees of freedom of the molecule. Using numerical simulations or the transfer integral method for the statistical mechanics, we investigte the formation of the large amplitude nonlinear excitations leading to denaturation. We show that a mechanism involving an energy localization analogous to self focusing may start the denaturation locally.

1. Introduction

The dynamics of DNA transcription is a fascinating but very difficult problem due to the complex roles played by RNA polymerases in the process. It is now well established[1] that it involves a local denaturation of DNA. Consequently, as a preliminary step for understanding the transcription, it is interesting to investigate the thermal denaturation of the double helix.

In the last few years the idea that nonlinear excitations could play a role in DNA has become increasingly popular. Englander *et al.*[2] suggested first a theory of soliton excitations as an explanation of the open states of DNA and then various nonlinear models were proposed. One major question to select the appropriate model is to determine among the many degrees of freedom in a DNA molecule the ones which are relevant.

Depending on their aim the models can be classified in three categories. First, the main variable can be chosen to be the sugar puckering which is associated with two configurations of the sugar molecule[3]. Such models are essentially designed to describe the transition between the A and B forms of DNA. The two other classes of models are intended to describe the dynamics of DNA opening. The models of the second category assume that the base rotation is the relevant variable and use a plane base-rotator model. The first one, presented by Yomosa[4] was further refined by Takeno and Homma[5] who introduced some discreteness effects, and by Zhang[6] who improved the model for base coupling. The third approach is more closely connected to experimental results and treats the stretching of the hydrogen bonds connecting two bases in a pair as the main variable. Vibrational normal modes analysis of infrared or Raman experiments suggested that local melting could be achieved through breathing modes[7] and a self-consistent phonon theory was proposed for the hydrogen bond melting[8] [9] [10]. However, in this theory, nonlinearity is treated as a perturbation in spite of the large motions involved in the denaturation.

Once the model is selected, another question is "what do we do with it?". Apart from the third approach developped to analyse experimental data, all the other models, in which nonlinearities are treated intrinsically, have been only investigated with the aim of finding solitary wave solutions and, in some cases checking their stability. Since such excitations are expected to play an important role in DNA denaturation, this is indeed interesting. However these approaches do not address the important question *how can these excitations be created at the rather low denaturation temperature?* The answer can be given by the *statistical mechanics* of the nonlinear models.

Finally, another important question is the role of inhomogeneities on the nonlinear dynamics of DNA. They are built in the system since the base sequence contains the genetic code and determine for instance the site selection for the initiation of the transcription[1].

This paper proposes preliminary answers to the questions listed above. In section 2, we investigate the formation of large amplitude nonlinear excitations leading to denaturation in a plane base-rotator model, without and with inhomogeneities. Section 3 presents a model similar to those used to analyze the infrared and Raman experiments in which nonlinearity is now treated intrinsically. We employ a transfer integral technique to investigate the statistical mechanics of the model and determine the inter-strand separation in the double helix as a funtion of temperature. Both approaches suggest that energy localization mechanisms could initiate the denaturation. This is discussed in section 4.

2. Dynamics of a plane base-rotator model.

In its simplest version, the plane base-rotator model reduces to a sine-Gordon model[4] [5] [6]. We have chosen this simple version to investigate the formation of the nonlinear excitations, but, instead of treating it in a continuum limit which may be a severe approximation for DNA[5], we keep the discrete character of the model in our analysis. In addition the model includes inhomogeneities in the stacking energy between adjacent bases. The hamiltonian is

$$H_1 = \sum_n \frac{1}{2} I \dot{\phi}_n^2 + \frac{1}{2} S(n) \left(\phi_n - \phi_{n-1}\right)^2 + B(1 - \cos\phi_n), \tag{2.1}$$

where ϕ_n design the rotation of base n, I its moment of inertia, $S(n)$ the coupling constant between adjacent bases, and the last term is the restoring potential provided by the hydrogen bonds connecting the two strands and by the dipolar interactions in the molecule.

We have investigated the *formation* of the large amplitude excitations leading to denaturation by a numerical simulation of the equations of motion resulting from the hamiltonian (2.1). In a typical calculation, the molecule contains 200 cells with periodic boundary conditions. The equations are solved with a fourth order Runge Kutta scheme with a time-step chosen to preserve energy to an accuracy better than 0.1 % during the whole calculation. The choice of "realistic" parameters is difficult since the model represents many degrees of freedom with a single variable ϕ. We have used an average between the values given by Yomosa[4] and Zhang[6] which are in rather good agreement. The moment of inertia is $I = MR^2$ with $M = 308m_p$ (proton mass) and $R = 4\mathring{A}$ and we have chosen $B = 0.15\,eV$ and $S = 0.03\,eV$ in the homogeneous case. The time unit is $1\,ps$, a value which is of the order of the characteristic times of macromolecular dynamics. We have considered two types of initial conditions, a sinusoidal plane wave and random velocities simulating thermal fluctuations.

Although it is not physical, the sinusoidal plane wave initial condition is useful to provide a clean situation to observe the build-up of the large amplitude motions. Figures 1 and 2 show typical results for a plane wave of amplitude $1\,rd$. Note that this value is high enough to allow large nonlinear effects but well below the value π corresponding to the instable point of a single oscillator.

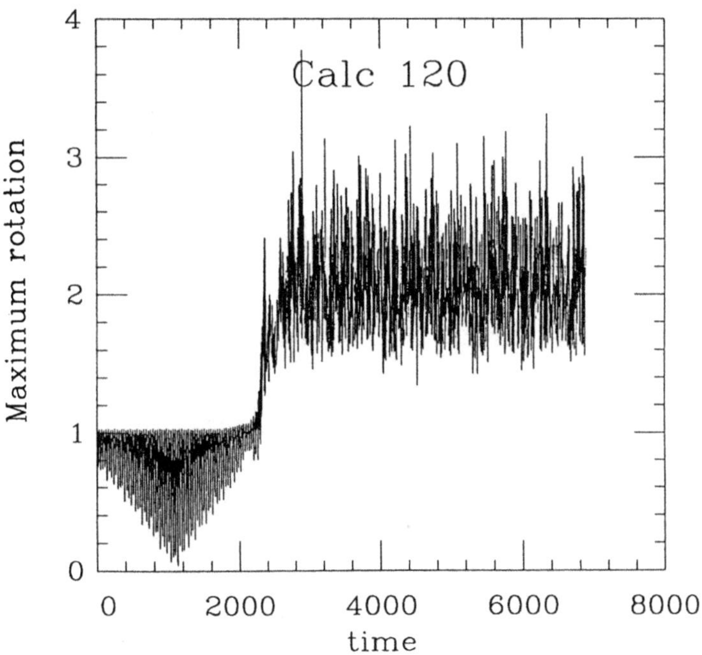

Figure 1 : Variation versus time of the maximum value of ϕ_n along the chain. The large oscillations below $1\,rd$ are due to a sampling effect in the calculation. Only the envelope joining the maxima is physically meaningful.

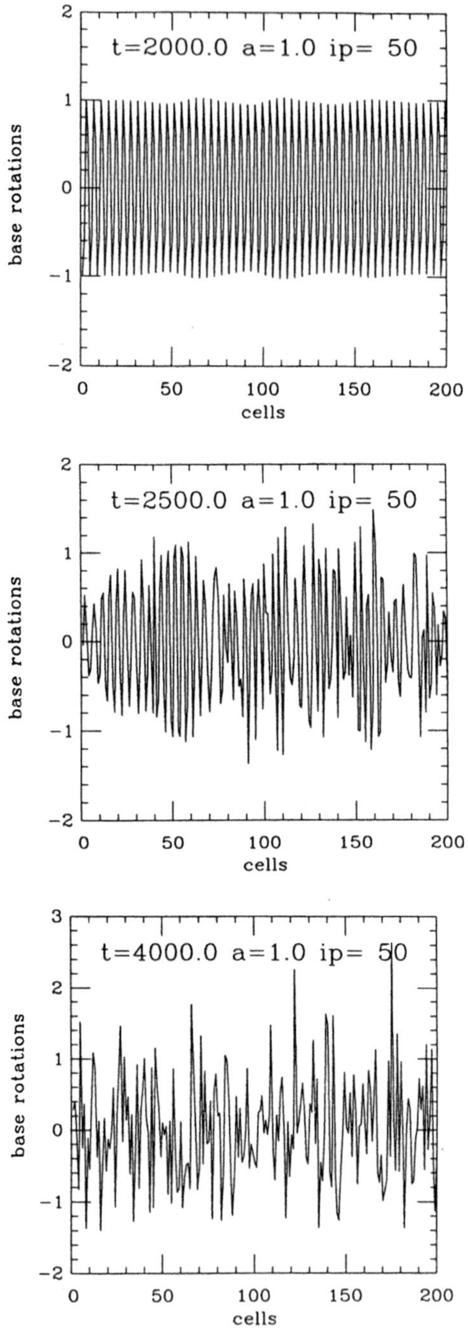

Figure 2 : Base rotations ϕ_n along the molecule at three times. (a) Upper figure: $t = 2000\,ps$. The initial plane wave is slightly modulated. (b) Central figure: $t = 2500\,ps$. The plane wave is disturbed by large fluctuations. (c) Lower figure: $t = 4000\,ps$. The plane wave has been completely destroyed and very large local rotations are observed (note the change in scale).

As shown in Fig. 1, the maximum value of ϕ_n exhibits a very interesting behavior. For about 2000 *ps*, it stays close to its initial value of $1\,rd$ with only a very slow increase. Then, in less than 100 *ps*, the maximum rotation doubles and exhibits large fluctuations. The mechanism of this evolution can be understood from Fig. 2 which shows the rotations of the bases along the molecule at different times. At time $t = 2000\,ps$, the motion is still a plane wave but a slight ***modulation*** is already visible. This modulation grows rapidly and at $t = 2500\,ps$ large irregular variations are found in the amplitude of the plane wave. At a later time ($t = 4000\,ps$), the plane wave structure has been completely destroyed and a chaotic-looking pattern is observed. In most of the sites, the rotation ϕ_n is now smaller than $1\,rd$ but in few places it is, on the contrary, as large as $2.5\,rd$. As a result the energy is no longer uniformly distributed along the chain as it was in the plane wave, but there are large peaks in the energy density as shown in Fig. 3.

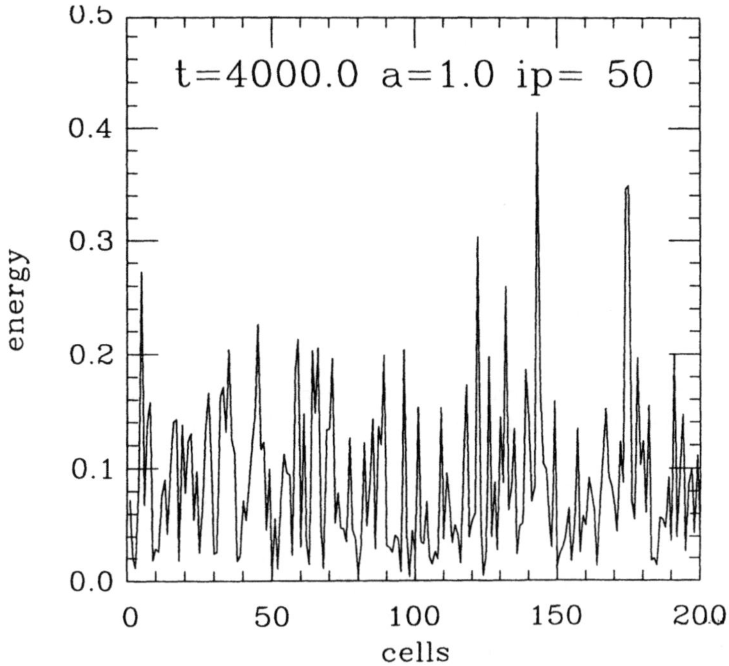

Figure 3 : Energy density along the molecule at $t = 4000\,ps$.

These results attest that ***energy tends to become localized in the molecule.*** In the case presented in Figs. 1-3, the initial energy was not high enough to produce a complete denaturation but the large short-lived rotations found in the simulation can probably be identified with the fluctuational openings that are observed experimentally in DNA because they allow fluorescent markers to bind to the molecule[1]. When the initial amplitude of the plane wave is increased, the same energy localization mechanism leads to local denaturation of the molecule. Figure 4 shows for instance the base rotations at $t = 1500\,ps$ for an initial amplitude of the plane wave equal to

1.5 rd. Although this amplitude is still well below π, the formation of several solitons is observed. As shown in Fig. 4, their width is of the order of the lattice spacing. This points out the importance of discreteness effects in the dynamics of DNA.

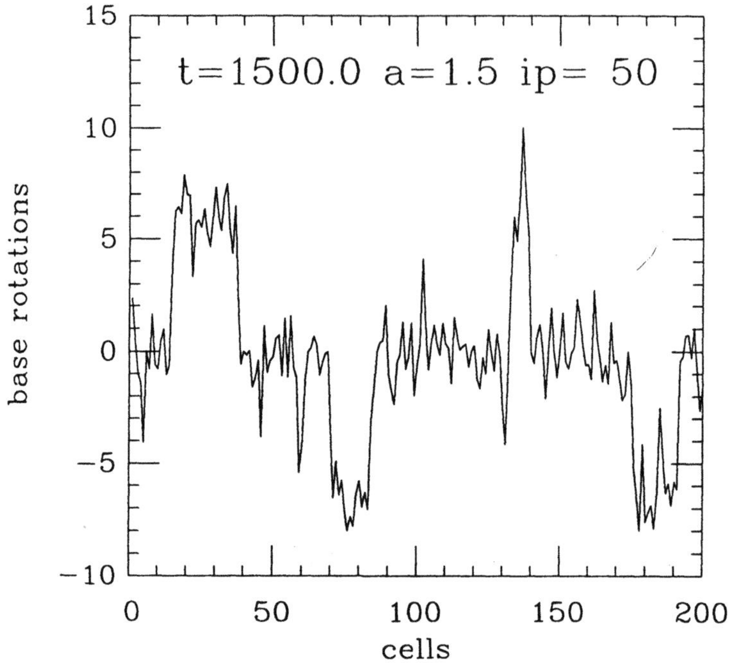

Figure 4: Base rotations ϕ_n along the molecule at $t = 1500\,ps$ for an initial plane wave amplitude of 1.5 rd. Note the formation of several 2π kinks.

In order to estimate the temperature required for thermal denaturation in the model the initial condition has been changed to a gaussian distribution of velocities $\dot{\phi}_n$. The initial condition contains only kinetic energy. Assuming that equipartition of energy is roughly valid to define temperature in the system, the variance of the distribution is set to $\sqrt{2k_BT/I}$. The results show that denaturation starts in the model only when k_BT approaches $0.05\,eV$, *i.e.* $T \approx 550\ K$. Although this is lower than the value of $B = 0.15\,eV$ indicating that the energy localization mechanism plays a significant role, this is far too high. As the various estimations for the parameters of the plane base-rotator model [4] [6] are in rather good agreement, this suggests that an important feature is missing in the model. *It can be the inhomogeneities in the molecule.* To test this idea, we have introduced in the calculation a modulation of the stacking energy $S(n)$. Figure 5 shows the results of a calculation in which $S(n)$ is reduced from $0.03\,eV$ to $0.012\,eV$ in the 10 central cells of the molecule. All the other model parameters are the same as in the calculation shown in Figs. 2-4. In the presence of inhomogeneities, a plane wave amplitude of $1\,rd$ is sufficient to initiate a denaturation. It is interesting to notice that, although the opening starts in the weakly coupled region, it does not stay confined in this region as shown in Fig. 5 where the opened region is not around the center of the molecule.

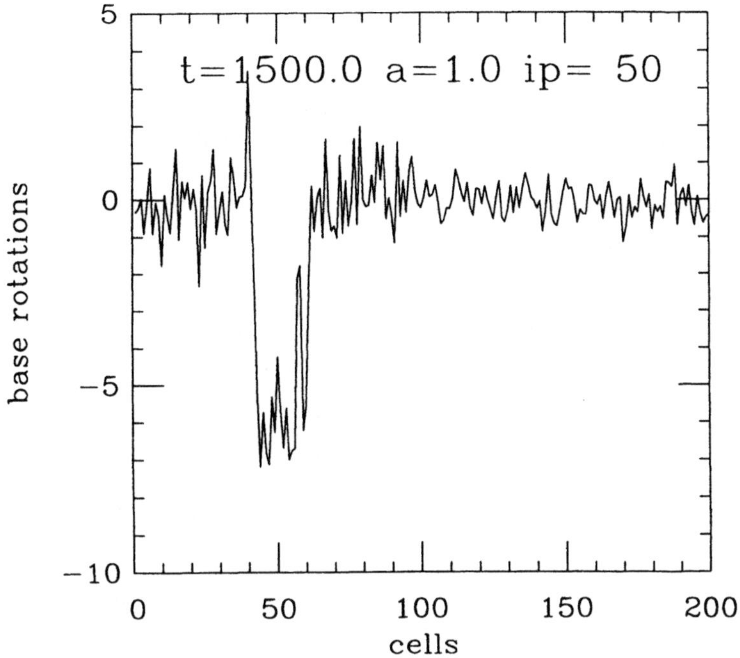

Figure 5: Base rotations ϕ_n along an inhomogeneous molecule at $t = 1500\,ps$ for an initial plane wave amplitude of $1.\,rd$.

Altough these results are only preliminary, they point out the role of inhomogeneities in the denaturation.

3. Model for the hydrogen bonds stretching.

In order to test a model of the third category for the dynamics of DNA, we have investigated the properties of a model directly derived from the models proposed to analyse infrared and Raman data[7] [8] [9]. For each base pair, it includes two degrees of freedom u_n and v_n which correspond to the displacements of the bases from their equilibrium positions along the direction of the hydrogen bonds that connect the two strands. Following the previous investigations on DNA[8] [9], the potential for the hydrogen bonds is approximated by a Morse potential. An harmonic coupling due to the stacking is assumed between neighboring bases so that the Hamiltonian of the model is:

$$H_2 = \sum_n \frac{1}{2} m \left(\dot{u}_n^2 + \dot{v}_n^2\right) + \frac{1}{2} k \left[\left(u_n - u_{n-1}\right)^2 + \left(v_n - v_{n-1}\right)^2\right] + V(u_n - v_n), \quad (3.1)$$

with

$$V(u_n - v_n) = D \left\{exp[-a(u_n - v_n)] - 1\right\}^2 . \quad (3.2)$$

In a first approach, the inhomogeneities due to the base sequence and the asymmetry of the two strands are again neglected so that we use a common mass m for the bases and the same coupling constant k along the two strands. The Morse potential V is an average potential representing the 2 or 3 bonds which connect the two bases in a pair. It can be estimated from the parameters obtained for the individual bonds by the analysis of the vibrational modes of DNA[9].

With these assumptions, the motions of the two strands can be described in terms of the variables

$$x_n = \frac{1}{\sqrt{2}}(u_n + v_n), \text{ and } y_n = \frac{1}{\sqrt{2}}(u_n - v_n) \tag{3.3}$$

which represent the in-phase and out-of-phase motions respectively. Only the out-of-phase displacements y_n stretch the hydrogen bonds. The hamiltonian (3.1) becomes:

$$\begin{aligned} H_2 = H(x) + H(y) = & \sum_n \left\{ \frac{p_n^2}{2m} + \frac{1}{2}k\left(x_n - x_{n-1}\right)^2 \right\} \\ & + \sum_n \left\{ \frac{q_n^2}{2m} + \frac{1}{2}k\left(y_n - y_{n-1}\right)^2 + D\left[exp(-a\sqrt{2}y_n) - 1\right]^2 \right\}. \end{aligned} \tag{3.4}$$

where $p_n = m\dot{x}_n$ and $q_n = m\dot{y}_n$. The classical partition function of a chain containing N base pairs

$$Z = \int_{-\infty}^{+\infty} \prod_{n=1}^{N} \left(dx_n dy_n dp_n dq_n e^{-\beta H_2(p_n, x_n, q_n, y_n)} \right). \tag{3.5}$$

factors into different parts

$$Z = Z_p \, Z_x \, Z_q \, Z_y. \tag{3.6}$$

The two momentum parts are readily integrated and, since the motions in x and y are decoupled, Z_x is simply the contribution of the potential energy in an hamonic chain. As the coupling involves only nearest neighbors interactions, Z_y can be expressed in the form

$$Z_y = \int_{-\infty}^{+\infty} \prod_{n=1}^{N} dy_n e^{-\beta f(y_n, y_{n-1})} \tag{3.7}$$

where f denotes the potential energy part in $H(y)$. This integral can be evaluated exactly in the limit of large systems using the eigenfunctions and eigenvalues of a transfer integral operator[11] [12]

$$\int dy_{n-1} e^{-\beta f(y_n, y_{n-1})} \varphi_i(y_{n-1}) = e^{-\beta \epsilon_i} \varphi_i(y_n). \tag{3.8}$$

The calculation is similar to the one performed by Krumhansl and Schrieffer[12] for the statistical mechanics of the ϕ^4 field. It yields

$$Z_y = e^{-N\beta\epsilon_0}, \tag{3.9}$$

where ϵ_0 is the lowest eigenvalue of a Schrödinger-like equation which determines the eigenfunctions of the transfer integral operator (3.8)

$$-\frac{1}{2\beta^2 k}\frac{\partial^2\varphi_i}{\partial y^2}+D\left(e^{-4ay}-2e^{-2ay}\right)\varphi_i(y)=\left(\epsilon_i-s_0-D\right)\varphi_i(y), \qquad (3.10)$$

with $s_0=(1/2\beta)\ln(\beta k/2\pi)$.

Eqn. (3.10) is formally identical to the Schrödinger equation for a particle in a Morse potential, so that it can be solved exactly[13] [14]. This equation has a discrete spectrum when $d=(\beta/a)\sqrt{kD}>1/2$ and the eigenvalue and the normalized eigenfunction for the ground state are then

$$\epsilon_0=\frac{1}{2\beta}\ln\left(\frac{\beta k}{2\pi}\right)+\frac{a}{\beta}\sqrt{\frac{D}{k}}-\frac{a^2}{4\beta^2 k} \qquad (3.11)$$

$$\varphi_0(y)=(\sqrt{2}a)^{1/2}\frac{(2d)^{d-1/2}}{\sqrt{\Gamma(2d-1)}}\exp\left(-de^{-\sqrt{2}ay}\right)\exp[-(d-1/2)\sqrt{2}ay]. \qquad (3.12)$$

These results can be used to compute the mean stretching $\overline{y}_m$ of the hydrogen bonds. It is given by

$$\overline{y}_m=\frac{1}{Z}\int\prod_{n=1}^{N}y_m e^{-\beta H_2}dx_n dy_n dp_n dq_n,$$

but, since H_2 separates into a sum over the x, y, p, q variables, $\overline{y}_m$ reduces to

$$\overline{y}_m=\frac{1}{Z_y}\int\prod_{i=1}^{N}y_m e^{-\beta f(y_n,y_{n-1})}dy_n. \qquad (3.13)$$

As the model is assumed to be homogeneous, the result does not depend on the particular site considered. The integral can again be calculated with the transfer integral method[11] and yields

$$\overline{y}=\frac{\sum_{i=1}^{N}\langle\varphi_i(y)|\,y\,|\varphi_i(y)\rangle\, e^{-N\beta\epsilon_i}}{\sum_{i=1}^{N}\langle\varphi_i(y)|\,\varphi_i(y)\rangle e^{-N\beta\epsilon_i}}. \qquad (3.14)$$

In the limit of large N, the result is again dominated by the lowest eigenvalue ϵ_0 so that $\overline{y}$ is given by

$$\overline{y}=\langle\varphi_0(y)|\,y\,|\varphi_0(y)\rangle=\int\varphi_0^2(y)y\,dy \qquad (3.15)$$

for a normalized eigenfunction $\varphi_0(y)$. This integral has been evaluated numerically with the expression (3.12) of $\varphi_0(y)$. The results are shown in Fig. 6 for three values of the coupling constant k. In the calculation, we have used $D=0.33\ eV$ and $a=1.8\ \mathring{A}^{-1}$ which corresponds to mean values for the $N-H\cdots N$ and $N-H\cdots O$ bonds in the $A-T$ and $G-C$ base pairs[9].

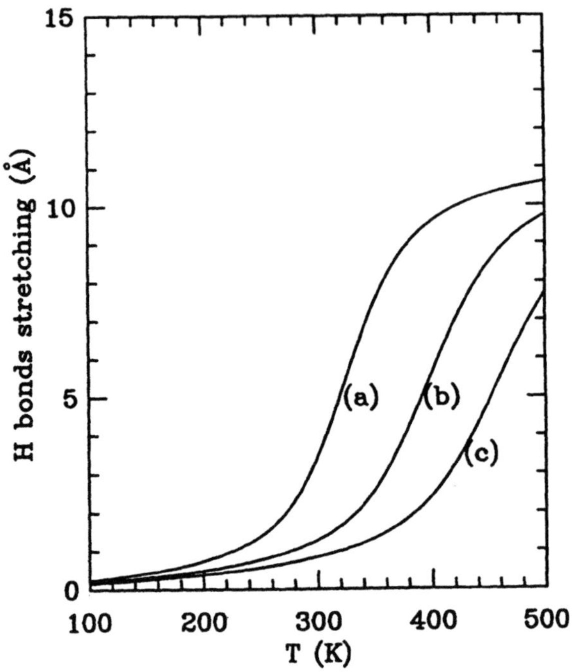

Figure 6: Variation of $\overline{y}$ as a function of temperature for three values of the coupling constant k : (a) $k = 2.0\ 10^{-3}\ ev/\AA^2$, (b) $k = 3.0\ 10^{-3}\ eV/\AA^2$ and (c) $k = 4.0\ 10^{-3}\ ev/\AA^2$.

This figure shows a rapid increase of $\overline{y}$ around a particular temperature which is a characteristic of DNA denaturation as observed for instance by measuring its absorbance of ultraviolet light at 260 nm[1]. The denaturation temperature is indeed sensitive to the parameters of the hydrogen bonds which bind the two strands, but fig. 6 shows that it is also very sensitive to the intra-strand interaction constant, a parameter which is no so well determined by experimental results. As k increases the denaturation temperature increases. This is similar to the results obtained with the plane base-rotator model and consistent with the increase observed experimentally in the presence of reagents that increase the hydrophobic interactions[1]. Our results indicate that, in the absence of inhomogeneities, k must be of the order of 3.0 $10^{-3}\ eV/\AA^2$ to get a reasonable denaturation temperature. This value characterizes a weak coupling between the transverse base motions in DNA which means that discreteness effects have to be taken into account in the models describing DNA denaturation[2] [4] [5] [6].

4. A possible mechanism for denaturation.

Although it indicates *at which temperature* the denaturation occurs, the calculation of $\overline{y}$ does not indicate *how* it occurs. The numerical simulations of the plane base-rotator model have already suggested that *modulational instability* could be responsible for the initiation of the denaturation. To understand this initiation in the model for hydrogen bonds stretching,, it is interesting to relate it to standard models involving nonlinear excitations. Since the nonlinearities appear in terms of the variable y, lets consider the equations of motion which derive from $H(y)$ (Eq. (3.4))

$$m\frac{\partial^2 y_n}{\partial t^2} = k(y_{n+1} + y_{n-1} - 2y_n) - 2\sqrt{2}Dae^{-\sqrt{2}ay_n}\left(e^{-\sqrt{2}ay_n} - 1\right) = 0 \qquad (4.1)$$

The phenomena precursor to denaturation can be investigated by expanding Eq. (4.1) for small y as

$$m\frac{\partial^2 y_n}{\partial t^2} = k(y_{n+1} + y_{n-1} - 2y_n) + 4Da^2 y_n - 6\sqrt{2}Da^3 y_n^2 + \frac{28}{3}Da^4 y_n^3 = 0. \qquad (4.2)$$

A solution of this equation can be obtained with a multiple-scale expansion[15] under the form

$$y_n = F_1(X_n, \tau)e^{i\theta_n} + c.c. + \varepsilon\left[F_0(X_n, \tau) + F_2(X_n, \tau)e^{2i\theta_n} + c.c.\right] \qquad (4.3)$$

with $X_n = \varepsilon x = \varepsilon nl$ and $\tau = \varepsilon t$, l being the lattice spacing *i.e.* the mean distance between ajacent bases. In this expression y is written as a modulated wave in which the carrier wave $\exp(i\theta_n) = \exp[i(qnl - \omega t)]$, with ω and q related by the dispersion relation of the lattice, includes the discreteness effect while the modulation factor is treated in the continuum limit (the so called *semi-discrete* approximation[16]). However, for a qualitative understanding of the denaturation, we can restrict ourselves to a continuum approximation for the carrier wave as well, which simplifies significantly the calculations. The d.c. and first harmonic terms in (4.3) , F_0 and F_2, are necessary because Eq. (4.2) contains an even power of y. The multiple-scale expansion yields a Non Linear Schrödinger (NLS) equation for $F_1(X, \tau)$[17]

$$iF_{1s} + PF_{1ZZ} + Q|F_1|^2 F_1 = 0, \qquad (4.4)$$

with $Z = X - V_g T$, $\quad s = \varepsilon\tau$ and

$$V_g = \frac{d\omega}{dq} = \frac{q}{\omega}\frac{k^2 l^2}{m}, \qquad P = \frac{(k^2 l^2/m) - V_g^2}{2\omega}, \qquad Q = \frac{4}{3}\frac{Da^4}{\omega}. \qquad (4.5)$$

The solitary waves in y that result from this equation are the breathing modes suggested by the infrared and Raman experiments[7].

The statistical mechanics of the NLS equation has been investigated recently by Lebowitz, Rose and Speer[18]. They show that the system can develop singularities in a finite time. Such singularities which correspond to self-focussing phenomena in plasmas may well be responsible for DNA denaturation because they occur when the L^2 norm of the field

$$N(F_1) = \int_0^L |F_1|^2 dx$$

(where L denotes the size of the system) is such that $N\beta$ exceeds some threshold. In our calculation, $N(F_1)$ is given at temperature T by $\overline{y^2}$ which can be calculated similarly to $\overline{y}$ and is given by an expression analogous to Eq. (3.15) :

$$\overline{y^2} = \langle \varphi_0(y) | \, y^2 \, | \varphi_0(y) \rangle = \int \varphi_0^2(y) y^2 dy \tag{4.6}$$

The calculation shows that $\overline{y^2}$ rises by several orders of magnitude in a small temperature range around DNA denaturation so that the threshold for energy localization could be reached. This suggests that the denaturation " bubble" which is observed experimentally at the begining of the denaturation process could be created by *energy localization due to nonlinear effects*. It is however difficult to provide a more quantitative analysis of this phenomenon within the framework of the present model because the coefficients of the NLS derived in the limit of small y depend on the frequency of the carrier wave introduced in the expansion. Although this frequency must lie within the frequencies of the phonon modes in the lattice, it is not well defined for the thermal fluctuations that we consider here.

Thus *both models suggest the possible existence of an energy localization mechanism in DNA*. If we are only interested in the precursor phenomena, *i.e.* in rather small amplitude motions, both models can be expanded to give an NLS equation and this explains certainly the common features in their behavior. The plane wave initial condition in the base-rotator model has shown that the localization can be considered as generated by modulational instability. Another important feature that emerges from our results is the *role of discreteness* in the dynamics of DNA. Both models give a reasonable denaturation temperature only if the coupling between adjacent bases is weak so that the energy can localize itself in regions containing only one or two base pairs to initiate the opening. Finally the simulations with the base-rotator model have shown that the inhomogeneities must not be ignored. They can act as extrinsic nucleation centers for the energy localization mechanism and lower significantly the denaturation temperature. A more complete investigation of this phenomenon will be performed.

ACKNOWLEDGEMENTS

One of us (M.P.) would like to thank D. K. Campbell for his hospitality in the Center for Nonlinear Studies of the Los Alamos National Laboratory, where part of this work has been performed. We would like to thank H. A. Rose, A. Garcia, M. Remoissenet and C. Reiss for helpful discussions.

REFERENCES

1. D. Freifelder, *Molecular Biology*, Jones and Bartlett Publishers, Boston 1987
2. S. W. Englander, N. R. Kallenbach, A. J. Heeger, J. A. Krumhansl and S. Litwin, Proc. Nat. Acad. Sci. USA **777**, 7222 (1980)
3. Krumhansl Lombdhal paper
4. S. Yomosa, Phys. Rev. A**27**, 2120 (1983) and A**30**, 474 (1984)
5. S. Takeno and S. Homma, Prog. Theor. Phys. **77**, 548 (1987)
6. Chun-Ting Zhang, Phys. Rev. A**35**, 886 (1987)
7. E. W. Prohofsky, K. C. Lu, L. L. Van Zandt and B. E. Putnam, Physics Letters **70A**, 492 (1979)
8. Y. Gao and E. W. Prohofsky, J. Chem. Phys. **80**, 2242 (1984)
9. Y. Gao, K. V. Devi-Prasad and E. W. Prohofsky, J. Chem. Phys. **80**, 6291 (1984)
10. Y. Kim, K. V. Devi-Prasad and E. W. Prohofsky, Phys. Rev. B**32**, 5185 (1985)
11. D. J. Scalapino, M. Sears and R. A. Ferrel, Phys. Rev. B**6**, 3409 (1972)
12. J. A. Krumhansl and J. R. Schrieffer, Phys. Rev. B**11**, 3535 (1975)
13. P. M. Morse, Phys. Rev **1**, 57 (1929)
14. G. Dana Brabson, J. of Chemical Education **50**, 397 (1973)
15. D. J. Kaup and A.C. Newell, Phys. Rev. B**18**, 5162 (1978)
16. Tt. Pnevmatikos, N. Flytzanis, M. Remoissenet, Phys. Rev. **33**, 2308 (1986)
17. M. Remoissenet, Phy. Rev. B**33**, 2386 (1986)
18. J. L. Lebowitz, H. A. Rose and E. R. Speer, J. Stat. Phys. **50**, 657 (1988)

NON LINEAR EXCITATIONS IN CHAINS OF HYDROGEN-BONDED MOLECULES : INCOHERENT NEUTRON SCATTERING IN ACETANILIDE AND DERIVATIVES

M.Barthes,R.Almairac,J.L.Sauvajol,J.Moret
G.D.P.C. - U.S.T.L. - 34060 Montpellier cedex - France.
R.Currat,J.Dianoux
I.L.L. - 156X - 38042 Grenoble cedex - France .

Cristalline acetanilide (ACN) is an anharmonic solid characterised by soft hydrogen-bonded chains of molecules. Anomalous behaviours in the I.R. and Raman mode spectra were observed at low temperature . The different theoretical interpretations of these anomalies involve alternatively, Davydov-like solitons,localized "polaronic" modes,topological defects or Fermi resonance.

We present incoherent neutron scattering measurements of the vibrational density of states, using samples with different specific deuterations. These new data :

- allow to identify a very intense maximum ,that we assign to the NH out-of-plane bending mode.
- display the specific behaviour of the methyl torsionnal modes , unidentified in previous optical investigations .
- confirm our first inelastic neutron scattering investigations, indicating no obvious anomalies in the range of frequency of the acoustic phonons .
- show the existence of thermally activated quasi-elastic scattering above 100 K, assigned to the random diffusive motion of the methyl protons.

These new results are discussed in the light of the available theoretical models .

Considerable theoretical and experimental interest is currently being focused on acetanilide ($C_6H_5NHCOCH_3$, or ACN) which is a very interesting crystal displaying chains of hydrogen-bonded amide groups, with bond distances close to those found in polypeptides. Thus ACN is considered as a model system for some biological molecules and proteins; it is also a potential candidate for the observation of biological solitons.

The anomalous temperature dependence of some infrared and Raman modes of ACN [1-4] has received several theoretical interpretations :

It has been assigned to a trapped Davydov-soliton state[2,5], to vibron solitons [6], or to a vibronic analog of a small Holstein polaron [7,8,9], due to a coupling of the unperturbed amide-I mode with both acoustic and optic longitudinally polarized lattice modes. The occurence of topological solitons has also been suggested [10].

All these models have been questionned by another explana tion involving a temperature tuning of a Fermi resonance between the amide -I mode and a accidentally degenerated combination level [11].

To get more informations about the nature of the non-linear excitations in this material, we have done some neutron scattering and optical investigations on ACN and some deuterated derivatives. Brillouin, Raman scattering[12] and infrared data will be published elsewhere. Our coherent inelastic neutron scattering study [13] has revealed an important anharmonicity, but no observable anomaly on the dispersion and temperature dependence of the acoustic modes, on the fully deuterated derivative, in contradiction with some expectations. The sound velocities and the corresponding elastic constants, calculated from the slopes of the acoustic dispersion curves, reflect a strong anisotropy of the bonding between ACN molecules.

We present here new results of an incoherent neutron scattering (INS) study of four specifically deuterated species of ACN :

/1/ $C_6H_5NHCOCH_3$ (or fully hydrogenated ACN)

/2/ $C_6H_5NDCOCH_3$

/3/ $C_6D_5NHCOCD_3$

/4/ $C_6D_5NDCOCD_3$ (or fully deuterated ACN),

initiated with the aim of measuring the temperature dependence of the phonon density of states and the quasi-elastic scattering.

MATERIALS AND INSTRUMENTATION

The synthesis of powdered samples of the four derivatives were achieved in our laboratory. The rate of deuteration of the benzene ring and of the methyl group is better than 99% and stable with time. The rate of deuteration of the proton of the amide group (-NHCO-) has been controlled by IR absorption and Raman scattering and is better than 90%. Samples N°/2/ and /4/ were always handled in dry nitrogen or helium atmosphere, in order to prevent the H/D exchange on the H bond.

All measurements have been carried out at the Institut Laue-Langevin in Grenoble, with IN6 time-of-flight spectrometer. The resolution was about 50 μeV at zero energy. The experiments were performed at different temperatures between 1.5 and 300K.

After subtracting the sample holder contribution and Bragg scattering, the experimental scattering law $S(Q,\omega)$ is treated by standard procedures [14,15], to extract the $P(\alpha,\beta)$ function (with $\alpha = \hbar^2 Q^2/2MkT$ and $\beta = \hbar\omega/kT = \varepsilon/kT$) and then the generalized frequency distribution, or vibrational density of states $G(\varepsilon)$. The data were corrected from the multiphonon scattering contribution with the help of an algorithm developped by A.J. Dianoux

RESULTS

a) The generalized frequency distributions, in the highest attainable energy range, are displayed in fig 1. The comparison of data from the four species indicates that a strong and very wide maximum exists at about 92-95 meV ($\sim$ 750-770 cm-1), in sample /1/ and /3/ only.

This fact, and the absence of the same maximum in the two other samples, unambiguously shows that the associated vibration involves the proton of the hydrogen-bond.

This maximum appears as the main difference between /1/ and /2/ (and between /3/ and /4/) INS spectra. (Subsequent investigations of I.R. spectra in the same range of frequencies have confirmed the presence of an absorption band at 750-770cm-1 in samples /1/ and /3/ only. An anomalous temperature dependence of its intensity has been observed [16], and at low temperature, the appearance of several multiplets .) It is therefore possible to assign this maximum to the N-H out-of-plane bending mode whose frequency is expected in this energy region, but never before identified. Fig 1b shows the temperature evolution of this wide maximum for sample /3/.

b) Fig 2 gives the experimental values of the phonon density of states at different temperatures, for the four selectively deuterated samples, in a smaller range of energy (0-40 meV,or 320 cm-1). Mainly, normal external modes, librations and acoustic phonons contribute to this part of the frequency distribution. However a detailed assignment of all the maxima is beyond the scope of the present paper. Only the anomalous features will be analysed, and their possible implication in a non -linear

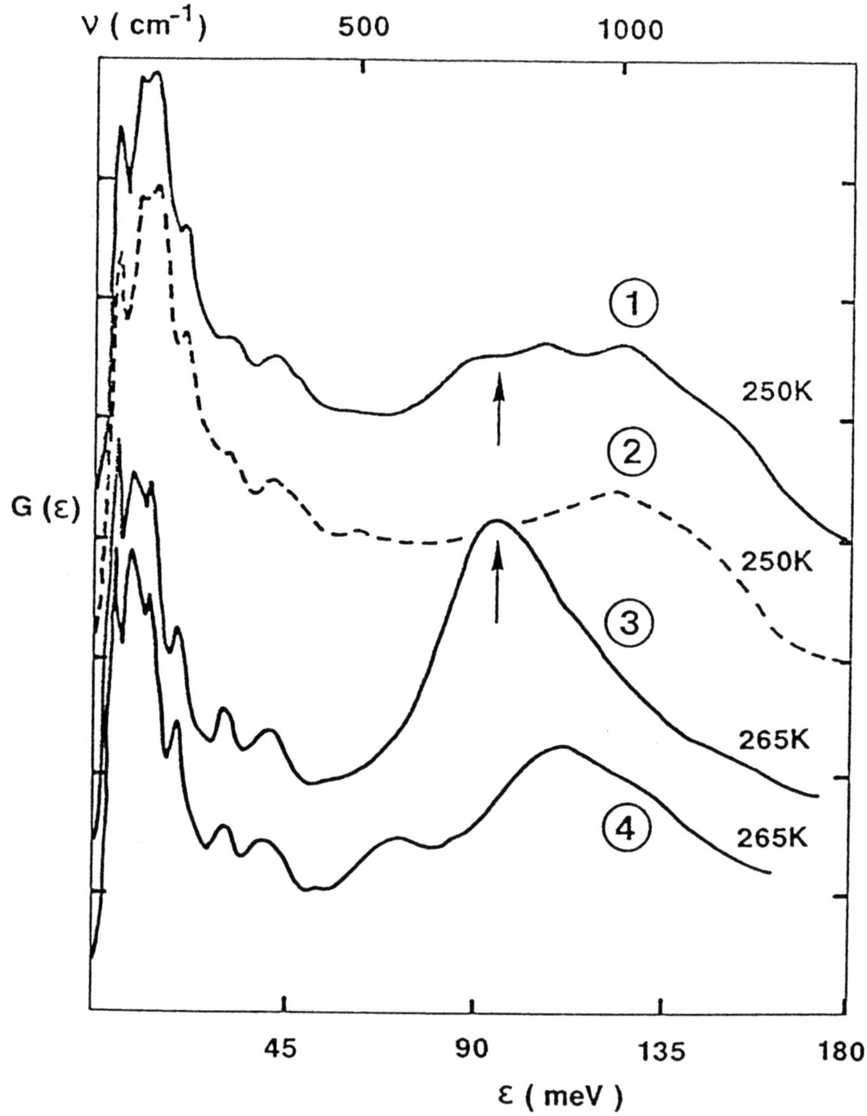

Fig. 1 - Generalized Frequency Distribution for the four ACN derivatives. The arrows indicate the maximum at 92 - 95 meV in samples /1/ and /3/.

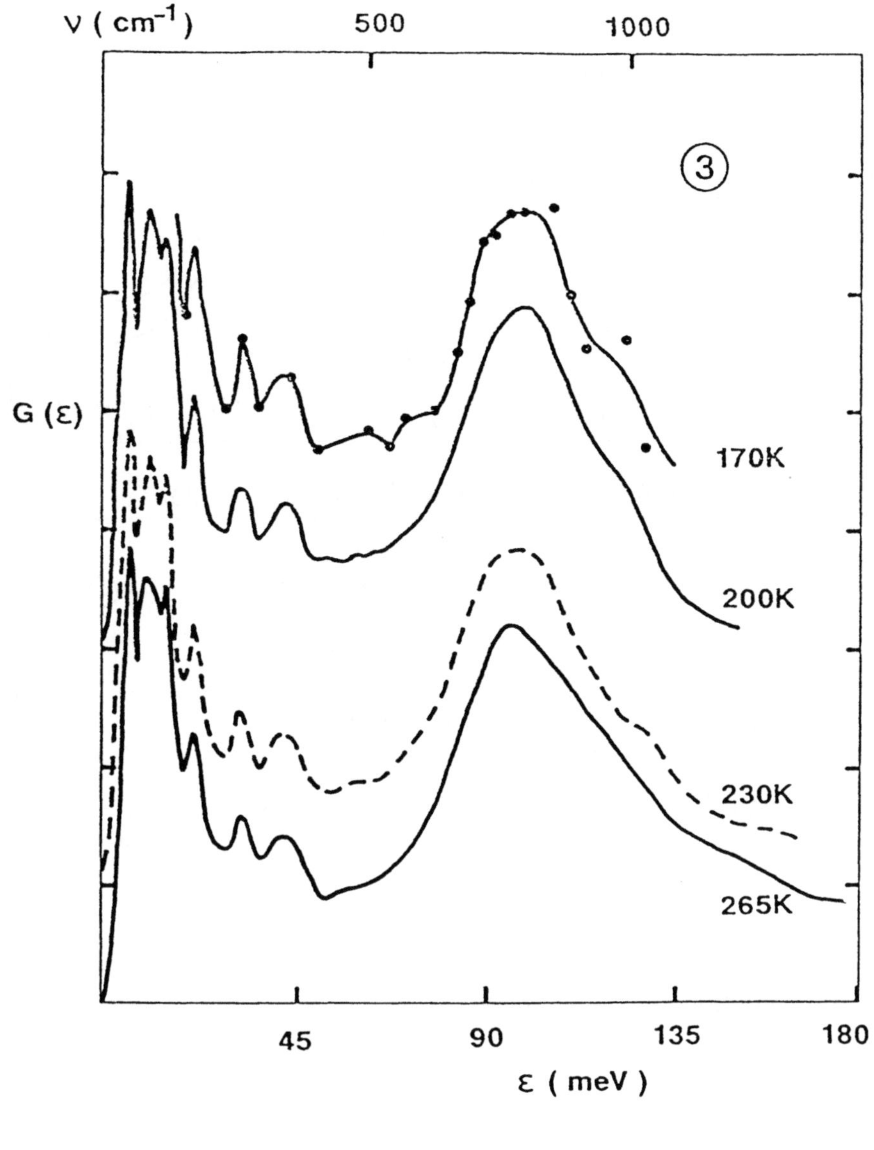

Fig. 1b - Temperature dependence of the wide maximum in G(ε) at 92 - 95 meV.

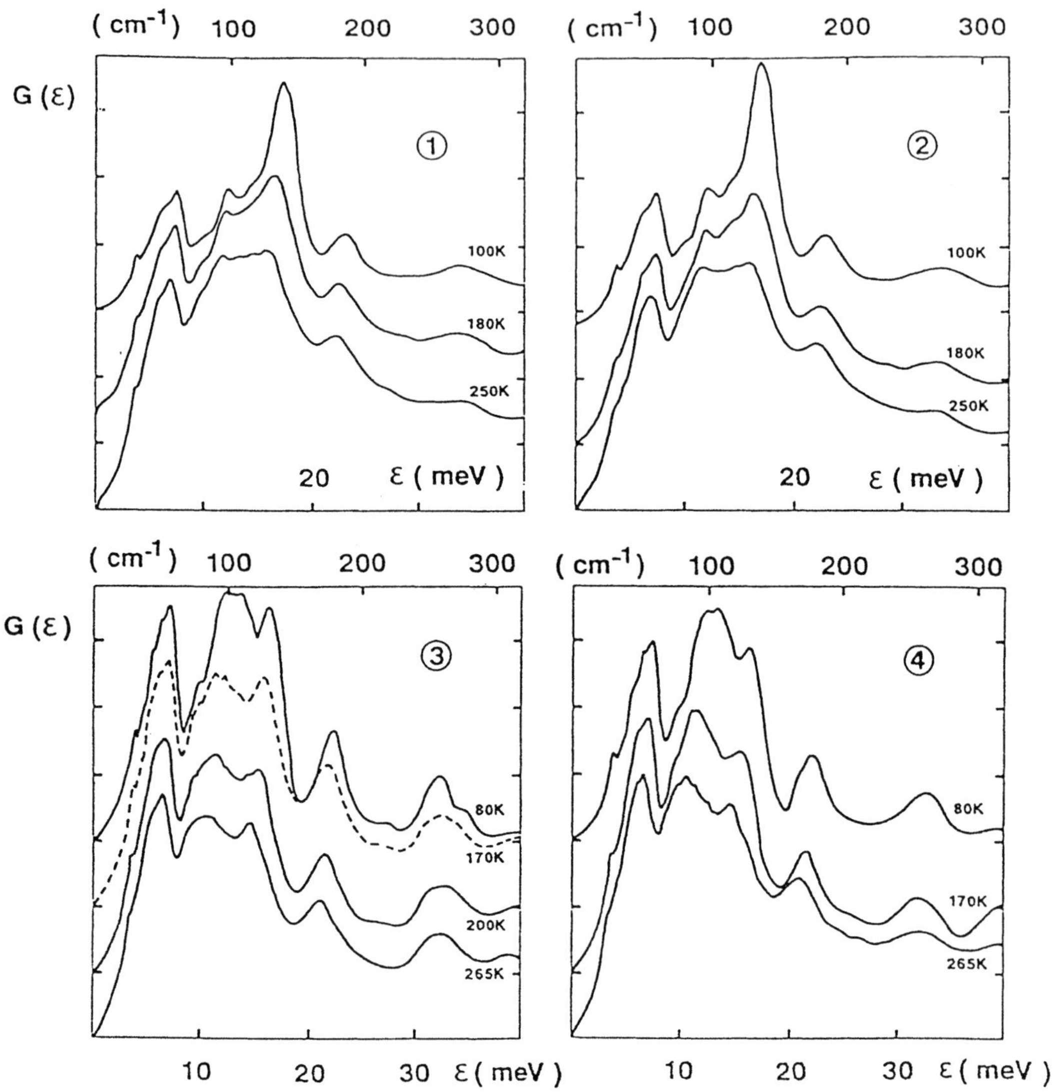

Fig. 2 - Vibrational Density of States of the four ACN specimens, at different temperatures ,between 0 and 40 meV.

coupling mechanism with higher frequency modes will be considered.

It is clear from fig 2 that, among the strong maxima, the peak situated at ~ 16-17 meV in samples /1/ and /2/ - and the maximum at ~ 10-12 meV in samples /3/ and /4/ - exhibit a different behaviour from others :

- larger energy shift at low temperature
- faster broadening with temperature than the other peaks in the phonon density of states
- significant intensity decrease by deuteration
- very large isotopic shift :

$$\Delta = \frac{\varepsilon_{/1/}}{\varepsilon_{/3/}} = \frac{\varepsilon_{/2/}}{\varepsilon_{/4/}} \simeq 1.3$$

These properties are precisely the same as the special characteristics of the methyl torsional modes [17,18]. In particular, the observed isotopic ratio, Δ , may be identified as follows :

$$\Delta \# \left(\frac{\theta_{CH_3}}{\theta_{CD_3}} \right)^{1/2}$$

where θ_i is the momentum of inertia of the methyl (and CD3) group. So, the energy of this motion scales with the mass of the hydrogen atom (and not with the mass of the whole molecule, or parts of it).

Moreover, the librationnal excitations have usually little dispersion, and therefore lead to a pronounced peak in the inelastic neutron scattering spectra. The strong intensity of the methyl torsionnal peaks is also a consequence of the large amplitude motion of the protons. The modes under consideration in ACN display also a significant intensity. It is thus reasonable to assign these peaks to the torsional transitions of the methyl, and CD3 groups, in consistency with their specific behaviour. It is obvious from the close similarities of their properties in samples /1/ and /2/ in one hand, in /3/ and /4/ in the other hand that they could not result from motions involving the amide-group (otherwise,there would be similarities between /1/ and /3/ on one hand,and between /2/ and /4/,on the other hand) .

Librations involve very small dipole moment change or polarizability ,and therefore are infrared and Raman inactive or weak. However in some cases, they may appear. The criterium of strong relaxation effect with increasing temperature and of large isotopic shift may be retained for the torsional band assignment. In the case of ACN, no unambiguous evidence of their activity in Raman scattering has been found[12], though the existence of a band at about the same frequency,and exhibiting a strong energy shift versus temperature,has been observed in a fully deuterated single crystal.

From the low energy part of the frequency distributions, it is also possible to deduce another physical information : no obvious anomaly is observable in the range of acoustic phonons, for none of the four species, thus confirming the previously acquiered data of our coherent inelastic scattering study [13].

C) The quasi-elastic scattering (QES) and its temperature dependence has also been studied. Samples /1/ and /2/ exhibit the same broadening and the same temperature evolution (fig.3). A weaker quasi-elastic broadening is also observed in samples/3/ and /4/, with again, the same temperature dependence for both. In the four species, the quasi-elastic broadening is thermally activated. It disappears below 100K in samples /3/ and /4/. Fig 4 shows a plot of the temperature dependence of the full width at half-maximum, Γ , estimated at each temperature after correction for the elastic peak [15], and fitting the QES with a lorentzian.

As in the preceding case, the similarities between /1/ and /2/ QES in one hand, the similarities between /3/ and /4/ QES in the other, demonstrate that the broadening may be attributed to some diffusive motion involving either the methyl group or the phenyl ring, but in no case the amide-group (because the activation energy is different for /1/ and /3/).

Though the accuracy on the Γ estimation is not very large, fig 4 however shows that the QES temperature dependence is consistent with a thermally-activated process, with an activation energy E_o of about 500K (43 meV, or ~ 350 cm-1) for samples /1/ and /2/, and about 630K (54 meV, or ~ 440 cm-1) for /3/ and /4/ .

As observed in other methylated compounds [15,18,19] above 50K, the quasi-elastic broadening reflects the thermally activated hopping of protons of the methyl group between adjacent wells of the rotational potential. The activation energy gives a measurement of the hindering barrier of the methyl rotation. The most commonly used rotational model for this motion involves random reorientations over the barrier by instantaneous jumps of 120° around the threefold axis. The time τ between jumps is related to the temperature by an Arrhenius law.

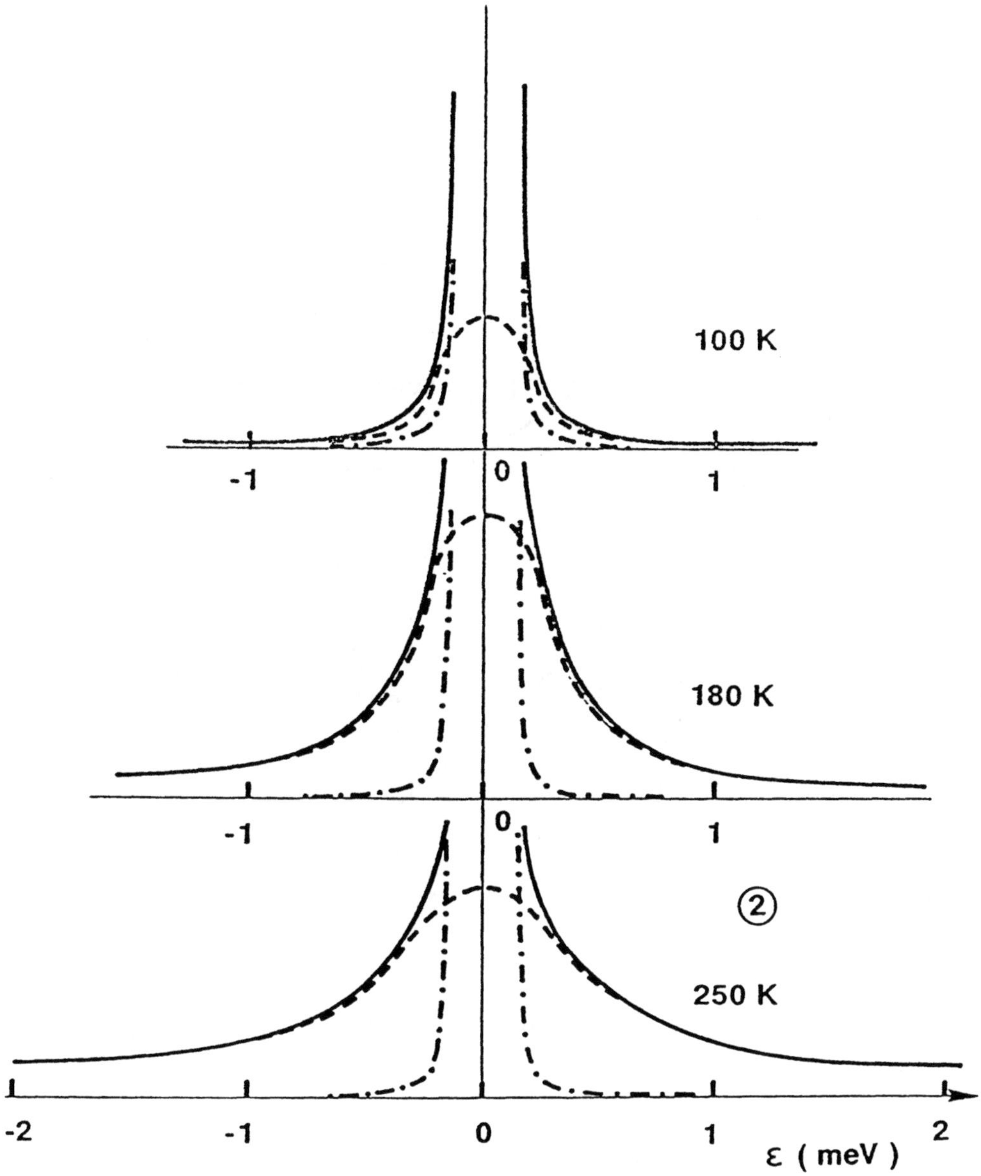

Fig. 3 - Quasi-elastic broadening measured in sample /2/ (dashed line). The Q.E.S. of sample /1/ is the same, and may be exactly superimposed (not shown). The solid line is $S(Q, \omega)$; the dash-dotted line represents the elastic contribution, measured by the scattering of vanadium .

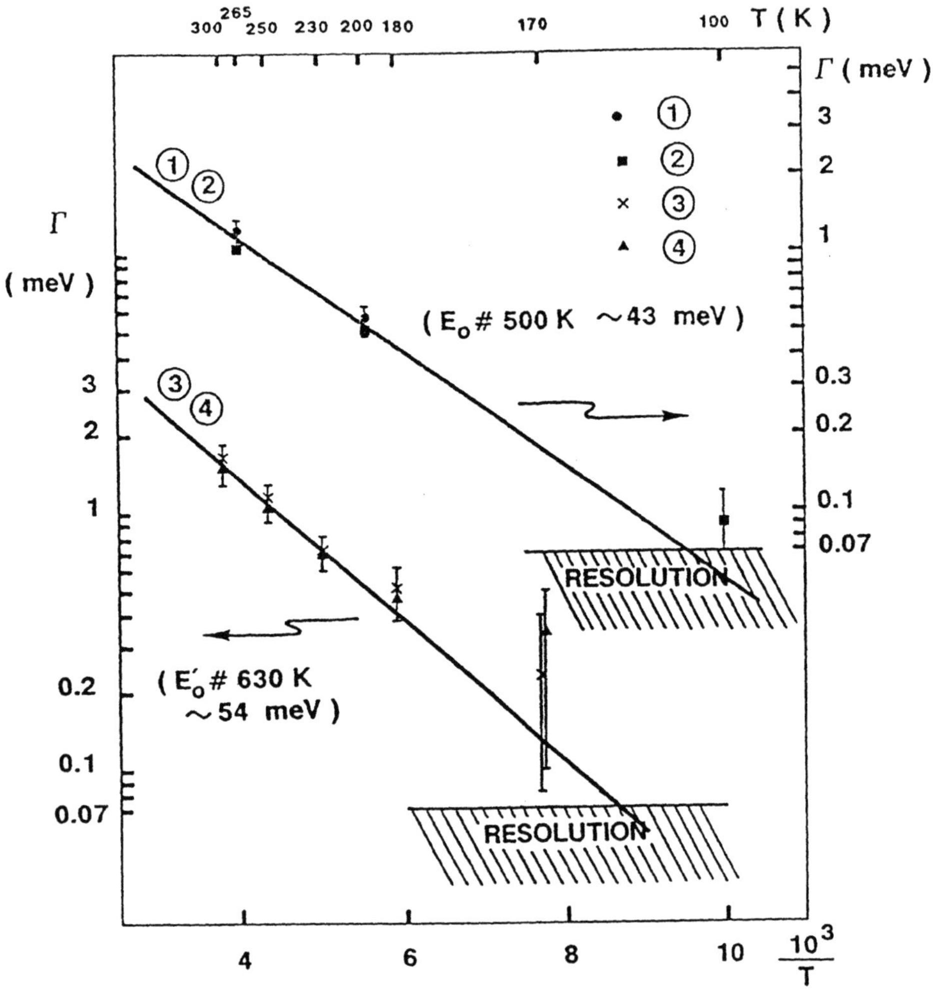

Fig. 4 - Temperature dependence of the full width at half maximum of the quasi elastic scattering,for the four samples.

If we assume that the observed QES in ACN has the same origin (and to get the unambiguous proof of this hypothesis,it will be necessary to do the same measurements with other selectively deuterated ACN : /5/ C6H5NHCOCD3 and /6/ C6D5NHCOCH3) we may estimate the time between two random jumps of the proton ($\tau \sim 10$ picoseconds). The high activation energies indicate a high hindering barrier and as a consequence, the tunnel splitting for the methyl protons is presumably very small, below the resolution of back scattering neutron spectrometers.

It is important to note that the activation energy E_0 is of the same order of magnitude as those deduced from the temperature dependence of the relaxation time T_1 of protons in NMR measurements.[19]

DISCUSSION

In these INS measurements,we have observed :

- a thermally activated quasi-elastic broadening above 100 K, identified as diffusive motions of the protons of the methyl group.The measured activation energy is related to the height of the hindering barrier of the methyl rotation .The large values of E_0 (40 to 50 meV) in the present case,indicate that at low temperature,the quantum tunneling of the methyl protons has a very low energy .

- the methyl torsional transitions,at 16-17 meV in /1/ and /2/ , at 10-12 meV in /3/ and /4/,unambiguously assigned,owing to their specific temperature dependence and isotopic shift .

- no obvious anomaly , in the energy range of the acoustic modes,for none of the four species, confirming the data of our previous inelastic scattering study on a fully deuterated single crystal.

- a new,very broad band,at 92-95 meV,only found in sample /1/ and /3/,assigned to a movement of the proton of the hydrogen bond,namely the N-H out-of-plane bending mode.This band is unusually wide (> 40 meV, or 300 cm-1) . Such an effect has been already observed in the case of KH_2PO_4 [20] .The existence of very broad hydrogen modes that appear prominently only in neutron scattering supports the picture that there exists an asymetric double minimum potential well where the proton of the hydrogen bond tunnels .However, in the case of ACN there is ,as yet, no direct experimental indication of quantum tunneling of this proton .High resolution neutron scattering investigations (below 50 μeV)have to be done to check this point .

The relationship of this mode at 92-95 meV with the other optical anomalies in ACN will be discussed in more details elsewhere[16]. However, it is possible to mention that, considering its anomalous intensity and energy dependence versus temperature, the question of the coupling of this mode with the methyl torsional transitions is open. An effect of the methyl deuteration on the intensity of the components of the γ- NH mode has recently been observed in the N-methylacetamide, in INS spectra [21].

Regarding the available theoretical models of the optical anomalies in ACN, these new results :

- confirm that a Davydov-like model involving a nonlinear coupling of the C=O stretching mode to some acoustic phonons is unlikely. These results are consistent, on this point, with the theory of localisation of the C=O vibrational energy involving polarons [9]. No experimental fact, as yet, contradicts the possibility of a nonlinear coupling between high energy internal modes and low lying optical phonons or methyl librations.

- bring a new argument against the explanation of anomalies by means of a Fermi resonance, in consistency with Raman experiments under pressure[22].

Indeed, if it is assumed that the anomalous infra-red band [16]at 92-95 meV results from a temperature tuning of an accidental Fermi resonance, it would probably not give rise to a strong maximum in the vibrational frequency distribution , specially at room temperature. It would also appear at slightly different frequencies in sample /1/ and /3/.

- will be more accurately analyzed by comparison with the corresponding infra-red data. Measurements under pressure ,and new selective deuterations will be used to characterize the coupling between the methyl librations and the out-of-plane NH bending.

REFERENCES

1 - G.CARERI,U.BUONTEMPO,F.CARTA,E.GRATTON and A.C.SCOTT - Phys.Rev. Lett. 51,304,1983
2 - G.CARERI,U.BUONTEMPO,F.GALLUZZI,A.C.SCOTT,E.GRATTON and E.SHYAMSUNDER - Phys.Rev.B 30,4689,1984.
3 - A.C.SCOTT,E.GRATTON,E.SHYAMSUNDER and G.CARERI - Phys.Rev.B 32, 5551,1985.
4 - G.GARERI, E.GRATTON AND E.SHYAMSUNDER - Phys.Rev.A 37, 4048, 1988.
5 - J.C.EILBECK, P.S.LOMDAHL AND A.C.SCOTT Phys.Rev. B 30, 4703, 1984.
6 - S. TAKENO - Prog.Theor.Phys. 75,1,1986.
7 - D.M. ALEXANDER -Phys.Rev.Lett.54,138,1985
8 - D.M.ALEXANDER and J.A.KRUMHANSL - Phys.Rev.B 33,7172,1986.
9 - A.C. SCOTT,I. J.BIGIO and C.T. JOHNSTON - Phys.Rev.B 39,15 june 1989
10 - G. BLANCHET and C. FINCHER - Phys.Rev.Lett. 54, 1310,01985.
11 - C.T.JOHNSTON, B.J.SWANSON - Chem.Phys.Lett. 114, 547,1985.
12 - J.L.SAUVAJOL, R.ALMAIRAC, M.BARTHES ,J.MORET and J.L.RIBET - J.of Raman spectroscopy , (in press)
13 - M.BARTHES ,R.ALMAIRAC, J.L.SAUVAJOL,R.CURRAT,J.MORET and J.L.RIBET - Europhysics Letters 7,55, 1988
14 - P.EGELSTAFF and P.SCHOFIELD - Nucl. Sci. Eng. ,12,260,1962.
15 - A.J. DIANOUX - "The Time Domain in Surface and Structural Dynamics" Ed. G.LONG and F.GRANDJEAN - Kluwer Acad.Publ. 1988
16- M.BARTHES et al (to be published). (Preliminary report on these IR measurements have been given in "Molecular Aspects of Systems of Biological Interest" J. of Molecular Liquids ,special issue) J.of Mol. Liquids - Special Issue -1989)
17 - M.PRAGER, J.STANISLAWSKI and W.HAUSLER - J.Chem. Phys. 89, 2563, 1987
18 - D.CAVAGNAT J.Chimie Phys.82, 239,1985
19 - G.GUSMAN , F.MASIN and P. BROEKAERT - 6th Int. Workshop on Nonlinear Coherent Struct. in Physics,Mechanics and Biolog. Systems - 1989 - Montpellier .
20 - "Molecular Spectroscopy with Neutrons" - J.BOUTIN and M.YIP , MIT Press, 1968.
21 - F.FILLAUX and J.TOMKINSON - Annual ISIS Report - 3,A140,1987-1988
22 - C.JOHNSTON ,unpubl. results - J.L. SAUVAJOL ,unpubl. results.

NMR STUDY OF NONLINEAR EXCITATIONS IN PREBIOLOGICAL SYSTEMS

F. Masin *, G. Gusman *, P. Broekaert + and C. Maerschalk x

Physique des Solides *, Résonance Magnétique + et Institut Bordet x

Université Libre de Bruxelles, Bd. du Triomphe, 1050 Bruxelles (Belgium)

1. Introduction

The study of the dynamical properties of crystalline acetanilide, (CH_3 CONH $C_6H_5)_n$ or shortly ACN, is an object of active work in physics for it provides an interesting model for a system which looks like a polypeptide, i. e. a system of biological interest. Indeed its structure has the form of one-dimensional chains of hydrogen bonded amide groups whose bond lengths have values close to those found in actual polypeptides.

During these last years, a lot of experimental studies have been done mainly by optical methods, namely infra-red and Raman spectroscopies and anomalous absorption properties have been reported. Among these works, Careri et al. (1) pointed out the occurrence of one new absorption band at 1650 cm^{-1} at low temperatures. These authors assigned this band to a new amide-I component, C=O stretching, red-shifted by 15 cm^{-1} from the usual 1665 cm^{-1} band. On the other hand, Blanchet and Fincher (2) discovered several new bands in the middle and far infra-red regions whose intensities are also temperature dependent, moreover multiple side-bands appear which are regularly spaced. They conclude that N-H stretching plays a main part in this behaviour and explain their results using the idea of zero-phonon lines transitions between two vibrational states, as a result they suggest that the one-dimensional lattice is twofold degenerate. They also report that this side-band structure already exists at room temperature (and even at higher temperatures up to melting temperature). We shall come back to this point later.

Anyway the main attribute of ACN is its one-dimensional character which is well-known today to enhance the effects of nonlinear excitations in solids. Several approaches have been proposed. Most of them use the concept of the so-called "Davydov soliton" originally introduced to explain the possibility of transporting energy on α-helix proteins. This idea has been used namely by Scott et al.[(3)] whose conclusions are that the existence of a strongly localized soliton is in agreement with experimental data. We shall not develop this theoretical aspect here but remind that soliton arise from a coupling between inner molecular vibrational modes of a coupled oscillator system , and the crystal itself through lattice-vibrations (the phonons playing the role of a heat bath) which in turn modify the oscillators energy.

Two main questions, we feel, remain unanswered :

1. What could be the influence of the experimental method itself on the existence of the soliton? Is this excitation induced by the laser irradiation?

2. Are they Davydov solitons (Carreri et al.) or are they topological defects (Blanchet et al.) ?

To try to give an answer at least to the first question, we have applied Nuclear Magnetic Resonance techniques (NMR) which due to the low energies involved, do not create any lattice excitations. This way we make use of the previous experience some of us have acquired during a study of solitons present in polyacetylene [(4)]. Although the nonlinear electron-lattice coupling is not supposed to work here as it does in polyacetylene : we do not expect to observe paramagnetic solitons, but we could reasonably assume that spin-lattice relaxation of the proton spin system after a radiofrequency excitation would be influenced by localized molecular vibrations if present in the system.

2. Experimental technique

We performed the experiments partim on a 200 MHz Brüker MSL spectrometer partim on a 42 MHz Brüker SXP. In both cases, our measurements range in a temperature domain from 4 K to 300 K. They consist in the recording and analysis of the ^1H proton and ^2H deuteron absorption spectra and in the determination of the evolution of the proton Zeeman spin-lattice relaxation times, T_{1Z}, by the use of the inversion-recovery pulse technique.The chemical synthesis of our crystals of ACN has been carefully undertaken to avoid any residual impurities nor any presence of water. Eventually in order to obtain informations separately on the influence of the position of the protons in the molecule on the spin-lattice relaxation, we have prepared, by isotopic substitution, various samples of ACN such that the three kinds of proton positions can be isolated and studied separately : the protons in the ring C_6H_5, the hydrogen-bonded protons in the chain, and the protons in the group CH_3. these samples are called H-H-H, D-H-D, H-H-D and D-D-H respectively.

3. **Experimental results**

In the temperature range extending from 300 K to 160 K, the NMR proton absorption spectrum looks like a "superposition" of two different lineshape components : a wide one and a very narrow one, with nearly same symmetries and centres. To illustrate this the lineshape at 166 K, illustrated in fig. 1, clearly shows the presence of the liquid-like component.

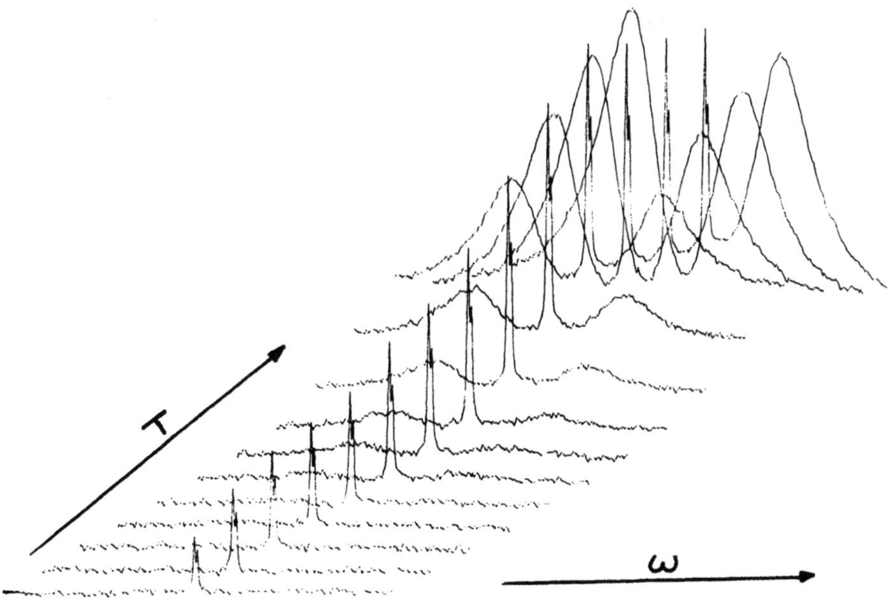

Figure 1 : Proton line shapes at 200 Mhz and at 166 K .

But we observe that the narrow line disappears below 160 K and this transition takes place on a few degrees. On the contrary, the absorption spectrum as seen at 144 K in fig. 2, does no more exhibit this narrow component.

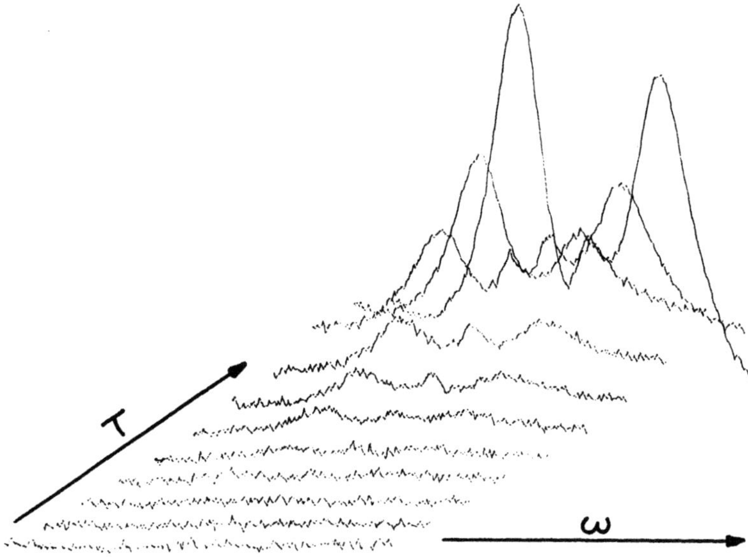

Figure 2 : Proton line shapes at 200 Mhz and at 144 K .

This would indicate that from room temperature to 160 K two kinds of protons or more precisely two different environments coexist in the system, some of them feel a local magnetic field due to the neighbouring nuclei, hence the broad line, as usual in solids. The narrow line, as usually met in liquids is concerned with spins whose interactions between neighbours are averaged by rapid motion. This motion stops suddenly at 160 K as if a phase transition has occurred.

First, we have studied **the broad line** to obtain the associated spin-lattice relaxation rate T_{1Z}^{-1} versus temperature. As shown in fig. 3, the shape of its temperature dependence reminds us of a typical thermal activated motion. It can be fitted by the following expression:

$$T_{1Z}^{-1} = \frac{2}{3} \gamma^2 h_0^2 \frac{\tau}{1 + \omega^2 \tau^2}$$

where γ is the magnetogyric ratio of the considered spin.

the local magnetic field value is $h_0 = 4\ 10^{-4}$ tesla

the characteristic time is $\tau = \tau_0 \exp\left(\frac{E}{T}\right)$

with $\tau_0 = 2\ 10^{-3}$ s ; E = 560 K

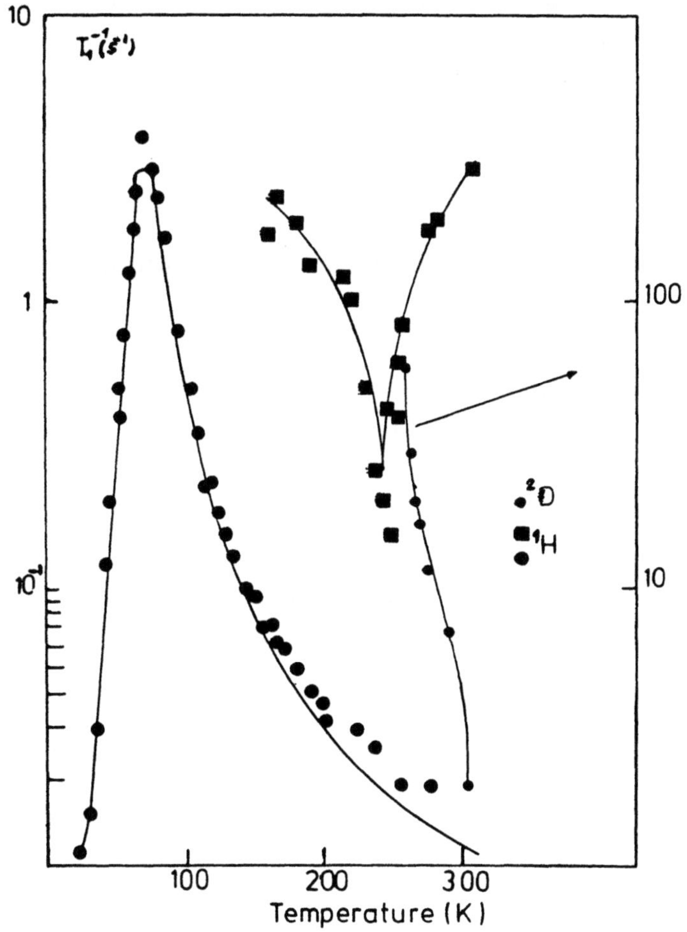

Figure 3 : Spin-lattice relaxation rate of ^1H and ^2H in ACN at 200 Mhz versus temperature (●: narrow line; ■: broad line)

On the other hand, the T_{1Z} associated to the narrow line depends on temperature in a very different way. Indeed it seems to follow a dependence usually encountered when ferroelectric transitions occur (in KH_2PO_4 e. g., thoroughly studied by Blinc(5)). We have thus tentatively fitted our data (at 200 MHz) by the following expression (see fig. 2):

$$T_{1Z}{}^{-1} \propto \sqrt{T - T_C}$$

$$\text{with} \quad T_C = 240 \text{ K}$$

It is interesting to remark that when we substitute ^2H (or D) to all ^1H in the sample the inverse behaviour is observed at a slightly higher temperature exactly like what happens in KD_2 PO_4 (Blinc(5)). In this latter system a soft mode is known to appear and accompanies a structural phase transition. It can be added that the narrow line suddenly spreads at 160 K exactly at the same temperature where an inflexion point is observed in the temperature dependence of the thermal conductivity (Carreri et al.).

To be able to compare both behaviours, let us insist on the large value of the spin-lattice relaxation time of the broad component which is of the same order of magnitude as T_{1Z} usually obtained in insulators.

To be precise, at room temperature and at 42 MHz, we obtain :

broad line : T_{1Z} = 106 s

narrow line : T_{1Z} = 50 ms

In order to attempt to identify those protons responsible of the narrow line, we have looked at various samples obtained by selective substitution of the three different kinds of ^1H by ^2H. In fig.3 various lineshapes are shown measured in these samples at room temperature and at Larmor frequency 42 MHz.

We also evaluate the ratio of the amplitude of the narrow line to the broad line, A_n/A_b, for various samples.

We have : in D-H-D	A_n/A_b = 0,169
in H-H-D	A_n/A_b = 0,169
in D-D-H	A_n/A_b = 0,158 and 0,185 (two different samples)

Obviously, this indicates that all types of protons do participate in the rapid motion. Moreover we have tried to make dynamic transfer from the narrow line to the wide one at 300 K and at 42 MHz without any success showing the absence of any efficient coupling between both.

In addition, to investigate the existence of much slower processes than those involved in the Zeeman spin-relaxation times, we initiated a study of the dipolar spin-lattice relaxation phenomenon. We have prepared a certain amount of dipolar order with the standard two pulses method (6) applied to the spins concerned with the broad line, and we have measured the T_{1D} (dipolar spin-lattice relaxation time). A first set of measurements on the protons, at room temperature and Larmor frequency 42 MHz gives a T_{1D} = 1,65 s. This value, two orders of magnitude shorter than the corresponding T1Z indicates the existence of a more active mechanism of relaxation in the range of frequencies lower than 1 MHz.

4. Concluding remarks

Our data show that, from a NMR point of view, ACN seems to have two subsystems of protons. Some do not look very different from those encountered in usual solids : the associated lineshape is broad and their spin-lattice relaxation time T_{1Z} is relatively long. This excludes the presence of electrons or relatively abundant paramagnetic impurities. The others obviously are subjected to a completely different influence : their lineshape is very narrow and their T_{1Z} is short as encountered when rapid motion exists. These results suggest that some NMR spin-lattice relaxation processes are induced by a localized and immobile excitation, its localized spatial character being in favour of its non-linear origin.

Acknowledgment

The authors are indebted to the BANQUE NATIONALE DE BELGIQUE and to the FONDS NATIONAL DE LA RECHERCHE SCIENTIFIQUE for their financial support during this work.

References

1. G. Careri, U. Buontempo, F. Carta; E. Gratton, A. C. Scott, Phys. Rev. Lett. **51**, 304 (1983).

2. Graciela B. Blanchet, C. R. Fincher Jr.,Phys. Rev. Lett. **54**, 1310 (1985).

3. A. C. Scott, E. Gratton, E. Shyamsunder, G. Careri, Phys. Rev.**B 32**, 5551 (1985).

4. F. Masin, G. Gusman, Phys. Rev. **B 36**, 2153 (1987).

5. R. Blinc, J. Slak, F. C. Sa Barreto, A. S. T. Pires, Phys. Rev. Lett. **42**, 1000 (1979).

6. J. Jeener, P. Broekaert, Phys. Rev. **157**, 232 (1967).

FINITE TEMPERATURE BEHAVIOUR OF THE EXCITON BOUND STATES IN THE COMPRESSIBLE MOLECULAR CHAIN

D.V. Kapor, M.J. Škrinjar and S.D. Stojanović
Institute of Physics, Faculty of Sciences, 21000 Novi Sad, Yugoslavia

Introduction

There exists an extensive literature on the theory of Davydov solitons [1-8] concerning theoretical as well as experimental aspects of the subject. The most important reason for studying the nonlinear excitations in a molecular chain is probably the fact that this can give a qualitative description of what is going on in some living tissues. There is still a lot of controversy on the validity of the model, and we wish to contribute to the subject by proposing an alternate model.

Our approach is based on the fact that Frenkel excitons are not really bosons (which is the common assumption), but they have more complicated nature. They are so called Paulions [9]. To be more precise, we shall consider a molecular chain with one molecule per site, and lattice constant a. We accept two-level scheme, which implies that all higher energy levels are sufficiently separated from the first excited state, so that they can be neglected. The separation from the ground state is Δ. Resonant energy of dipole-dipole interaction between the nearest neighbours is I.

The system is described by the Hamiltonian:

$$\hat{H} = \hat{H}_{ex} + \hat{H}_{ph} + \hat{H}_{int} \tag{1}$$

$$\hat{H}_{ex} = \Delta \sum_n \hat{P}^+_n \hat{P}_n - I \sum_n (\hat{P}^+_n \hat{P}_{n+1} + \hat{P}^+_{n+1} \hat{P}_n) \tag{2}$$

Operator $\hat{P}^+_n$ creates an exciton in the first excited state at the site n. The nature of the operators can be seen from their commutation relations:

$$[\hat{P}_n, \hat{P}_m] = 0 \qquad \hat{P}^2_n = \hat{P}^{+2}_n = 0$$

$$[\hat{P}_n, \hat{P}^+_m] = \delta_{n,m}(1 - 2\hat{P}^+_n \hat{P}_n) \tag{3}$$

It is quite common to use also the operators of the pseudo-spin 1/2, assigned in the following way:

$$\hat{P}^+_n \to \hat{S}_n \qquad \hat{P}_n \to \hat{S}^+_n, \quad \frac{1}{2} - \hat{P}^+_n \hat{P}_n \to \hat{S}^z_n$$

We have to take into account also lattice vibrations and we shall study only the longitudinal acoustic phonons

$$\hat{H}_{ph} = \sum_q \hbar\omega_q \hat{b}^+_q \hat{b}_q \qquad \omega_q^2 = 4\omega_o^2 \sin \frac{aq}{2} \qquad \omega_o = \frac{v_{sound}}{a} \tag{4}$$

$\hat{b}^*_q$ creates a phonon with the wave-vetor q.

There exist proposed several forms of the exciton-phonon coupling, and we use [1]:

$$\hat{H}_{int} = \sum_{n,q} \chi_n(q)\, \hat{P}^+_n \hat{P}_n (\hat{b}^+_{-q} + \hat{b}_q) \tag{5}$$

$$\chi_n(q) = 2i\chi \left(\frac{\hbar}{2MN\omega_q}\right)^{1/2} \sin (q\ a)\, e^{iqna} \tag{6}$$

χ is the exciton-phonon coupling constant and M is the mass of the molecule. Spring constant $\mathcal{K}$ is defined through $Mv_s^2 = \mathcal{K}a^2$.

Effective Hamiltonian

It is not possible to decouple completely exciton and phonons, but we shall try to do it approximately by performing the unitary transformation:

$$\hat{H}_{eq} = e^{i\hat{S}}\, \hat{H}\, e^{-i\hat{S}} \qquad |\psi\rangle_{eq} = e^{i\hat{S}}|\psi\rangle \tag{7}$$

with

$$\hat{S} = \sum_{q,m} F_m(q) \left(\frac{1}{2} - \hat{S}^z_m\right)(\hat{b}_q - \hat{b}^+_{-q}) \equiv \sum_{q,m} F_m(q)\, \hat{P}^+_m \hat{P}_m (\hat{b}_q - \hat{b}^+_{-q}) \tag{8}$$

The following choice

$$F_m(q) = -\frac{i\chi_m(q)}{\hbar\omega_q} = \frac{2\chi}{\hbar\omega_q}\left(\frac{\hbar}{2MN\omega_q}\right)^{1/2} \sin(q\ a)\, e^{iqma} \tag{9}$$

leads to an equivalent Hamiltonian of the form:

$$\hat{H}_{eq} = \left(\Delta - \frac{2\chi^2}{M\omega_o^2}\right) \sum_n \left(\frac{1}{2} - \hat{S}^z_n\right) - I \sum_n \left(\hat{T}_n \hat{S}^-_n S^+_{n+1} + h.c.\right) -$$

$$- \frac{\chi^2}{M\omega_o^2} \sum_n \left(\hat{S}^z_n \hat{S}^z_{n+1} - \frac{1}{4}\right) + \sum_q \hbar\omega_q \hat{b}^+_q \hat{b}_q \tag{10}$$

where

$$\hat{T}_n = \exp\left(\sum_q \gamma_n(q)(\hat{b}_q - \hat{b}^+_{-q})\right) \tag{11.a}$$

$$\gamma_n(q) = \gamma(q)\, e^{iqna} = iF_n(q)\,(e^{iqa} - 1) \tag{11.b}$$

One can notice that for I=0, phonons are decoupled from "renormalized" excitons whose exciton-exciton interaction is induced by exciton-phonon interaction.

If we assume that exciton radiative lifetime is much larger than the relaxation time in the exciton-phonon system, (which is fulfilled in typical molecular crystals [9]) we can suppose that excitons and phonons are in the thermal equilibrium at the temperature of the crystal $T = \Theta/k_B$.

The effective exciton Hamiltonian corresponding to the given temperature T will be obtained by averaging $\hat{H}_{eq}$ over the phonon ensamble, taking into account that phonons are "nearly decoupled" from "renormalized" excitons.

Introducing

$$\hat{\rho}_{ph} = \frac{e^{-\hat{H}_{ph}/\Theta}}{\mathrm{Tr}\left(e^{-H_{ph}/\Theta}\right)} \tag{10}$$

we define

$$\hat{H}_{eff} = \mathrm{Tr}\left[\hat{\rho}_{ph}(\hat{H}_{eq} - \hat{H}_{ph})\right] = \sum_n \left(\Delta - \frac{2\chi^2}{M\omega_0^2}\right)\left(\frac{1}{2} - \hat{S}^z_n\right) -$$

$$- I\, e^{-W(\Theta)} \sum_n \left(\hat{S}^-_n \hat{S}^+_{n+1} + \hat{S}^-_{n+1}\hat{S}^+_n\right) - \frac{\chi^2}{M\omega_0^2}\sum_n \left(\hat{S}^z_n \hat{S}^z_{n+1} - \frac{1}{4}\right) \tag{11}$$

where

$$e^{-W(\Theta)} = \mathrm{Tr}(\hat{\rho}_{ph}\hat{T}_n) \tag{12}$$

$$W(\Theta) = \frac{1}{2}\sum_q |\gamma_q|^2 \coth\frac{\hbar\omega_q}{2\Theta} = \frac{2\chi^2}{\hbar MN}\sum_q \frac{\sin^2 qa(1 - \cos qa)}{\omega_q^3}\coth\frac{\hbar\omega_q}{2\Theta} \tag{13}$$

The system described by $\tilde{H}_{eff}$ (11) can be treated as an anisotropic ferromagnetic chain using Bethe-Ansatz [10,11] in order to obtain the spectrum of multimagnon bound states [12]. We shall refer later to some of the result, but a numerical study using particular data for α - helix indicates that it is sufficient to use the results for the isotropic chain.

Classical solitons

In order to study the solitons in this system, we shall use spin coherent states (SCS) [13,14] which are defined in the following way for $S=\frac{1}{2}$:

$$|\alpha\rangle = \prod_m |\alpha_m\rangle \tag{14.a}$$

$$|\alpha_m\rangle = \frac{1}{\sqrt{1+|\alpha_m|^2}}\, e^{\alpha_m \hat{S}_m^-} |0\rangle_m = \cos\frac{\Theta_m}{2}|0\rangle_m + e^{i\phi_m}\sin\frac{\Theta_m}{2}|1\rangle_m$$

$$\text{where}\quad \alpha_m = \mathrm{tg}\frac{\Theta_m}{2}\, e^{i\phi_m} \quad \text{and} \quad |0\rangle_m = |\uparrow\rangle \qquad |1\rangle = |\downarrow\rangle \tag{14.c}$$

We define the classical Lagrangian of the system [15]:

$$L = \langle\alpha|\, i\hbar \overleftrightarrow{\frac{\partial}{\partial t}}\, |\alpha\rangle - \langle\alpha\,|\,H_{eff}\,|\alpha\rangle \tag{15}$$

and write down the equations of motion for the variables $(\phi_n, \frac{1}{2}\cos\Theta_m)$. The solution can be expressed in terms of the following integrals of motion in the continuum approximation [16,17]:

a) "Magnetization" = number of excited molecules

$$M = \sum_n \left(\frac{1}{2} - S_n^z\right) = \frac{1}{2a}\int(1-\cos\Theta)dx \tag{16}$$

b) momentum of the system

$$P = \frac{\hbar}{a}\int \frac{1}{2}(1-\cos\Theta)\,\phi_x\, dx \tag{17}$$

The solution is [16,17]:

$$\cos\Theta = 1 - \frac{2\varphi_o^2}{\mathrm{ch}^2\frac{2\varphi_o^2}{Ma}(x-vt)} \tag{18}$$

with

$$\varphi_o^2 = \sin^2\frac{Pa}{2\hbar} \tag{19}$$

The energy of the system becomes:

$$E_s = \langle\alpha|H_{eff}|\alpha\rangle = \frac{2J}{M}\sin^2\frac{Pq}{2\hbar} + M\varepsilon_o \qquad \varepsilon_o = \Delta - \frac{2\chi^2}{M\omega_o^2} \qquad J = 2I e^{-W(\Theta)} \tag{20}$$

and group velocity is

$$v = \frac{\partial E}{\partial P} = \frac{Ja}{M\hbar} \sin \frac{Pa}{\hbar} \tag{21}$$

It is important to notice that the comparison with the results of Bethe - Ansatz calculation indicates that this soliton solution coincides with multiexciton bound state, only if we perform the semi-classical quantisation M=m, m=1,2,... .

One can also notice that following the standard definition, from (18) we can calculate the soliton dimension as

$$l_o = \frac{Ma}{\varphi_o^2} = \frac{Ma}{\sin^2 \frac{Pa}{2\hbar}} \tag{22}$$

which equals to Ma at the Brillouin zone boundary and diverges as we close the zone center (excitation becomes delocalized).

Let us now calculate the induced relative displacement

$$\hat{\rho}_n = - \frac{1}{a} \left[\hat{U}_{n+1} - \hat{U}_n\right] = - \frac{1}{a} \sum_q \left[\frac{\hbar}{2MN\omega_q}\right]^{1/2} e^{iqna}\left[e^{iqa} - 1\right]\left[\hat{b}_q + \hat{b}^+_q\right] \tag{23}$$

Following our approximation, we shall define the average relative induced displacement as

$$\rho_n = \langle\hat{\rho}_n\rangle = \mathrm{Tr}\ (\hat{\rho}_{ph}\langle\alpha|\ e^{i\hat{S}}\hat{\rho}_n e^{-i\hat{S}}|\alpha\rangle) \tag{24}$$

Taking into account

$$e^{i\hat{S}}\hat{b}_q e^{-i\hat{S}} = \hat{b}_q - i \sum_m F^*(q)\ \hat{P}^+_m\hat{P}_m$$

we obtain

$$\rho_n = - \frac{4a\chi}{Mv_s^2} \left[\langle\alpha|\hat{P}^+_n\hat{P}_n|\alpha\rangle + \langle\alpha|\hat{P}^+_{n+1}\hat{P}_{n+1}|\alpha\rangle\right] \tag{25}$$

In the continuum approximation $\langle\hat{P}^+_{n+1},\hat{P}_{n+1}\rangle \approx \langle\hat{P}^+_n\hat{P}_n\rangle = \frac{1}{2}(1 - \cos\Theta_n)$ leading finally to

$$\rho(x,t) = - \frac{4a\chi}{Mv_s^2} \left[1 - \cos\Theta(x,t)\right] = \frac{8a\chi}{Mv_s^2} \frac{\varphi_o^2}{\mathrm{ch}^2 \frac{2}{l_o}(x-vt)} = \rho_o \frac{1}{\mathrm{ch}^2 \frac{2}{l_o}(x-vt)} \tag{26}$$

and

$$u(x,t) = - \int \rho(x',t)dx' = u_o \mathrm{th} \frac{2}{l_o}(x-vt); \quad u_o = \frac{4a\chi}{Mv_s^2}\ ma \tag{27}$$

where m is the number of excited molecules (=M).

$\rho(x,t)$ and $u(x,t)$ are induced by the excitation of the molecule due to exciton-phonon interaction since $\mathrm{Tr}(\rho_{ph}\hat{\rho}_n) = 0$ if $\langle\hat{P}^+_n,P_n\rangle = 0$.

Analyzing the above expression for the soliton solution, and its consequencies, we see that only the soliton energy and velocity depend on the temperature through the renormalized transfer integral $J = 2Ie^{-W(\Theta)}$. This change is rather small in the range 0 - 300 K, so the soliton solution is well defined in the whole range.

It is much more interesting to study the amplitude of the thermal fluctuations of the molecule:

$$\Delta\rho_n = \sqrt{Tr(\hat{\rho}_{ph}(\hat{\rho}_n)^2)} \qquad (\langle\hat{\rho}_n\rangle = 0) \tag{28}$$

and compare it with the amplitude ρ_0 of the induced relative displacement.

Using (23), we obtain:

$$\langle\hat{\rho}_n^2\rangle = Tr(\hat{\rho}_{ph}\hat{\rho}_n^2) = \frac{2\hbar}{MNa^2}\sum_q \frac{\sin^2\frac{qa}{2}}{\omega_q}\coth\frac{\hbar\omega_q}{2\Theta} \tag{29}$$

Let us perform an estimate for the temperatures $\Theta \geq \hbar\omega_D \geq \hbar\omega_q$:

$$\langle\hat{\rho}_n^2\rangle = \frac{k_B T}{\mathcal{K}a^2} \tag{30}$$

Using values for α-helix [5,8]: $a = 5.4\cdot10^{-10}$m, $\mathcal{K} = 13$ N/m, $K_B = 1.38\cdot10^{-23}$J/K, we estimate for T = 300 K:

$$\langle\hat{\rho}_n^2\rangle \cong 10^{-3} \qquad \Delta\rho_n \cong 0.03$$

One can perform the same calculation for T=0 K [12] corresponding to quantum fluctuation and obtain $\Delta\rho_n \approx 0.01$. On the other hand, for $\chi \geq 6.2\cdot10^{-11}$ N, we have

$$\rho_0 = \frac{8a\chi}{Mv_s^2} = \frac{8\chi}{\mathcal{K}a} \approx 0.1$$

We can see that at low temperatures, ρ_0 is one order of magnitude higher from $\Delta\rho_n$, while for T = 300 K, it is about 3 times higher. This implies that lattice energy localized at a site, $\sim \rho_0^2$ is about 10 times higher then the energy of thermal fluctuatins $(\sim\Theta \simeq \mathcal{K}a^2\langle\tilde{\rho}_n^2\rangle)$. For this reasson, we conclude that solitons of Davydov type can exist even at biologically relevant temperatures, which is contrary to previous conclusions of Lomdahl and Ker [18].

Here we must state our result more precisely. The treatment of excitons as Paulions introduces an important change with respect to boson approach: bound states are formed irrespective to the value of coupling constant χ. Yet, the threshold value appears $(\chi_{th} \sim 6.2\cdot10^{-11})$, above which the whole concept of soliton becomes reasonable, because for smaller values of χ, fluctuations destroy the excitation.

Analyzing the above expression for the soliton solution, and its consequencies, we see that only the soliton energy and velocity depend on the temperature through the renormalized transfer integral $J = 2Ie^{-W(\Theta)}$. This change is rather small in the range 0 - 300 K, so the soliton solution is well defined in the whole range.

It is much more interesting to study the amplitude of the thermal fluctuations of the molecule:

$$\Delta\rho_n = \sqrt{\mathrm{Tr}(\hat{\rho}_{ph}(\hat{\rho}_n)^2)} \qquad (\langle\hat{\rho}_n\rangle = 0) \tag{28}$$

and compare it with the amplitude ρ_0 of the induced relative displacement.

Using (23), we obtain:

$$\langle\hat{\rho}_n^2\rangle = \mathrm{Tr}(\hat{\rho}_{ph}\hat{\rho}_n^2) = \frac{2\hbar}{MNa^2}\sum_q \frac{\sin^2\frac{qa}{2}}{\omega_q}\coth\frac{\hbar\omega_q}{2\Theta} \tag{29}$$

Let us perform an estimate for the temperatures $\Theta \geq \hbar\omega_D \geq \hbar\omega_q$:

$$\langle\hat{\rho}_n^2\rangle = \frac{k_B T}{\varkappa a^2} \tag{30}$$

Using values for α-helix [5,8]: $a = 5.4\cdot10^{-10}$m, $\varkappa = 13$ N/m, $K_B = 1.38\cdot10^{-23}$J/K, we estimate for T = 300 K:

$\langle\hat{\rho}_n^2\rangle \cong 10^{-3} \qquad \Delta\rho_n \cong 0.03$

One can perform the same calculation for T=0 K [12] corresponding to quantum fluctuation and obtain $\Delta\rho_n \approx 0.01$. On the other hand, for $\chi \geq 6.2\cdot10^{-11}$ N, we have

$$\rho_0 = \frac{8a\chi}{Mv_s^2} = \frac{8\chi}{\varkappa a} \approx 0.1$$

We can see that at low temperatures, ρ_0 is one order of magnitude higher from $\Delta\rho_n$, while for T = 300 K, it is about 3 times higher. This implies that lattice energy localized at a site, $\sim \rho_0^2$ is about 10 times higher then the energy of thermal fluctuatins $(\sim\Theta \simeq \varkappa a^2\langle\tilde{\rho}_n^2\rangle)$. For this reasson, we conclude that solitons of Davydov type can exist even at biologically relevant temperatures, which is contrary to previous conclusions of Lomdahl and Ker [18].

Here we must state our result more precisely. The treatment of excitons as Paulions introduces an important change with respect to boson approach: bound states are formed irrespective to the value of coupling constant χ. Yet, the threshold value appears $(\chi_{th} \sim 6.2\cdot10^{-11})$, above which the whole concept of soliton becomes reasonable, because for smaller values of χ, fluctuations destroy the excitation.

We have mentioned that semiclassical quantization enables us to establish the correspondence between solitons and multi-exciton bound states. The state with two bound excitons whose energy is $\varepsilon_2 \approx 2\Delta$ is of particular importance. Namely, if we suppose that Δ equals to the energy of amide I bond $\varepsilon_I = 0.203$ eV, it follows that $\varepsilon_2 \simeq 2\varepsilon_I$ is nearly equal to the energy released through the hydrolysis of ATP into ADP (~ 0.43 eV), indicating to a possible role of these excitations in the biological systems [1,19].

Finally, let us mention that all numerical estimates are based on the data for α-helix, because the experiments concerning ACN indicate that its structure must be taken into account, and the model becomes much more complicated.

References

[1] A.S. Davydov: Solitons in Molecular Crystals, Reidel (Dordrecht) (1985).

[2] G. Carreri, U. Buontempo, E. Gratton and A.C. Scott: Phys. Rev. Lett. **51**, 304 (1984).

[3] G. Carreri, U. Buontempo, F. Galluzzi, A.C. Scott, E. Gratton and E. Shyamsunder: Phys. Rev. **B30**, 4689 (1984).

[4] G. Carreri, E. Gratton and E. Shyamsunder: Phys. Rev. **A37**, 4048, (1988).

[5] A.C. Scott: Phys. Rev. **A26**, 578 (1982).

[6] Mac Neil and A.C. Scott: Phys. Ser. 29, 284 (1984).

[7] W.C. Kerr and P.S. Lomdahl: Phys. Rev. **B35**, 3629 (1987).

[8] D.W. Brown, K. Lindenberg and X. Wang: to appear in Singular Behaviour and Nonlinear Dynamics, Ed. F.H. Chionh, World Scientific (Singapore), (1989).

[9] V.M. Agrenovich and M.D. Galanin: Electronic Excitation Energy Transfer in Condensed Matter, North-Holland, P.C. (Amsterdam), (1982).

[10] J.D. Johnson and J.C. Bonner: Phys. Rev. Lett. **44**, 616 (1980).

[11] R.P. Hodgson and J.B. Parkinson: J. Phys.**C17**, 3223 (1984).

[12] M.J. Škrinjar, D.V. Kapor and S.D. Stojanović: Submitted in J.Phys. Cond. Matt.

[13] J.M. Radcliffe: J. Phys. **A4**, 313 (1971).

[14] A. Perelomov: Generalized Coherent States and Their Application, Springer (Berlin) (1985).

[15] L.R. Mead and N. Papanicolaon: Phys. Rev. B28, 1633 (1983).

[16] J. Tjon and J. Wright: Phys. Rev. B15, 3470 (1977).

[17] F.M. Haldane: J. Phys. C15, L 83, (1982).

[18] P.S. Lomdahl and W.C. Kerr: Phys. Rev. Lett. 55 1235 (1985).

[19] H. Bolterauer and R.D. Henkel: Phys. Scr. 13, 314 (1988).

Solitons on a diatomic chain with alternating interactions

F.G. Mertens and D. Hochstraßer

Physics Institute, University of Bayreuth,

D-8580 Bayreuth, F.R. Germany

We search for pulse-like solitary waves on a one-dimensional lattice with two alternating masses and two alternating interactions. The equations of motion in Fourier space are treated by an iterative method which takes into account the discreteness effects. We present numerical results for the case that one of the interactions is linear (this is an excellent approximation in the case of a generalized Yomosa model for the energy transport in α-helical proteins). Moreover we discuss the finite lifetime of the solitons due to the emission of optical phonons.

1. Introduction

The motivation for this work is to generalize Yomosa's model [1] for the energy transport in muscle proteins with a double α-helical structure. In each helix there are three chains of peptide groups coupled by hydrogen bonds, which are strongly nonlinear. Yomosa considered a single chain of rigid peptide groups and represented the nonlinear interactions V betweeen them e.g. by a Toda potential. In this model, energy can be transported by lattice solitons, in contrast to the solitons of Davydov's model [2]. Ref. [3] discusses several advantages of modelling the energy transport by lattice solitons.

We generalize Yomosa's model in two ways: (1) The peptide groups are no longer rigid; for simplicity we represent them by two masses M_1 and M_2 coupled by an interaction W. The generalization to more internal degrees of freedom of the peptide groups will turn out to be straightforward. (2)

Contrary to Yomosa, we do not work in the continuum limit but apply an iterative method [4], where the discreteness effects are taken into account more and more accurately.

2. Equations of motion

The Lagrangian of a diatomic chain with alternating interactions is

$$L = \sum_n \{ \tfrac{1}{2} M_1 \dot{A}_n^2 + \tfrac{1}{2} M_2 \dot{B}_n^2 - V(A_{n+1} - B_n) - W(B_n - A_n)\} \tag{2.1}$$

Introducing relative displacements (in units of the lattice constant)

$$\varphi_n = A_{n+1} - B_n \; , \qquad \Delta_n = B_n - A_n \tag{2.2}$$

we obtain the equations of motion

$$\ddot{\varphi}_n + \alpha \frac{\delta V}{\delta \varphi_n} - \frac{1}{1+\mu} \{ \frac{\delta W}{\delta \Delta_{n+1}} + \mu \frac{\delta W}{\delta \Delta_n} \} = 0 \tag{2.3a}$$

$$\ddot{\Delta}_n + \frac{\delta W}{\delta \Delta_n} - \frac{\alpha}{1+\mu} \{ \mu \frac{\delta V}{\delta \varphi_n} + \frac{\delta V}{\delta \varphi_{n-1}} \} = 0 \; . \tag{2.3b}$$

μ is the mass ratio M_1/M_2. The different strength of the two interactions is taken into account by α. In this way V and W can be written with the same prefactor, here we choose the units such that a linearization yields $\delta V/\delta\varphi_n \simeq \varphi_n$ and $\delta W/\delta\Delta_n \simeq \Delta_n$.

The dispersion curves of acoustic and optical phonons are

$$\omega_\pm^2(q) = \frac{1+\alpha}{2} \{ 1 \pm [1 - \frac{16\alpha}{(1+\alpha)^2} c_m^2 \sin^2 \frac{q}{2}]^{1/2} \} \tag{2.4}$$

The sound velocity c_s is

$$c_s^2 = \frac{\alpha}{1+\alpha} c_m^2 \; , \qquad c_m^2 = \frac{\mu}{(1+\mu)^2} = \frac{m}{M} \; , \tag{2.5}$$

where m and M are the reduced mass and the total mass.

3. Solitary waves

We are interested in pulse-like solitary waves with velocity c

$$
\begin{aligned}
\varphi_n(t) &= \varphi(n - ct) = \varphi(z) \\
\Delta_n(t) &= \Delta(n - ct) = \Delta(z) ,
\end{aligned}
\tag{3.1}
$$

satisfying decaying boundary conditions such that the Fourier transforms $\tilde{\varphi}(q)$ and $\tilde{\Delta}(q)$, as well as $\tilde{f}(q)$ and $\tilde{g}(q)$ exist, here f and g denote the nonlinear parts of the forces

$$
\frac{\delta V}{\delta \varphi} = \varphi(z) + f[\varphi(z)] \ ; \qquad \frac{\delta W}{\delta \Delta} = \Delta(z) + g[\Delta(z)] \ . \tag{3.2}
$$

The equations of motion (2.3) now read

$$
R \begin{pmatrix} \tilde{\varphi} \\ \tilde{\Delta} \end{pmatrix} + S \begin{pmatrix} \tilde{f} \\ \tilde{g} \end{pmatrix} = 0 \tag{3.3}
$$

with

$$
R = \begin{pmatrix} \alpha - c^2q^2 & -\frac{1}{1+\mu}(\mu + e^{iq}) \\ -\frac{\alpha}{1+\mu}(\mu + e^{-iq}) & 1 - c^2q^2 \end{pmatrix} \tag{3.4}
$$

$$
S = \begin{pmatrix} \alpha & -\frac{1}{1+\mu}(\mu + e^{iq}) \\ -\frac{\alpha}{1+\mu}(\mu + e^{-iq}) & 1 \end{pmatrix} \tag{3.5}
$$

Equ. (3.3) is formally solved by

$$
\begin{pmatrix} \tilde{\varphi} \\ \tilde{\Delta} \end{pmatrix} = \frac{1}{[c^2q^2 - \omega_+^2(q)][c^2q^2 - \omega_-^2(q)]} \, T \begin{pmatrix} \tilde{f} \\ \tilde{g} \end{pmatrix} \tag{3.6}
$$

with

$$T = \begin{pmatrix} \alpha[c^2q^2 - \omega_m^2(q)] & -\frac{c^2q^2}{1+\mu}(\mu + e^{iq}) \\ -\alpha\frac{c^2q^2}{1+\mu}(\mu + e^{-iq}) & c^2q^2 - \alpha\omega_m^2(q) \end{pmatrix} \tag{3.7}$$

where

$$\omega_m(q) = 2\, c_m |\sin q/2| \; . \tag{3.8}$$

However, $\tilde{f}$ and $\tilde{g}$ implicitely depend on $\tilde{\varphi}$ and $\tilde{\Delta}$, since f and g are functionals of φ and Δ, see (3.2).

In order to solve (3.6) we generalize an <u>iterative</u> method which has been developed recently for the monoatomic chain [4]. In this method the velocity c must be released in order to prevent a convergence to the trivial solution. We demonstrate this procedure for the special case that one of the interactions, e.g. $W(\Delta)$ is chosen to be harmonic, i.e. $\tilde{g}(q) \equiv 0$. For our peptide model this is an excellent approximation since the displacements Δ of the strong covalent bonds within the peptide groups are much smaller than the displacements φ of the hydrogen bonds between the peptide groups.

We now apply the following iteration procedure to the first equation in (3.6)

$$\tilde{\varphi}_{i+1}(q) = \frac{\alpha[c_i^2q^2 - \omega_m^2(q)]}{[c_i^2q^2 - \omega_+^2(q)][c_i^2q^2 - \omega_-^2(q)]}\, \tilde{f}_i(q) \tag{3.9}$$

where $\tilde{f}_i$ is the Fourier transform of $f[\varphi_i(z)]$. We have replaced the constant velocity in (3.6) by

$$c_i^2 = c_s^2 \frac{\tilde{\varphi}_i(0) + \tilde{f}_i(0)}{\tilde{\varphi}_i(0) + c_s^2/c_m^2 \cdot \tilde{f}_i(0)} \tag{3.10}$$

which is obtained by demanding $\tilde{\varphi}_{i+1}(0) = \tilde{\varphi}_i(0)$. Thus the iteration does not select a solution with a given velocity c, but a solution with a given integrated amplitude

$$\tilde{\varphi}_1(0) = \int_{-\infty}^{\infty} dz \; \varphi_1(z) \quad . \tag{3.11}$$

In order to get a starting function $\varphi_1(z)$ we use the quasi-continuum approach in q-space [4,3].

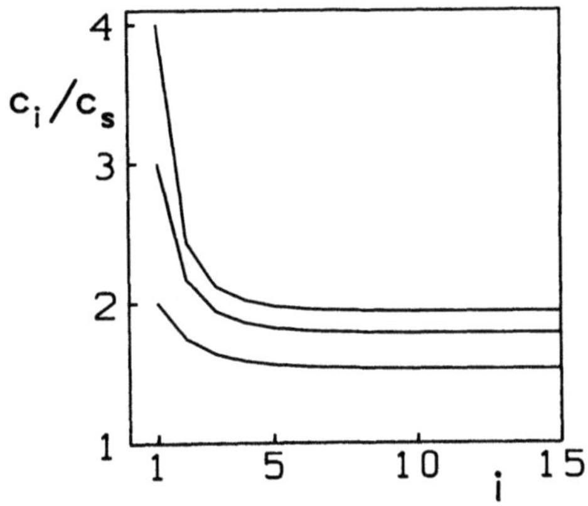

Fig. 1. Velocities c_i during the iteration, for 3 different initial velocities c_1; using a Toda potential $\alpha/\beta^2(\exp(-\beta\varphi) + \beta\varphi - 1)$ with $\alpha = 0.00492$ and $\beta = 18$ from ref. [1], mass ratio $\mu = 1$.

The c_i change during the iteration and the final value usually is considerably lower than the initial one (Fig. 1). Equ. (3.10) implies $c_s < c_i < c_m$, thus there is an upper velocity limit, in addition to the lower velocity limit of the monoatomic chain. Because of (2.5) the range of allowed velocities becomes larger for decreasing α. For our peptide model we have $\alpha \ll 1$, since the hydrogen bonds are considerably weaker than the covalent bonds of the peptides.

We have used Toda and Lennard-Jones potentials for the hydrogen bonds and a sufficient convergence is achieved after at most 8 iterations. The results were tested by computer experiments, an example for a very narrow soliton is shown in Fig. 2.

However, a complete convergence cannot be achieved. Computer simulations and theoretical investigations [5] have shown that a solitary wave with velocity c in principle always looses energy due to the emission of phonons, if the line cq has an intersection with one of the phonon branches.

In our case this intersection occurs for the optical branch. The problem shows up technically by a pole in (3.9). In each step of the iteration a cut-off is necessary, which makes sense only if $\tilde{f}(q)$ and $\tilde{\varphi}(q)$ are negligible for $|q| \geq q_o$ where q_o is the intersection point. For the parameters we have been using, this point is far outside of the first Brillouin zone. Thus even for very narrow solitons (in the order of the lattice constant) the energy loss is negligible, in fact in Fig. 2 a phonon emission is not observed.

Though our model is not expected to be integrable, the solitary waves practically behave like real solitons, even if the width is very small: In a scattering experiment no change of the shape of the pulses is observed on the scale of Fig. 3, and the emission of phonons can hardly be seen.

We have only shown results for the mass ratio $\mu = 1$ because there is no qualitative change of the results for $\mu \neq 1$, i.e. the essential feature of the model consists in the alternating linear and nonlinear interactions.

Numerical results for the case that both interactions are nonlinear will be presented in a forthcoming paper.

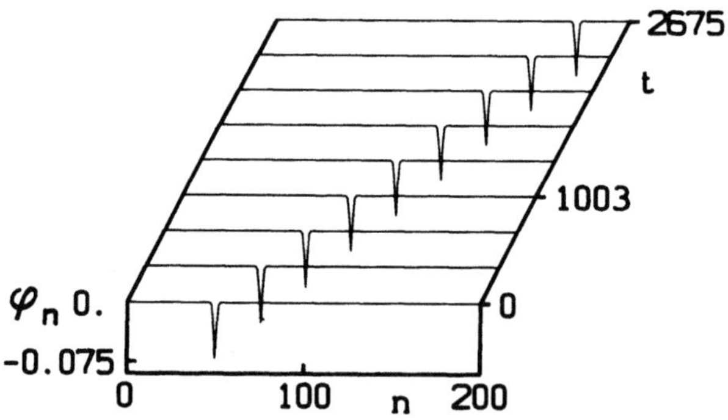

Fig. 2. Computer simulation on a chain of 200 unit cells; initial condition is an iterative solution of (3.9) with $c = 1.28\ c_s$. Same potential and parameters as in Fig. 1.

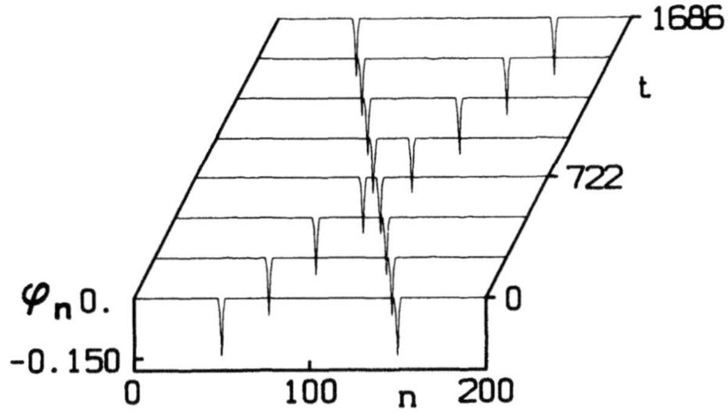

Fig. 3. Scattering experiment for 2 very narrow solitary waves with $c = 1.9\ c_s$. Same potential and parameters as in Fig. 1.

Acknowledgement

This work is financed by DFG (Deutsche Forschungsgemeinschaft), project D 7, SFB 213.

References

1. S. Yomosa, J. Phys. Soc. Japan 53, 3692 (1984); Phys. Rev. A 32, 1752 (1985)
2. A.S. Davydov, J. Theor. Biol. 38, 559 (1973); Stud. Biophys. 47, 221 (1984); Phys. Scr. 20, 387 (1979), Biology and Quantum Mechanics (Pergamon, New York, 1982)
3. D. Hochstraßer, F.G. Mertens, and H. Büttner, Phys. Rev. A, in press
4. D. Hochstraßer, F.G. Mertens, and H. Büttner, Physica D 35, 259 (1989)
5. M. Peyrard, St. Pnevmatikos, and N. Flytzanis, Physica 19 D, 268 (1986)

LONGITUDINAL AND TRANSVERSE PROTON COLLECTIVE DYNAMICS ON A TWO-DIMENSIONAL MULTISTABLE SUBSTRATE

St. Pnevmatikos[1], A.V. Savin[2] and A.V. Zolotaryuk[3]

[1]Research Center of Crete, P.O. Box 1527, 711 10 Heraklio, Crete, Greece
[2]Institute for Physico-Technical Problems, 119034 Moscow, USSR
[3]Institute for Theoretical Physics, UkrSSR Academy of Sciences, 252130 Kiev, USSR

ABSTRACT

The longitudinal and transverse collective dynamics of protons is studied in a zig-zag hydrogen-bonded model. Protons are interacting harmonically with their first proton neighbors and with Morse forces with the nine nearest negative ion neighbors that are considered rigid (frozen). Ionic and bonding defect solutions of soliton type are obtained numerically and their collision properties are analysed.

INTRODUCTION

A great deal of activity has been devoted to the understanding of the physical and electrochemical processes that are responsible for the anomalously high proton mobility of many materials that are characterised by long zig-zag hydrogen-bonded networks [1]. Onsager associated the conductivity in ice, which is not electronic but protonic in nature, to a hopping mechanism that allows the protons of the hydrogen bonds to move along hydrogen bonded atomic channels [2]. Experimental evidence strongly indicates that charge transport proceeds via the motion of two types of defects that can be created in the network, viz. the ionic defects and the bonding (or orientational or Bjerrum) defects [3]. The former involves an intrabond motion of the (unique) binding proton of the hydrogen bond, whereas the latter results from interbond transfer of the protons due to rotations of the water molecules. These ideas are extended to other hydrogen-bonded systems [1,3].

The aforementioned problem has been reconsidered by many authors introducing the idea of a collective transition of the interacting proton subsystem that could explain qualitatively the ionic and bonding defect creation and propagation along the crystal [4]. In some recent works both ionic and bonding defects are shown to be approximate soliton solutions of some well known nonlinear partial differential equations derived in the continuum limit by doubly periodic on-site potentials [5,6]. In addition, interesting properties have been discovered when the negative ion atomic environment was considered to interact with the proton sublattice [7,8].

In the above works the theoretical and numerical approaches are based on some one-dimensional models where protons and negative ions (if included) are allowed to move only longitudinally under the influence of a doubly periodic on-site potential. In the present paper, for the first time, many of these restrictions are removed by the introduction of a new atomic model where protons move both longitudinaly and transversely interacting on the one hand with their first proton neighbors harmonically and on the other hand with their nine closest negative neighbors anharmonically (for instance with Morse pair interactions). The former interaction is responsible for the collective nature of the proton dynamics, whereas the latter, while of pair type, is resulting to the formation of a multistable two-dimensional substrate in the vallleys of which protons move along the hydrogen-bonded zig-zag structure.

THE MODEL

Let us consider a standard zig-zag hydrogen bonded chain (see for instance hydrogen halides [9]) that is represented in figure 1. Here l is the equilibrium length of the X-X bridge between two neighbor negative ions, α is the angle between two such adjacent bridges, and $l_x = l\sin(\alpha/2)$, $l_y = l\cos(\alpha/2)$.

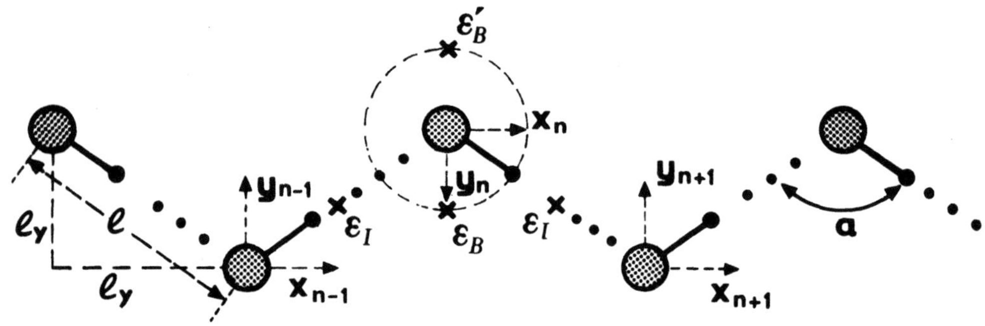

Figure 1 : Schematic representation of the zig-zag hydrogen-bonded chain. Large and small circles denote frozen negative ions and moving protons respectively; x marks represent ionic (ε_I) and bonding (ε_B) local potential maxima.

In a first crude approximation we assume all negative ions frozen and we focus our attention to the two-dimensional dynamics of protons. The coordinates (x_n, y_n) of protons are measured from the corresponding n-th frozen negative ions. For the purposes of this paper, we introduce the following Hamiltonian :

$$H = \sum_n \left\{ \frac{m}{2} \left[\left(\frac{dx_n}{dt}\right)^2 + \left(\frac{dy_n}{dt}\right)^2 \right] + \frac{K_x}{2} (x_{n+1}-x_n)^2 + \frac{K_y}{2} (y_{n+1}-y_n)^2 + U(x_n, y_n) \right\} \quad (1)$$

where m is the proton mass and K_x, K_y are the longitudinal and transverse elasticity constants which can be evaluated from the expansion of the dipole-dipole interaction energy. The first term in (1) represents the kinetic energy of protons analyzed in the x and y directions. The next two terms represent the proton-proton interaction which is limited to the first proton neighbors and it is assumed to be harmonic. The last term is a long range pair potential denoting the interaction of each proton with some of the closest frozen negative neighbors. We consider the following Morse form for this term :

$$U(x_n,y_n) = D \sum_{i=-4}^{4} [\, 1 - e^{-b(r_{ni}-r_o)} \,]^2 \qquad (2)$$

where

$$r_{ni} = [\, (x_n - il_x)^2 + (y_n - \frac{(-1)^{i+1} + 1}{2} l_y)^2 \,]^{1/2} \qquad (3)$$

and D is the binding energy of the X-H covalent bond in an isolated molecule with r_o the length of this bond in equilibrium. It can be shown that nine nearest negative neighbors are enough in order to give to the potential function $U(x_n,y_n)$ the right physical structure of the hydrogen-bonded network (see figure 2).

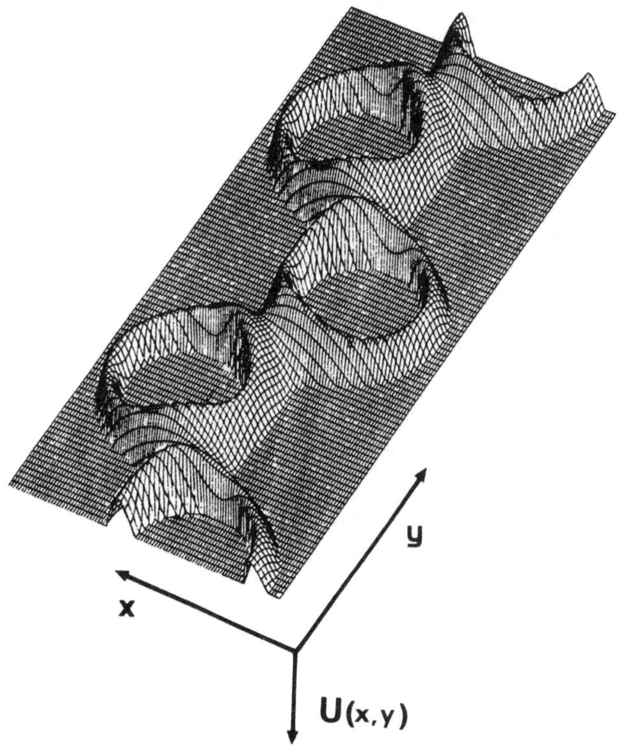

Figure 2 : The potential relief -U(x,y) which results by the interaction of one proton with its nine nearest negative ions. The standing out valleys correspond to the proton channels of low potential energy. For convenience we plot the -U(x,y) and we cut the plot to certain level by a plane.

In such a structure we have a valey of low energy around each negative ion which have two absolute degenerate minima

$$U(x_o,y_o) = U(-x_o,y_o) = \min_{x,y} U(x,y) \qquad (4)$$

corresponding to possible proton equilibrium positions. Besides these two stable equilibrium positions there are also four unstable equilibrium positions (local maxima) around every negative atom defined by the following relations :

$$x = \pm l_x/2 \qquad , y = l_y/2 \qquad (5a,b)$$

$$x = 0 \ , y = y_1 \ , \quad U(0,y_1) = \min_{y>0} U(0,y) \qquad (5c)$$

$$x = 0 \ , y = y_2 \ , \quad U(0,y_2) = \min_{y<0} U\ (0,y) \qquad (5d)$$

The transfer of protons over the barrier :

$$\varepsilon_I = U(l_x/2,\ l_y/2) - U(x_o,y_o) \qquad (6)$$

generates what we call ionic defect. The transfer of protons over the barrier :

$$\varepsilon_B = U(0,y_1) - U(x_o,y_o) \qquad (7)$$

generates what we call bonding (or Bjerrum) defect corresponding to an appropriate rotation of the dipole X-H around the negative frozen center. Our model allows the generation of bonding defects following an unphysical path over the barrier :

$$\varepsilon_B' = U(0,y_2) - U(x_o,y_o) \qquad (8)$$

This possibility may be removed with slight modification of the model [10].

Knowing the value of ε_I the phenomenological parameter b in the Morse potential (2) can be determined uniquely since the length r_o of the X-H bond is also fixed. For instance for the next nearest neighbors case in (2) the parameter b is given by

$$b = \ln \{ \frac{1}{2} [1 - (\varepsilon_I/D)^{1/2}]\} \ / \ (r_o-1/2) \qquad (9)$$

From the Hamiltonian (1) we can derive the following equations of motion :

$$m \frac{d^2x_n}{dt^2} = K_x\ (x_{n+1} - 2x_n + x_{n-1}) - \frac{\partial}{\partial x_n} U(x_n,y_n) \qquad (10a)$$

$$m \frac{d^2y_n}{dt^2} = K_y (y_{n+1} - 2y_n + y_{n-1}) - \frac{\partial}{\partial y_n} U(x_n,y_n) \qquad (10b)$$

If the following conditions are satisfied :

$$K_x - m\upsilon^2/l_x^2 < 0 \qquad \text{and} \qquad K_y - m\upsilon^2/l_x^2 < 0 \qquad (11)$$

then the transfer of protons over the barriers is produced smoothly and the dynamics of defects is clearly collective. In this case looking for travelling wave solutions where

$$x_n(t) = x(nl_x-\upsilon t) \qquad \text{and} \qquad y_n(t) = y(nl_x-\upsilon t) \qquad (12)$$

we can use a convenient numerical scheme [10,11] and solve the following minimization problem :

$$\sum_n [\frac{1}{2} (K_x-m\upsilon^2/l_x^2) (x_{n+1}-x_n)^2 + \frac{1}{2} (K_y-m\upsilon^2/l_x^2) (y_{n+1}-y_n)^2 + U(x_n,y_n)] \rightarrow \min \qquad (13)$$

Using this procedure we can obtain exactly the coordinates $\{x_n,y_n\}$ for some forms of travelling wave solutions of the discrete Hamiltonian (1). Then, we can approximately obtain their particle velocities $\{dx_n/dt , dy_n/dt\}$ by the expressions :

$$dx_n/dt = -\upsilon dx_n/dx = -\frac{\upsilon}{2l_x} (x_{n+1}-x_n)_1 \qquad (14a)$$

$$dy_n/dt = -\upsilon dy_n/dx = -\frac{\upsilon}{2l_x} (y_{n+1}-y_n)_1 \qquad (14b)$$

Therefore a set of initial conditions can be produced representing "exact" solutions of the ionic and bonding defect type. In figure 3, we plot the longitudinal and transverse components with their derivatives for three of such soliton solutions. We note that, both components of an ionic defect have the form of kinks. The bonding defect has a kink type x-component and a pulse type y-component. Finally, the components of the unphysical long-path bonding defect are also plotted.

NUMERICAL RESULTS

Let us introduce, as an example, the following set of parameters corresponding to the Ih ice structure :

l=2.76 A , r_o=0.97A, α=120° , D=5.11eV

so that l_x=2.39A and l_y=1.38A. Also

$$\varepsilon_I = 0.76 \text{ eV} \quad , \ \varepsilon_B = 0.68 \text{ eV}$$

so that b=2.878 .... x 10^{10} m^{-1}. Finally,

$$K_x = 1000 \text{ N/m} \quad , \ K_y = 500 \text{ N/m}$$

Using some combinations of initial conditions, produced with the previous minimization scheme, we can study various interaction properties of the defects of both types.

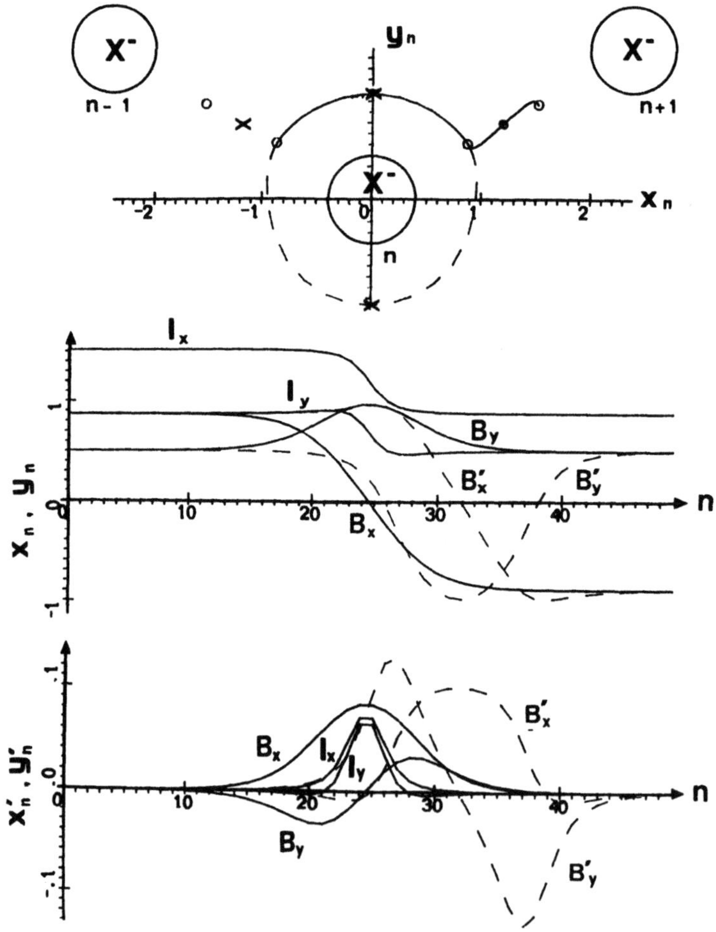

Figure 3 : a) Possible paths of proton transfer. Large and small circles denote negative frozen ions and protons respectively. b) Longitudinal (I_x, B_x, B_x') and transverse (I_y, B_y, B_y') ionic, bonding and long path bonding respectively displacements of protons. c) The corresponding longitudinal and transverse proton velocities.

The collision of opposed charged ionic defects can be resulted either to an inelastic reflection (figure 4) for fast incident waves or to a breather formation (figure 5) for slow incident waves. Even with no damping the breather is not long-lived as for the one-dimensional case [12]. This is due to the coupling between the x and y components which forces the breather to radiate permanently. In some cases we may even observe the direct recombination of the two incident waves which results to their transformation to phonons.

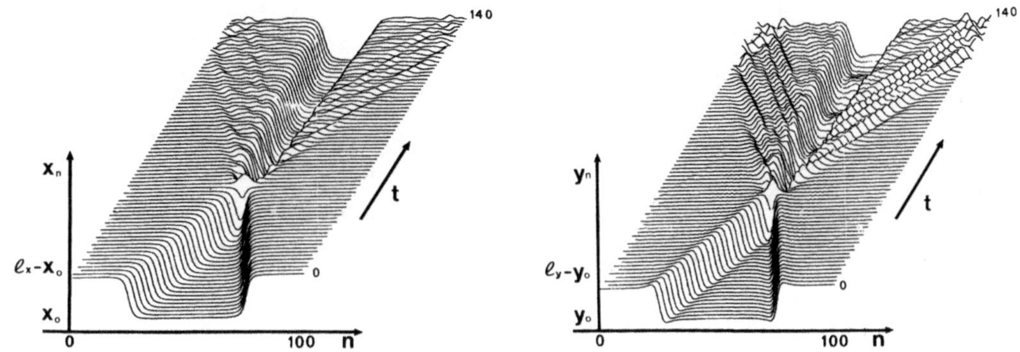

Figure 4 : Reflection of a positive (antikink) and a negative (kink) ionic defects. Incident velocities : $v_+ = -v_- = 0.5$. Velocities of reflected waves : $v_+ = -v_- = 0.07$. Left: longitudinal coordinates. Right: transverse coordinates.

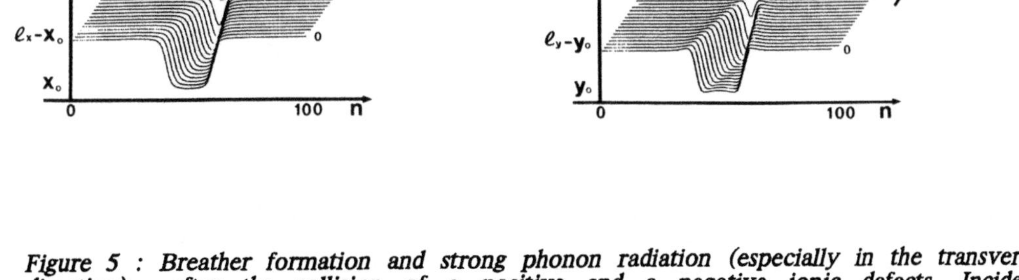

Figure 5 : Breather formation and strong phonon radiation (especially in the transverse direction) after the collision of a positive and a negative ionic defects. Incident velocities : $v_+ = -v_- = 0.3$. Boundary conditions : absorbing ends.

The energy of bonding defects exceeds in our case the energy of ionic defects. Therefore, the collision of two opposed charged bonding defects results to the conversion into two ionic defects (see figure 6). In this case a strong radiation of phonons is observed (mostly in the y-direction).

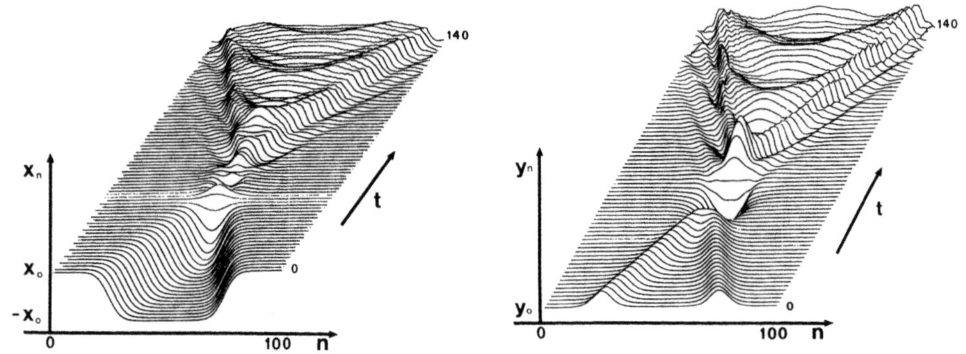

Figure 6 : Conversion of a pair of bonding defects (incident velocities : $v_+=-v_-=0.5$) to a pair of ionic defects accompanied by strong radiation especially in the transverse component.

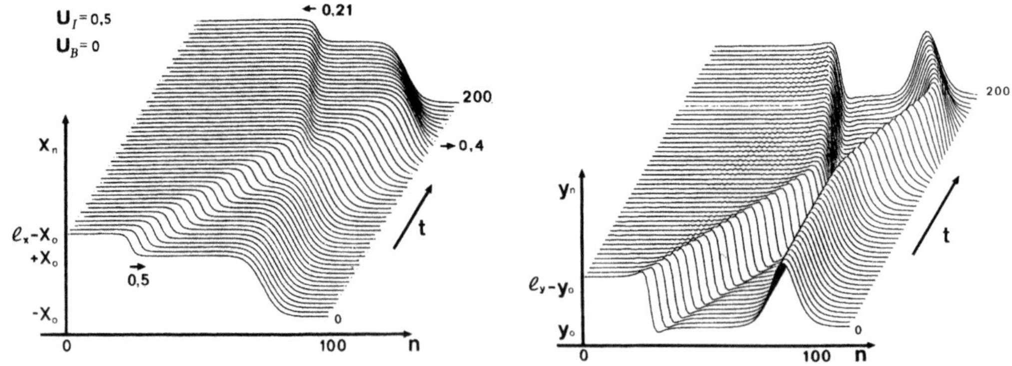

Figure 7 : Collision of a positive ionic defect with a positive bonding defect. Velocities before the collision : $v_I=0.5$, $v_B=0$ and after the collision : $v_I=-0.212$, $v_B=0.404$. (fixed boundary conditions). Left: longitudinal, right: transverse coordinates.

In the last figure (see figure 7) we present the collision between a positive ionic and a positive bonding defects (antikinks). The collision (reflection) is inelastic but both waves conserve their integrity. The unphysical long-path bonding defect is not studied.

CONCLUSIONS

This short paper is a first report of our results on the two-dimensional proton dynamics in a hydrogen-bonded zig-zag chain. We introduced here, for the first time, a new family of models where protons are allowed to move on the plane following both a longitudinal and transverse dynamics. Each proton interacts on the one hand with its first proton neighbors through harmonic forces and on the other hand with nine nearest negative ion neighbors through Morse type forces. Therefore only pair type interactions are considered in this model. However, the combination of nine Morse potential functions results to the construction of a multistable potential support which defines very naturally the proton channels for charge, mass and energy transport. In the present approach the negative ions are considered rigid (frozen) so that our study to be focused on the dynamics of protons. From previous studies [5-8], we know that the inclusion of the heavy sublattice dynamics will produce secondary but interesting effects on the proton dynamics. These effects are supposed to be studied in a subsequent paper where the model will be slightly modified in order to eliminate the long path bonding defect which is not expected to be favorized in the real system.

Since this model is hardly tractable analytically, we produced soliton solutions of the ionic and bonding defect type using a steepest descent minimization numerical scheme [10,11]. The numerical simulation of these solitons on the discrete lattice showed a meaningful stability with standard collision properties. However, the stability of these solitons is observed to be more sensitive in the transverse direction where the radiation appeared to be much stronger than in the longitudinal component [10].

Acknowledgment : One of us (A.V.Z.) is grateful to the hospitality of the Research Center of Crete where part of this work is completed.

REFERENCES

1. L. Glasser, Chem. Rev. 75, 21, (1975)
2. L. Onsager, Science 156, 541, (1967); 166, 1359, (1969)
3. J.F. Nagle and S. Tristram-Nagle, J. Membrane Biol. 74, 1, (1983)
4. J.H. Weiner and A. Askar, Nature 226, 842, (1970); S. Yomosa, J. Phys.Soc.Jpn, 51, 3318, (1982); O.E. Yanovitskii and E.S. Kryachko, Phys.Stat.Sol. (b), 147, 69,

(1988); L.N. Kristoforov and A.V. Zolotaryuk, Phys.Stat.Sol. (b), 146, 487, (1988); A. Gordon, Physica B, 150, 319, (1988)

5. St. Pnevmatikos, Phys.Rev.Lett, 60, 1534, (1988)
6. A.V. Zolotaryuk, in "Biophysical aspects of cancer", ed. J. Fiala and J. Pokorny, (Charles University Press, Prague 1987)
7. S. Yomosa, J.Phys.Soc.Jpn, 52, 1866, (1983); V.Ya. Antonchenko, A.S. Davydov and A.V. Zolotaryuk, Phys. Stat. Sol (b) 115, 631, (1983); St. Pnevmatikos, Phys. Lett. A, 122, 249, (1987)
8. M. Peyrard, St. Pnevmatikos and N. Flytzanis, Phys.Rev.A, 36, 903, (1987)
9. M. Springborg, Phys.Rev.B 38, 1438, (1988-I); R.W. Jansen et al, Phys.Rev.B 35, 9830, (1987-II)
10. A.V. Savin, A.V. Zolotaryuk and St. Pnevmatikos, unpublished, (1989)
11. St. Pnevmatikos, A.V. Savin, I. Stylianou and A.V. Zolotaryuk, in preparation, (1989)
12. D.K. Campbell, M. Peyrard and P. Sodano, Physica 19D, 165, (1986).

TODA LATTICE WITH TRANSVERSE DEGREE OF FREEDOM

P.L. Christiansen*),
Center for Nonlinear Studies
Los Alamos National Laboratory
Los Alamos, NM 87545, USA

and

Dipartimento di Matematica e Applicazioni
Università degli Studi di Napoli
Via Mezzocannone 8, I–80314 Napoli, Italy

P.S. Lomdahl and V. Muto
Center for Nonlinear Studies and Theoretical Division
Los Alamos National Laboratory
Los Alamos, NM 87545, USA

ABSTRACT

A transverse degree of freedom is introduced in the Toda lattice. The corresponding continuum approximations are discussed.

INTRODUCTION

One of the few integrable discrete systems is the nonlinear spring and mass chain introduced by Toda [1]. Its integrability was demonstrated by Flaschka [2] and effective analytical technique based on the spectral transform was developed subsequently [3]. Many theoretical and numerical studies of perturbed Toda lattices have been reported, see [4] e.g.. Recently, the Toda lattice has been applied to model the propagation of longitudinal waves along DNA [5]. Solitons were found to form spontaneously at physiological temperatures. In a more realistic model of DNA also transverse degrees of freedom must be taken into account. Thus a two chain model of DNA was

*) Permanent address:
Laboratory of Applied Mathematical Physics
The Technical University of Denmark
DK–2800 Lyngby, Denmark

treated by statistical mechanics methods in [6] with Morse potentials representing the H–bonds in the base pairs. Here, as a first step, we formulate the Toda lattice model for one strand with a transverse degree of freedom. The continuum approximations of the resulting equations and their solutions are investigated. Computer studies of the lattice dynamics will be presented elsewhere.

FORMULATION OF THE MODEL

We consider a one-dimensional lattice with lattice constant ℓ and N lattice points. At each lattice point we place identical masses (base pairs), m. The longitudinal and transverse displacements from the equilibrium positions are given by $y_1, y_2, \cdots, y_N$ and $v_1, v_2, \cdots, v_N$, respectively. In the case of a circular arrangement of the N masses (corresponding to a circular DNA molecule) we get the periodic boundary conditions for the displacements as function of time, t,

$$y_{n+N}(t) = y_n(t), \quad v_{n+N}(t) = v_n(t), \quad n = 1, 2, \cdots, N \quad . \tag{1}$$

The elongation (or compression) of the spring connecting the n'th and (n+1)'th masses is given by

$$r_n = [(\ell + y_{n+1} - y_n)^2 + (v_{n+1} - v_n)^2]^{\frac{1}{2}} - \ell \quad . \tag{2}$$

$r_n = 0$ when the length of the spring is equal to the lattice constant. Note that $r_0 = r_N$ is the elongation of the spring connecting the 1st and the N'th masses. The Toda potential [1] is given by

$$V(r_n) = \frac{a}{b} [\exp(-br_n) + 1] + ar_n \quad , \tag{3}$$

where a and b are constants. The Hamiltonian for the Toda chain becomes

$$H = \sum_{n=1}^{N} \frac{1}{2} m (\dot{y}_n^2 + \dot{v}_n^2) + V(r_n) \quad , \tag{4}$$

where dot denotes differentiation with respect to time. The dynamical equations become

$$m\ddot{y}_n = - V'(r_n) \frac{\partial r_n}{\partial y_n} - V'(r_{n-1}) \frac{\partial r_{n-1}}{\partial y_n} \tag{5a}$$

and

$$m\ddot{v}_n = - V'(r_n) \frac{\partial r_n}{\partial v_n} - V'(r_{n-1}) \frac{\partial r_{n-1}}{\partial v_n} \quad . \tag{5b}$$

There are two characteristic lengths in the model, ℓ and $1/b$, and we shall denote their ratio $\beta = \ell b$. Furthermore, we introduce the mass density $\rho = m/\ell$. A characteristic time is $\ell \sqrt{\rho/a}$. Parameter values for DNA [5] are given in Table 1.

$\ell = 3.4 \times 10^{-10}$ m	$\beta = \ell b = 21$
$a = 5.13 \times 10^{-10}$ N	$\rho = m/\ell = 3.77 \times 10^{-15}$ kg/m
$b = 6.18 \times 10^{10}$ m^{-1}	$\ell\sqrt{\rho/a} = 9.2 \times 10^{-13}$ s
$m = 1.282 \times 10^{-24}$ kg	

Table 1. Parameter values for DNA [5].

It is convenient to introduce the longitudinal and transverse strains

$$\epsilon^{\alpha} u_n = (y_{n+1} - y_n)/\ell \tag{6a}$$

and

$$\epsilon^{\gamma} w_n = (v_{n+1} - v_n)/\ell \quad , \tag{6b}$$

where ϵ is an indicator of smallness. We shall consider asymptotic expansions of Eq. (5) for small elongations in four cases:

i) longitudinal strain larger than transverse strain ($\alpha = 1,\ \gamma = 2$)

ii) longitudinal and transverse strain of same order of magnitude ($\alpha = \gamma = 1$)

iii) longitudinal strain smaller than transverse strain ($\alpha = 2,\ \gamma = 1$)

iv) longitudinal strain much smaller than transverse strain ($\alpha = 3,\ \gamma = 1$).

The resulting equations have the form

$$\epsilon^{\alpha} \frac{\ell^2 \rho}{a} \ddot{u}_n = U_{n+1} - 2U_n + U_{n-1} \tag{7a}$$

and

$$\epsilon^{\gamma} \frac{\ell^2 \rho}{a} \ddot{w}_n = W_{n+1} - 2W_n + W_{n-1} \tag{7b}$$

with

$$U_n \simeq (1 - \exp(-br_n))\,(1 - \epsilon^{2\gamma} w_n^2/2) \tag{7c}$$

and

$$W_n \simeq \epsilon^{\alpha}(1 - \exp(-br_n))\, w_n \quad . \tag{7d}$$

In the case where not only $r_n/\ell < 1$, but also $r_n b = \beta r_n/\ell < 1$ we may truncate the Toda potential to

$$V(r_n) \simeq \frac{a}{b}\left[\frac{(br_n)^2}{2} - \frac{(br_n)^3}{6}\right] \tag{8}$$

and arrive at the Boussinesq approximations given in Table 2.

Case	α,γ	U_n	W_n
i	1,2	$\epsilon\beta u_n - \epsilon^2 \frac{\beta^2}{2} u_n^2$	$\epsilon^3 \beta u_n w_n$
ii	1,1	$\epsilon\beta u_n - \epsilon^2 \left[\frac{\beta^2}{2} u_n^2 - \frac{\beta}{2} w_n^2\right]$	$\epsilon^2 \beta u_n w_n$
iii	2,1	$\epsilon^2\beta (u_n + \frac{1}{2} w_n^2)$	$\epsilon^3 \beta(u_n + \frac{1}{2}w_n^2) w_n$
		$- \epsilon^4\left[\frac{\beta^2}{2} u_n + \beta(1 + \frac{\beta}{2}) u_n w_n^2 + \frac{\beta}{8} (3 + \beta) w_n^4\right]$	
iv	3,1	$\epsilon^2 \frac{\beta}{2} w_n^2 + \epsilon^3 \beta u_n$	$\epsilon^3 \frac{\beta}{2} w_n^3$

Table 2. Asymptotic expansions of U_n and W_n (Eq. (7)) for small longitudinal and transverse excitations.

CONTINUUM APPROXIMATIONS

In order to derive continuum approximations for the lattice equations (7–8) we follow Collins [7] and Rosenau and Hyman [8]. These authors developed ideas by Kruskal and Zabusky [9] and showed that

$$T(f_{n+1}) - 2T(f_n) + T(f_{n-1}) \to \left[1 - \frac{\ell^2}{12}\frac{\partial^2}{\partial x^2}\right]^{-1} \ell^2 \frac{\partial^2}{\partial x^2} T(f) \quad . \tag{9}$$

Here T is a nonlinear function of $f_n(t) \to f(x,t)$ with $x = n\ell$ in the continuum limit $n \to \infty$, $\ell \to 0$. Identifying $T(f_n)$ with U_n and W_n in Eq. (7) obtain the following continuum approximations for the longitudinal and transverse strains, u(x,t) and w(x,t), in the four cases:

i) $$\frac{\rho}{a} u_{tt} = \beta u_{xx} - \frac{\beta^2}{2} (u^2)_{xx} + \frac{\rho}{a}\frac{\ell^2}{12} u_{xxtt} \tag{10a}$$

$$\frac{\rho}{a} w_{tt} = \frac{\rho}{a}\frac{\ell^2}{12} w_{xxtt} \tag{10b}$$

ii) $$\frac{\rho}{a} u_{tt} = \beta u_{xx} - \frac{\beta^2}{2} (u^2)_{xx} + \frac{\beta}{2} (w^2)_{xx} + \frac{\rho}{a}\frac{\ell^2}{12} u_{xxtt} \tag{11a}$$

$$\frac{\rho}{a} w_{tt} = \beta(uw)_{xx} + \frac{\rho}{a}\frac{\ell^2}{12} w_{xxtt} \tag{11b}$$

iii) $$\frac{\rho}{a} u_{tt} = \beta u_{xx} + \frac{\beta}{2} (w^2)_{xx} + \frac{\rho}{a}\frac{\ell^2}{12} u_{xxtt} \tag{12a}$$

$$\frac{\rho}{a} w_{tt} = \frac{\rho}{a}\frac{\ell^2}{12} w_{xxtt} \tag{12b}$$

and

$$\text{iv)} \quad \frac{\rho}{a} u_{tt} = \frac{\beta}{2} (w^2)_{xx} + \beta\, u_{xx} + \frac{\rho}{a} \frac{\ell^2}{12} u_{xxtt} \tag{13a}$$

$$\frac{\rho}{a} w_{tt} = \frac{\beta}{2} (w^3)_{xx} + \frac{\rho}{a} \frac{\ell^2}{12} w_{xxtt} \quad . \tag{13b}$$

Here we have kept terms of order ϵ and ϵ^2 in Eqs. (10–12), and terms of order ϵ, ϵ^2, and ϵ^3 in Eq. (13), and then omitted the ϵ's. In all four cases w(x,t) can be replaced by −w(x,t). The longitudinal strain, u(x,t), does not have this symmetry property.

ON THE SOLUTIONS

In case i) the longitudinal field and the transverse fields decouple. u(x,t) satisfies (10a) which is the improved Boussinesq equation [9]. In the solitonic limit (of infinite periodicity interval and finite velocity) the travelling wave solution to this equation becomes

$$u(x,t) = -\frac{3}{\beta} (s^2-1)\, \mathrm{sech}^2 \frac{\sqrt{3(s^2-1)}}{s\ell} \left[x - s\sqrt{\frac{\beta a}{\rho}}\, t - x_o \right] \quad . \tag{14}$$

This compressional travels with the supersonic velocity, $s\sqrt{\beta a/\rho}$ ($s > 1$), $\sqrt{\beta a/\rho}$ being the sound velocity ($= 1.69 \times 10^3$ m/s for DNA).

Eq. (10b) with the periodicity condition

$$w(x,t) = w(x + jL, t) \qquad j = 1, 2, \cdots \quad , \tag{15}$$

when $L = N \cdot \ell$, has the trivial solution

$$w(x,t) = \cosh \left[\frac{\sqrt{12}}{\ell} \left[x - \frac{L}{2} \right] \right] T(t) \tag{16}$$

for $j = 1$. Here T(t) is an arbitrary function of time. Thus the transverse strain is a standing wave.

In case ii) where the longitudinal and transverse strains are of the same order of magnitude we investigate the ansatz

$$w = Au + B \quad \text{or} \quad u = w/A - B/A \tag{17a,b}$$

where A and B are constants. Eqs. (11a) and (11b) become identical if we choose

$$A = \pm \sqrt{2+\beta} \text{ and } B = \mp \sqrt{2+\beta}/(1+\beta) \quad . \tag{18a,b}$$

The resulting equation is

$$\frac{\rho}{a} u_{tt} = -\frac{\beta}{1+\beta} u_{xx} + \beta (u^2)_{xx} + \frac{\rho}{a} \frac{\ell^2}{12} u_{xxtt} \tag{19a}$$

or

$$\frac{\rho}{a} w_{tt} = \frac{\beta}{1+\beta} \pm \frac{\beta}{\sqrt{2+\beta}} (w^2)_{xx} + \frac{\rho}{a} \frac{\ell^2}{12} w_{xxtt} \quad . \tag{19b}$$

For infinite periodicity interval we find the travelling wave solution

$$u(x,t) = \frac{1}{1+\beta} \left[1 + \frac{3}{2} (s^2-1) \operatorname{sech}^2 \frac{\sqrt{3(s^2-1)}}{s\ell} \left[x - s\sqrt{\frac{\beta a}{(1+\beta)\rho}}\, t - x_o \right] \right] \tag{20a}$$

or

$$w(x,t) = \pm \frac{3}{2} \frac{\sqrt{2+\beta}}{1+\beta} (s^2-1) \operatorname{sech}^2 \frac{\sqrt{3(s^2-1)}}{s\ell} \left[x - s\sqrt{\frac{\beta a}{(1+\beta)\rho}}\, t - x_o \right] \quad . \tag{20b}$$

The longitudinal solitary wave (20a) differs from the uncoupled solitary wave (14) by being elongated instead of compressional, by existing as a superposition to the constant elongation $(1+\beta)^{-1}$ (like a "dark soliton" in the terminology of optical solitons), and by having a velocity, $s\sqrt{\beta a/(1+\beta)\rho}$, which may be smaller than the sound velocity (for $1 < s < 1.024$ in the case of DNA where $\beta = 21$). Thus the ansatz (17) that the longitudinal and transverse strains travel together (with the same velocity) makes this hybrid wave travel slower than the uncoupled longitudinal strain wave by a factor $\sqrt{1+\beta}$ and faster than the uncoupled transverse strain wave which is a standing wave with zero velocity.

In case iii) the transverse strain is again a standing wave (solution to Eq. (12b) which is identical to (10b)). This wave acts as a source term in the linear dispersive wave equation (12a) for the longitudinal strain which can be solved by separation of variables.

In case iv) the wave equation for the longitudinal strain (13a) is identical to Eq. (12a). However, the transverse strain in the source term is given by the nonlinear equation (13b) for which we have obtained solutions by separation of variables

$$w(x,t) = X\left[\frac{\sqrt{12}}{\ell} x\right] T\left[\frac{1}{\ell}\sqrt{\frac{6\beta a}{\rho}}\, t\right] \quad , \tag{21}$$

yielding the differential equations

$$X'' - C = - \Lambda(x^3)'' \tag{22a}$$

and

$$T^3 = \Lambda T'' \tag{22b}$$

for the functions X and T. Here prime denotes differentiation with respect to the arguments, $\frac{\sqrt{12}}{\ell} x$

and $\frac{1}{l}\sqrt{\frac{6\beta a}{\rho}}$ t, and Λ is the separation constant. Phase plane analysis of the integrated systems.

$$\left.\begin{aligned} X' &= \frac{\pm\sqrt{\frac{3\Lambda}{2}X^4 + X^2 + C_1}}{3\Lambda X^2 + 1} \\ X'' &= \frac{X(1 - 6\ (X')^2}{3\Lambda X^2 + 1} \end{aligned}\right\} \qquad (23a)$$

and

$$\left.\begin{aligned} T' &= \pm\sqrt{\frac{1}{2\Lambda}}(T^4 - C_2) \\ T'' &= \frac{T^3}{\Lambda} \end{aligned}\right\} \qquad (23b)$$

shows blow up for $\Lambda > 0$. (Example: For the integration constants $(C_1, C_2) = \left[\frac{1}{6\Lambda}, 0\right]$ we get the rational solution

$$w(x,t) = \pm\sqrt{\frac{2\rho}{3\beta a}}\,\frac{x-x_o}{t-t_o} \qquad (24)\ )\quad .$$

For $\Lambda < 0$ bounded solutions to Eq. (13b) are found.

CONCLUSION

For the Toda lattice with a transverse degree of freedom the dynamical equations are derived for spring elongations, which are small compared to the lattice constant, in cases of different orders of magnitude of the ratio between the longitudinal field and the transverse field. Continuum approximation leads to partial differential equations of improved Boussinesq type when the spring elongation is small also compared to the Toda length parameter (1/b). When the longitudinal field is small compared to the transverse field the fields decouple into a supersonic longitudinal solitary wave and a standing transverse wave. When the two fields are of the same order of magnitude a hybrid wave which may be subsonic is found. (The longitudinal wave then has the character of a "dark" compressional solitary wave). The stability of the hybrid wave is presently under investigation. When the transverse field is larger than the longitudinal field, we find a standing transverse wave acting as a source for linear dispersive longitudinal waves. The standing transverse wave may be nonlinear and has been found by separation of variables.

ACKNOWLEDGEMENTS

David Campbell and Salvatore Rionero are thanked for their helpful comments in discussion of this work. The studies were performed under the auspices of the US Department of Energy. One of us (PLC) expresses thanks for the warm hospitality of the Center for Nonlinear Studies, Los Alamos National Laboratory and the Department of Mathematics and Applications, The University of Naples and acknowledges support from Julie Damms Studiefond, the Danish Technical Research Council, and Consiglio Nazionale delle Ricerche, Rome, Italy.

REFERENCES

[1] M. Toda, Phys. Rep. 18 (1975) 1.

[2] H. Flaschka, Phys. Rev. B 9 (1974) 1924; Prog. Theor. Phys. 51 (1974) 703.

[3] H. Flaschka and D.W. McLaughlin, Prog. Theor. Phys. 55 (1976) 438; W.E. Ferguson, H. Flaschka, and D.W. McLaughlin, J. Comput. Phys. 45 (1982) 157.

[4] B.L. Holian, H. Flaschka, and D.W. McLaughlin, Phys. Rev. A 24 (1981) 2595.

[5] V. Muto, A.C. Scott, and P.L. Christiansen, Phys. Lett. A 136 (1989) 33.

[6] M. Peyrard and A.R. Bishop, Phys. Rev. Lett. 62 (1989) 2755.

[7] M.A. Collins, Chem. Phys. Lett. 77 (1981) 342.

[8] P. Rosenau, Phys. Lett. A 118 (1986) 222; J.M. Hyman and P. Rosenau, Phys. Lett. A 124 (1987) 287.

[9] M.P. Soerensen, P.L. Christiansen, and P.S. Lomdahl, J. Acoust. Soc. Am. 76 (1984) 871; M.P. Soerensen, P.L. Christiansen, P.S. Lomdahl, and O. Skovgaard, J. Acoust. Soc. Am. 81 (1987) 1718.

PART II

SOLID-STATE PHYSICS

CHARGE DENSITY WAVE AND SUPERCONDUCTIVITY:
LOCALIZED BIPOLARONS VERSUS EXTENDED COOPER PAIRS?

Serge AUBRY and Gilles ABRAMOVICI, Denis FEINBERG[1],
Pascal QUEMERAIS[2] and Jean-Luc RAIMBAULT[2]
Laboratoire Léon Brillouin (Laboratoire commun CEA-CNRS)
CEN Saclay 91191-Gif-sur-Yvette Cédex (France)

The electron-phonon coupling in conducting materials may produce at low temperature either a charge density wave (CDW) or standard superconductivity (SC). We analyse the limits of validity of these two kinds of theories which up to now are not well-defined on the particular HOLSTEIN model (where the non-interacting electrons are described by a tight binding model and are coupled to a dispersionless optic mode by on-site electron-phonon coupling). In the adiabatic limit (no quantum lattice fluctuations), we prove rigorously at any dimension that when the electron-phonon coupling is large enough, the ground-state and the first excited states can be described as a superlattice of bipolarons pinned to the lattice which forms the CDW. Then, the low temperature thermodynamics of the electron-phonon system can be described by an Ising spin Hamiltonian (lattice gas model). When the quantum lattice fluctuations are taken into account, this Hamiltonian becomes a quantum model with spins 1/2. For smaller electron-phonon coupling, we argue that the CDW with gapless phasons (or FROHLICH modes usually considered in the litterature) are unstable against quantum lattice fluctuations. A superconducting phase may then appear. The conjectured phase diagram at 0K thus involves either superconducting phase or pinned bipolaronic structures (CDW).

1-Introduction

At low temperature, the electron-phonon coupling is known to be responsible of very different physical phenomena in systems with conducting electrons, such as standard Superconductivity (SC) and Charge Density Waves (CDW). The standard theory of superconductivity (BCS)[1] assumes first that the electron-phonon coupling which is treated as a perturbation, can be eliminated by a canonical transformation and replaced by an effective attractive electron-electron interaction. Next, it is shown that the collective pairing of electrons with opposite spins takes place and forms the BCS superconducting state through an optimized variational form excluding other types of instabilities. The quantum character of the phonons is essential in this theory because the electron-electron attraction originates by the exchange of one or few quanta of phonon energy only.

---

[1] Permanent Address:
LEPES, CNRS 25 Avenue des Martyrs 166X 38042-GRENOBLE Cedex (France)

[2] Laboratoire de Physique Cristalline, Institut de Physique et Chimie des Matériaux
2 chemin de la Houssinière 44072-NANTES Cédex 03 (France)

By contrast, the standard theories of CDW[2] neglect the quantum character of the phonons. Then, it is shown that in some conditions which are favored by the low dimensionality of the systems, the energy cost due to a periodic lattice distortion which raises the degeneracy of the electrons at the Fermi surface ("nesting") may become negative. The system then exhibits a CDW associated to a periodic lattice distortion (PLD).

These two theories are based on different assumptions which seems to be incompatible. Thus, in order to get a better understanding of the competition between CDW and SC, it is useful to first focus our attention on one of the simplest models (Holstein model). We analyse the stability of the CDW structures against quantum fluctuations of the lattice. The aim of this talk is to shed a new light on this problem through a new approach based on previous works on incommensurate and chaotic structures in PEIERLS models[4] which is currently under development.

2-Definition of the HOLSTEIN Hamiltonian

The Holstein Hamiltonian is the sum of three terms

(1) $$H = H_k + H_{ep} + H_p$$

H_k is the Hamiltonian of a single band of non-interacting electrons (within a tight binding approximation)

(2-a) $$H_k = -T \sum_{<i,j>,\sigma} c^+_{i,\sigma} c_{j,\sigma}$$

where T is the exchange constant between neighboring sites <i,j> on a d-dimensional square lattice and σ is the electron spin $\pm\frac{1}{2}$ noted $\uparrow$ or $\downarrow$. $c^+_{i,\sigma}$ and $c_{i,\sigma}$ are the creation and annihilation Fermions operators of an electron at site i with spin σ respectively. The band width is thus 4T d.

H_p is the Hamiltonian of quantum phonons corresponding to a dispersionless optical branch

(2-b) $$H_p = \sum_i \hbar \omega_0 \left(a^+_i a_i + \frac{1}{2} \right)$$

where a^+_i and a_i are the creation and annihilation boson operators of phonons at site i respectively. The on-site electron-phonon interaction with constant g is represented by the Hamiltonian

(2-c) $$H_{ep} = g \sum_i n_i \left(a^+_i + a_i \right)$$

the electronic density operator at site i is

(2-d) $n_i = c^+_{i,\uparrow} c_{i,\uparrow} + c^+_{i,\downarrow} c_{i,\downarrow}$

In this model, two dimensionless independant parameters can be defined. They are:

(3-a) $$\gamma = \frac{\hbar\omega_0}{T}$$

which measures the "quantum character" of the phonons and

(3-b) $$k = \frac{2g}{\sqrt{\hbar\omega_0 T}}$$

which is the reduced electron-phonon coupling constant in the classical phonon limit. By setting, the position operator

(4-a) $$u_i = \frac{\sqrt{\gamma}}{2}\,(a^+_i + a_i)$$

and its conjugate operator (the commutator is $[u_i, p_i] = i$)

(4-b) $$p_i = \frac{i}{\sqrt{\gamma}}\,(a^+_i - a_i)$$

the initial Hamiltonian (1) becomes in unit of energy 2T:

(5) $$\hat{H} = \frac{H}{2T} = -\frac{1}{2}\sum_{<i,j>\,\sigma} c^+_{i,\sigma} c_{j,\sigma} + \frac{k}{2}\sum_i n_i u_i + \frac{1}{2}\sum_i \left(u_i^2 + \frac{\gamma^2}{4} p_i^2\right)$$

It is useful first, to analyse the limit cases γ small (adiabatic limit) and γ large (antiadiabatic limit) and next to try to understand the crossover between these two limits.

3-Adiabatic Limit

For most real systems, the phonon energies with typical energies $h\,\omega_0$ are of the order of magnitude of several hundred degrees K. These energies are much smaller than the typical band energies T by factors 100 or more.The adiabatic limit is obtained by setting γ=0 in (5). Standard theories of CDW use this adiabatic approximation. As we shall see in the next, this approximation is different of the Born-Oppenheimer approximation which takes partially the quantum lattice fluctuations into account. Thus, the adiabatic solution may be *qualitatively* modified by quantum lattice fluctuations even for very small values of γ.

For $\gamma=0$, $\{u_i\}$ can be considered as scalar variables in (5) instead of being an operator. Then, the creation operator of an electron in the eigenstates μ and spin σ is:

$$\text{(6-a)} \qquad c^{+}_{\mu,\sigma} = \sum_i \Psi^{i}_{\mu} \; c^{+}_{i,\sigma}$$

where $\{\Psi^{i}_{\mu}\}$ fulfills the eigen equation

$$\text{(6-b)} \qquad -\sum_{j:<i,j>} \Psi^{j}_{\mu} + k\, u_i\, \Psi^{i}_{\mu} = E_{\mu}(\{u_i\})\, \Psi^{i}_{\mu}$$

depends on the variables $\{u_i\}$. (The sum in (6-b) $\sum_{j:<i,j>}$ is done for the nearest neighbouring sites j of i). For the ground-state and the low-lying excitations of this model, there are two electrons with opposite spins which occupy the eigenstates with energy smaller than the Fermi energy E_F. Then the energy (5) is a function $\Phi(\{u_i\}))$ of the scalar variables $\{u_i\}$

$$\text{(6-c)} \qquad \Phi(\{u_i\})) = \frac{1}{2}\sum_i u_i^2 + \sum_{E_\mu<E_F} E_{\mu}(\{u_i\})$$

The ground-state of the system is obtained by minimizing this variational form. When this functional has many local minima with energy close to the energy of the ground state, these states which are metastable, corresponds to the low-lying excitations which determine the low-temperature behavior of the system. In one dimension, this model has been studied and it has been shown that the ground-state exhibits a Transition by Breaking of Analyticity (TBA).

4-The Transition by Breaking of Analyticity in One-dimensional Peierls Models

The ground-state of (6-c) exhibits a periodic lattice distortion (PLD) with wave-vector $2k_F$ (k_F is the Fermi wave-vector) associated with a CDW. When the electronic band filling (number of electron pairs per site) is an irrational number ζ, the CDW-PLD state is degenerate with respect to the variation of its phase α. Then, it has been observed that this model exhibits at zero K, a transition by breaking of analyticity (TBA) as a function of the electron-phonon coupling constant k. This TBA was found to be perfectly similar to those of the Frenkel-Kontorowa (FK) model (with the same critical behavior). But, this last model has the advantage that the existence of the TBA and of many of its characteristics can be analysed *rigorously* [4,9].

Let us briefly recall here, the main characteristic of the TBA in this Holstein model. When k is smaller than some critical value $k_c(\zeta)$, the coordinates of the incommensurate CDW-PLD ground-state depend analytically on the phase ("analytic state"), and in addition this state is phase *undefectible* (or charge undefectible). This means that there exists no local minima of (6-c) with the same band filling ζ, not belonging to the family of incommensurate ground-states with arbitrary phase α. For example, if one adds to the system a *localized spinless* pair of electrons, it cannot remain localized in a metastable state and extends over the whole system. In that regime, the CDW phonon excitations which correspond to phase fluctuation are gapless and form the well-known phason branch.

(However, let us remark that taking into account the pair breaking of the electrons, the CDW-PLD ground-state is spin-defectible, which means that that this state can accept a localized defect with a spin $\pm\frac{1}{2}$ as in the well-known SSH model for polyacethylene. But in that case, there is no chaotic states with spins when the system is charge undefectible).

When k is larger than the critical value $k_c(\zeta)$, the coordinates of the incommensurate CDW-PLD depend discontinuously on the phase ("non-analytic state"), and then this state is phase *defectible* (or charge defectible). The functional energy (6-c) has infinitely many metastable states (local minima) with the same band filling ζ, other than the incommensurate ground-states. Most of these states are chaotic with a finite entropy. In that situation, the ground-state accepts *localized charge defects* and these configurational excitations are *gapless* while the phonon phase excitations have a finite gap (unlike the "analytic" regime). In this regime, the CDW and its metastable states can be interpreted as an incommensurate order of *localized bipolarons* and as glass of bipolarons respectively. (We extend the concept of a bipolaron which is usually well defined only when it is single). Then, it corresponds to a pair of electrons with opposite spins localized in a potential well due to the lattice distortion created self-consistently by the electron-phonon coupling. In section 6, we shall show that we can still define explicitly the bipolaronic lattice distortion of a single bipolaron for a many bipolarons system) .

Charge undefectibility characterizes experimentally the conducting materials while charge defectibility characterizes the insulating materials. Thus the TBA is a metal-insulator transition at zero degree K. More details about this transition were given in refs.3,4 .

Standard theories of CDW assume implicitely that the CDW is always in the conducting "analytic regime" ($k<k_c(\zeta)$). Up to now, our arguments for proving the existence of a TBA in a CDW model which is in contradiction with these standard theories, were essentially numerical and thus could be controversed. We can now provide a theorem valid at any dimension for the adiabatic Holstein model, which gives a *rigorous* fundation to this conjecture.

5- Bipolaronic States in the Adiabatic Holstein Model

When the electron-phonon coupling is large enough in the adiabatic Holstein model *at any dimension* , we prove the following theorem:

Theorem (1989)[6]

Let us choose an arbitrary configuration of pseudo-spins $\{\sigma_n = 0 \text{ ou } 1\}$ *such that* :

$\sum_n \sigma_n = r$ *be the number of electron pairs of a finite system with s sites with for example periodic boundary conditions (s is arbitrarily large and* $n \in Z^d$ *is a site of the d dimensional square lattice). Then, for*

(7-a) $$k > 2\sqrt{5d}$$

there exists one and only one local minima $\{u_n\}$ *of the functional energy* (6-c) *of the adiabatic Holstein model, such that*

(7-b) $$u_n = -k \ \rho_n$$

where ρ_n *is the electronic density of the* CDW *at site n and such that*

(7-c) $$|\rho_n - \sigma_n| \leq \frac{1}{4}$$

These metastable states have the following properties:

1-The electronic eigenstates exhibits a finite gap in energy with a lower band occupied and an upper band empty.

2- The phonon excitations exhibits a finite gap in frequency

3- The amplitude of a local perturbation at site n obtained for example by fixing an atom i *at a position different of its equilibrium position , decreases faster than a exponential of the distance between site i and n.*

The detailed proof of this theorem is too long for being described here and will be given elsewhere[6]. Let us just give the basic strategy which we used for proving this theorem:

1-We define an non-linear operator $\mathcal{O}$ in a Banach space for which the fixed points are the extrema of the variational form (6-c)

2- For each pseudospin configuration, we associate a domain $\mathcal{D}(\{\sigma_n\})$ by condition (7-c). When k is large enough, we prove that this domain $\mathcal{D}(\{\sigma_n\})$ is invariant by this operator $\mathcal{O}$ and

3- that the spectral norm of the linearized operator $\nabla\mathcal{O}$ is strictly smaller than 1 inside this domain..

Consequently, this operator contracts the metric distance in the domain so that the Banach fixed point theorem (1921) can be applied. This theorem asserts the existence of a unique fixed point inside each domain $\mathcal{D}(\{\sigma_n\})$ which thus proves the main part of the theorem. Let us note that the same strategy applied to the FK model allows one to prove the existence of chaotic states but in that case, the proof is much easier and shorter. Extensions and variations of this theorem and of its methods of proof to other adiabatic models and related results seem to be possible and are currently under study (but there will be slight changes in the theorems, for example when the phonons are not dispersionless).

Let us discuss some physical implications of this theorem. In this specific Holstein model, for any pseudo-spin configuration $\{\sigma_n\}$, there exists a metastable state of the variational energy form (6-c) such that the r maxima and the s-r minima of the associate electronic density $\{\rho_n\}$ corresponds to the sites where $\sigma_n=1$ and $\sigma_n=0$ respectively. All the occupied electronic states are below an energy gap while the unoccupied electronic states are above. In other words, one can choose arbitrarily the location of the peaks of the electronic density which physically means the location of the bipolarons. More precisely, we can say that site n is occupied by a localized bipolaron when the pseudospin σ_n is 1 and that this site n is empty when $\sigma_n=0$. (However, it is clear that the bipolarons are not strictly localized on single sites but extend more or less around the occupied sites.)

Athough the bipolarons can be considered as localized in the real space (because one can choose arbitrarily their location), the electronic eigenstates are not necessarily localized. (Nevertheless, the proof of this theorem works whatever the electronic states are). For example if the configuration of pseudo spins $\{\sigma_n\}$ is chosen periodic, the electronic states are extended Bloch eigenstates. If this configuration is random, they might be exponentially localized. In fact, in this model, the concept of localization of the electrons is physically meaningless when the (undistinguishable) electrons *are not independant particles.* The excitations of these metastable states are global excitations of the whole system of electrons coupled with the phonons which thus involve self-consistently the electronic density and the potential created by the lattice deformation. In such a situation, *the concept of defectibility replaces and generalizes the concept of localization.* Then, it is clear that these bipolaronic states are insulating at zero degree.

In the adiabatic approximation, the phonon frequencies ω_ν which correspond to small motions of the atoms (which are assumed to have a unit mass) are determined by the eigen equation

$$(8) \qquad \omega_\nu^2 \, u_n^\nu = \sum_m \frac{\partial^2 \Phi(\{u_i\})}{\partial u_n \partial u_m} u_m^\nu$$

We have proven for this theorem, that the set of eigenvalues ω_ν^2 of the quadratic form associated with the second order expansion of the adiabatic energy are strictly larger than a finite positive value ω_G^2 which implies the absence of a gapless phason branch. In addition, the local perturbation created for example by an impurity at some site n decays faster than an exponential far away from this impurity. In fact, numerical studies in 1-d have shown that the decay is exponential and the corresponding characteristic length is the coherence length ξ. (However ξ depends on k and on the pseudospin configuration $\{\sigma_n\}$).

Although we know that there exist many other metastable states involving atomic distortions with smaller amplitudes, for large enough k, the ground-state presumably belongs to the set of metastable states defined by this theorem, but for a special choice of the pseudo-spins $\{\sigma_n\}$. We have not proven this conjecture but we have numerical evidence and a strong physical intuition that it should be true.

As for the FK model[7], for large enough electron phonon coupling k, these chaotic states are those which are relevant for studying the thermodynamics of the system in 1,2 or 3 dimensions at low temperature. Since the configuration of the sytem are represented by a pseudo-spin configuration, the initial Hamiltonian can be expanded on this basis as a lattice gas model. This is the representation which has been also found by Alexandrov et al [8] within a completely different approach. In ref [4], we found the beginning of the expansion of this Hamiltonian as a function of $\frac{1}{k}$. In one-dimension, this Hamiltonian has indeed an incommensurate ground-state with the expected modulation wave-vector $2k_F$. But let us emphasize that for two or three dimensional models, the ground-state which corresponds to a particular arrangement of the bipolarons is unknown. They could interpret observed CDWs in 2-d systems.

As in the one dimensional model, when the electron-phonon coupling k becomes smaller, it can reasonably be expected that the bipolaron size also diverges in d-dimensional models at some critical value of k (which depends on the band filling) Then all the metastable states disappear after complex cascades of inverse bifurcations as for the FK model! . But unlike the 1-d adiabatic model which is always unstable against a CDW-PLD, the adiabatic ground-state should present for smaller k another transition to a state with no lattice distortion.

6- Effective Bipolaronic Shape in the Adiabatic 1-d Holstein Model

Since we have noted the similarity of the 1-d Holstein model and of the FK model, it is tempting to check if the "decomposition theorem" obtained for the FK model applies here. For the FK model, this theorem asserts that the "non-analytic" incommensurate ground-states can be decomposed as a *linear superposition of effective discommensurations.* The shape of these discommensurations is not those of a single one but is effective because it depends on the presence of the other discommensurations. The size of this discommensuration is the coherence length ξ and diverges when k approach $k_c(\zeta)$ from above. We found numerically the same result in the adiabatic Holstein model (and hope to prove it rigorously).More precisely, the PLD of the groundstate can be written as

$$\text{(9-a)} \qquad u_n = -\sum_i \sigma_i b_{n-i}$$

with a sequence of pseudospins $\{\sigma_i\}$ defined as

$$\text{(9-b)} \qquad \sigma_i = \chi(i\zeta+\alpha)$$

where function $\chi(x)$ has period 1, is equal to 1 for $0<x\leq\zeta$ and to zero for $\zeta<x\leq 1$. α is the arbitrary phase. The sequence $\{b_i\}$ is shown for two examples, figure1. This result suggests that the bipolaron are well-defined objects at least for the "non-analytic" incommensurate ground-states and its excited states. It size diverges at $k_c(\zeta)$ and it does not exist below $k_c(\zeta)$ when the ground-state is incommensurate and "analytic".

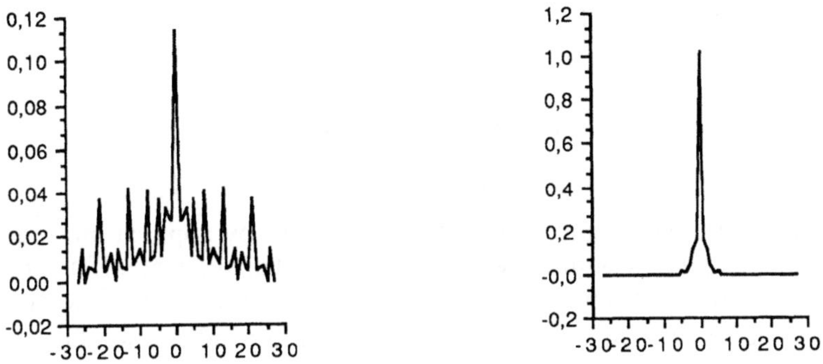

Figure 1: *Bipolaron shape* $\{b_i\}$ *versus* i *in the 1-d adiabatic Holstein model for* $\zeta=\frac{\sqrt{5}-1}{2}$ *for* $k=1.58\approx k_c(\zeta)$ *(left) and* $k=1.7$ *(right). This bipolaron is well localized far above the TBA and diverges when approaching* $k_c(\zeta)$ *from above. It is undefined for* $k<k_c(\zeta)$.

Let us now examine, the effect of quantum lattice fluctuations on the adiabatic solutions.

7- Quantum lattice fluctuations for small γ

The Born-Oppenheimer approximation consists of quantizing the atomic motion in the adiabatic potential with the Hamiltonian

(10-a) $$\mathbf{H} = \frac{\gamma^2}{4}\sum_i \frac{1}{2}p_i^2 + \Phi(\{u_i\})$$

Within this approximation, the electrons are always in adiabatic equilibrium with the atomic configuration. Clearly, they cannot order by themselves into a superconducting order. The global wave function of the system has the form

(10-b) $$\Psi = \Psi(\{u_i\}) \prod_{\mu\ \mathrm{occ},\sigma} c^+_{\mu,\sigma}(\{u_i\})\ |\mathrm{vacuum}\rangle$$

where $c^{+}_{\mu,\sigma}(\{u_i\})$ is defined by (6-a) and (6-b). In fact, this approximation is in general very hard to handle explicitly because the adiabatic energy $\Phi(\{u_i\})$ is a highly non-linear functional as suggested by the above theorem. In practice, one can use quadratic expansions around the local minima of $\Phi(\{u_i\})$.

When the bipolaronic structure is "non-analytic" (large k), we can perform a more suitable approach. The adiabatic states are represented as coherent states

$$(11\text{-a}) \qquad |\{u_i\}> = \prod_{\mu\ \mathrm{occ}} c^{+}_{\mu,\uparrow} \prod_{\mu\ \mathrm{occ}} c^{+}_{\mu,\downarrow} \times \exp\left(\sum_i \frac{u_i}{\sqrt{\gamma}} (a^{+}_i - a_i) \right) | \mathrm{vacuum} >$$

which consists of creating the PLD onto the vacuum by application of an appropriate distortion operator together with the electrons in the adiabatic state. Since there are many metastable states, one can calculate the overlaps between them. One find after some calculations, the exact overlaps between two arbitrary adiabatic states $|\{u_i\}>$ and $|\{u'_i\}>$ which are

$$(11\text{-b}) \qquad <\{u'_i\}|\{u_i\}> = (\det \bar{\bar{A}})^2 \exp\left(-\frac{1}{2\gamma} \sum_{i=1}^{S} (u_i - u'_i)^2 \right) \quad \text{and}$$

$$(11\text{-c}) \qquad <\{u'_i\}|\hat{H}|\{u_i\}> = \left(S\frac{\gamma}{4} + \frac{\Phi(\{u_i\}) + \Phi(\{u'_i\})}{2} - \frac{1}{4}\sum_{i=1}^{S} (u'_i - u_i)^2 \right) <\{u'_i\}|\{u_i\}>$$

($\det \bar{\bar{A}}$ is the determinant of the $r \times r$ matrix $\bar{\bar{A}} = \left(A_{\mu',\mu}\right)$ defined as

$$(11\text{-d}) \qquad A_{\mu',\mu} = < \psi'^{\mu'} | \psi^{\mu} > = \sum_i \psi'^{\mu' *}_i \psi^{\mu}_i$$

where $|\psi^{\mu}>$ and $|\psi'^{\mu'}>$ are the occupied electronic states given by (6-b) for the adiabatic state $|\{u_i\}>$ and $|\{u'_i\}>$ respectively.

For $k \gg k_c(\zeta)$ and small γ, it is easy to check that these adiabatic states $|\{u_i\}>$ are orthogonal one with each other because of the exponential in (11-b). For example, for two adiabatic states which differs only by the location of a bipolaron, one easily find the estimation

$$(12) \qquad <\{u'_i\} \mid \{u_i\}> \approx \exp(-\frac{k^2}{\gamma}).$$

For other adiabatic states which differs by the location of more and more pseudospins, the overlap becomes smaller and smaller. To fix the ideas, k=2 and g=10^{-2} yields an overlap of e^{-400} which clearly is physically neglegible. Clearly, the adiabatic metastable configurations are practically unaffected by the existence of small quantum lattice fluctuations.

When k approach $k_c(\zeta)$ from above, we noted that the bipolaronic size diverges, which implies that the overlap between close adiabatic states goes to 1.More precisely, the overlap calculated in (12) becomes roughly proportional to $\exp(-\frac{E_{PN}}{\gamma})$ where E_{PN} is the Peierls-Nabarro energy barrier that the height of the barrier which has to be overcome for moving a bipolaron of one lattice site. This feature is a proof that the above quantized adiabatic states (11-a) becomes very bad quantum states at the TBA.

In the bipolaronic regime $k>k_c(\zeta)$, one can consider the quantum lattice fluctuations as a correction to the lattice gas Hamiltonian of bipolarons. The extra terms appears as transverse spin operators which allows the tunnelling of bipolarons from one occupied site to an empty site. Setting

$$(13\text{-a}) \qquad S_i^z = \sigma_i - \frac{1}{2}$$

the spin Hamiltonian expands with the general form

$$(13\text{-b}) \qquad \mathbf{H} = \sum_i h\, S_i^z + \sum_{i,j} J_{i-j}\; S_i^z S_j^z + \sum_{i,j,k} K_{i-k,j-k}\, S_i^z S_j^z S_k^z + \ldots - \sum_{i,j} \Gamma_{i-j}\, (S_i^x S_j^x + S_i^y S_j^y) + \ldots$$

Thus for small γ, one can explicitly calculate the first terms of this expansion (see ref.4). At the lowest order, one finds again the same Hamiltonian as obtained by ALEXANDROV et al 1986 for describing bipolaronic superconductivity in this specific model. In their representation, superconductivity is obtained if the average of the transverse component $<S_i^x>$ is non zero. Howevér close to the adiabatic limit and the quantum terms in (13-b) which fulfills[4]

$$(13\text{-c}) \qquad \Gamma_{i-j} \approx J_{i-j} \exp(-\frac{4\, E_{PN}}{\gamma})$$

$$\approx J_{i-j} \exp(-\frac{k^2}{\gamma}) \qquad \text{for large k,}$$

are negligible unless k becomes close above to $k_c(\zeta)$ so that $E_{PN} \to 0$. Then the bipolaronic model collapses because due to the divergence of the bipolaron size, the series in (13-b) tends to diverge.

Among the arguments which we developped in ref.6 for proving the unstability of "analytic" CDW under quantum lattice fluctuations, we suggested that for $k<k_c(\zeta)$, when the phase fluctuations of the incommensurate "analytic" CDW extends over a large enough distance (which is about the size of a Cooper pair!), then it gains more energy by quantum tunnelling than it looses by the elastic deformation.

We are currently performing more sophisticated expansions with respect to the quantum parameter γ. At the present stage of our work, all attempts confirm that the adiabatic "analytic" CDW are far from a good quantum state of the Hamiltonian. Although, we have no rigorous proof, our present conjecture remains that the existence of a gapless phason branch for a CDW structure makes this structure unstable with respect to quantum lattice fluctuation whatever γ is small (but non zero)!

On the other hand, as we demonstrate above, bipolaronic states should remain stable providing that the quantum parameter be not too large $\gamma \ll k^2$.

8- The antiadiabatic limit $\hbar\omega_0 \gg T$

This regime is physically unusual but is plausible when the electrons are not coupled to phonons but to other excitations of the solid with much higher energies (plasmons, excitons?). In that regime, the Lang-Firsov unitary transformation $\exp(iS_{LF})$ (e.g. see ref. 8) transforms the Hamiltonian (1) into

(14-a) $$\hat{H} = \exp(iS_{LF})\, H \exp(-iS_{LF})$$

with

(14-b) $$i\, S_{LF} = \frac{g}{\hbar\omega_0} \sum_i \left(\sum_\sigma c^+_{i,\sigma} c_{i,\sigma}\right)(a^+_i - a_i)$$

One obtains

(14-c) $$\hat{H} = T\left[- \sum_{\langle i,j\rangle,\sigma} \hat{t}_{i,j}\, c^+_{i,\sigma} c_{j,\sigma} - \frac{1}{4} k^2 \sum_i \left(\sum_\sigma c^+_{i,\sigma} c_{i,\sigma}\right)^2 + \gamma \sum_i \left(a^+_i a_i + \frac{1}{2}\right) \right]$$

where

(14-d) $$\hat{t}_{i,j} = \exp\left(\frac{1}{2} \frac{k}{\sqrt{\gamma}} \left((a^+_i - a_i) - (a^+_j - a_j) \right)\right)$$

When

(15-a) $$k^2 \ll \gamma$$

or equivalently $g \ll \hbar\omega_0$ (this is the regime opposite to the regime of stability of the bipolaronic structures against quantum fluctuations)

(15-b) $\hat{t}_{i,j} = 1$

Then, one obtain an Hamiltonian where the electrons and the phonons *exactly decouple*. An electron-electron interaction is left which is on-site with the coefficient

(15-c) $-\frac{Tk^2}{2} = -U$

This situation allows one to try the standard BCS form for superconductivity

(16-a) $$\prod_q (u_q + v_q\, c^+_{q,\uparrow}\, c^+_{-q,\downarrow}) \mid \text{vacuum}>$$

where c^+_q is now the creation operator of an delocalized electron with wave vector q and spin $\uparrow$. The modulation u(q) and v(q) now is in the reciprocal space. This specificity allows the existence of quantum states (under magnetic field) of the whole system with a permanent current and thus of superconductivity.

(16-b) $$\prod_q (u_q + v_q\, c^+_{q+\delta,\uparrow}\, c^+_{-q+\delta,\downarrow}) \mid \text{vacuum} >$$

NOZIERES et SCHMITT-RINK[10] have studied this negative U Hubbard model.

9-Conclusion:

As we suggested above, the "analytic" incommensurate CDW should be unstable with respect to quantum lattice fluctuations. Thus, we conjecture that the phase diagram at zero degree K of the Holstein model should be shared into two main regions according to the role of the quantum lattice fluctuations (cf. fig. 2):

1- The bipolaronic region which starts from the adiabatic line γ=0 and from the critical point of the TBA when the band filling ζ is irrational. In that region, sufficiently far away from the critical line (according to (13-c) and (15-a), $\gamma \approx 4E_{PN} \approx k^2$ for large k), the quantum lattice fluctuations have a small effect . (For the commensurate band filling, this line reaches the origin in 1-d but should converge to the incommensurate line in the limit of high order commensurability.)

2- The superconducting region in the remainding part of the phase diagram, where the quantum lattice fluctuations always prevent the formation of a structure in real space.

It can be reasonably expected for models in more than 1 dimension, that there exist many other critical lines (perhaps infinitely many) inside the bipolaronic region which determine regions with different bipolaronic arrangements. In addition, superconductivity should behave very differently depending on the region of the phase diagram where it is supposed to exist (e.g. close to the adiabatic line for irrational ζ opposed to the region with both large k and $\gamma >> k^2$).

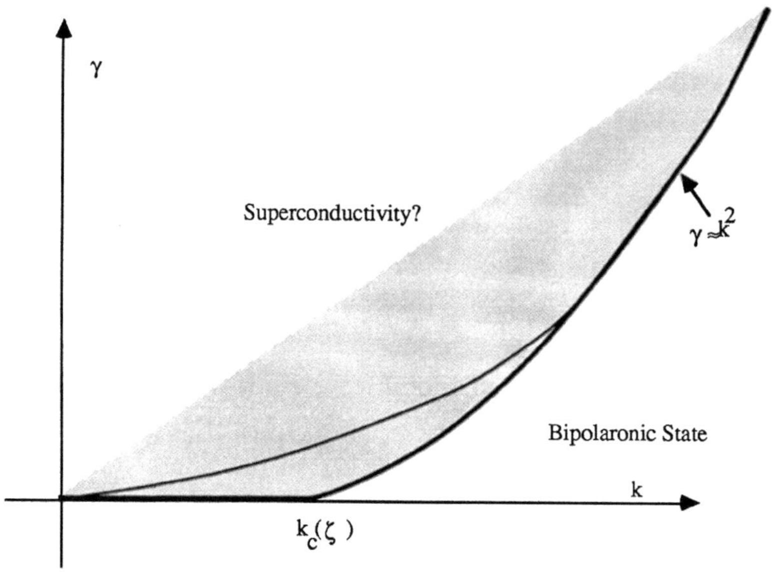

Figure 2 *Scheme of the conjectured phase diagram in* k *and* γ *of the Holstein model at zero degree K. A line separates the bipolaronic states from the superconducting region . For large* k, *this line follows approximately* $\gamma \approx k^2$ *but depends on the band filling for small* k. *For irrational band filling, it should reach (with an exponent?) the* k *axis at* $k_c(\zeta)$.

[1] e.g see C. KITTEL *Quantum Theory of Solids* Eds John WILEY & SONS (1976)
[2] P.A. LEE, T.M. RICE and P.W. ANDERSON *Phys.Rev.Lett.* **31** 462 (1973)
[3] P.Y. LE DAERON P.Y. and S. AUBRY J. Phys. **C16** 4827-4838 (1983) and J.Physique (Paris) **C3** 1573-1577 (1983)
[4] S.AUBRY and P.QUEMERAIS in *Low-Dimensional Electronic Properties of Molybdenium Bronzes and Oxides* 295-405 Ed. C.Schlenker, Kluwer Academic Publishers Group. (1989)
[5] S.AUBRY and P. QUEMERAIS "Lattice Pinning of a CDW and Quantum Fluctuations" in " Singular Behaviour & Nonlinear Dynamics" Ed. St. Pnevmatikos, T. Bountis & Sp. Pnevmatikos, in press (1989)
[6] S.AUBRY, G.ABRAMOVICI and J.L. RAIMBAULT, P. QUEMERAIS in preparation (1989)
[7] F.VALLET, R. SCHILLING, S. AUBRY Europhysics Letters **2** p.815-822 (1986) and J. Phys. **C21** 67-105 (1988)
[8] A.S. ALEXANDROV, J. RANNINGER and S. ROBASZKIEWICZ *Phys.Rev.* **B33** 4526-4542 (1986)
[9] S. AUBRY S. "Structures Incommensurables et Brisure de la Symétrie de Translation" in "Structure et Instabilité" Ed. C.Godrèche Edition de Physique (1986) pp.73-194
[10] P. NOZIERES and S. SCHMITT-RINK J. of Low Temp. Phys. **59** 195 (1985)

MODULATED PHASES IN CRYSTALS : SYMMETRIES OF WALLS AND WALL LATTICES . EXAMPLE OF QUARTZ

P. Saint-Grégoire (*) and V. Janovec (**)

(*) GDPC (URA CNRS n°233), USTL,case 026, 34060 MONTPELLIER CEDEX, F
(**) Institute of Physics, CsAV, Na Slovance 2, 18040 PRAGUE 8, CSSR

I . INTRODUCTION

The atomic displacements in modulated phases of real crystals obey non-linear equations which do not have in general exact solutions. A possible approach of the problem is to simplify the equations in a drastic way, and, e.g., to treat the system as a linear chain. Another approach is just to keep the maximum number of informations on the concrete system and to try to deduce other properties. In this respect, the tools offered by the theory of symmetry are very important, since they allow to treat the problem without simplifying assumptions.

The aim of this paper is to present such an approach to modulated phases which can be described by a lattice of domain walls, called also discommensurations. The incommensurate phase of quartz-type is used as an illustrative example.

II . STRUCTURE OF INCOMMENSURATE PHASES. EXAMPLE OF QUARTZ

a) General properties of incommensurate phases

Most of incommensurate phases are observed in a temperature region between two periodic crystalline phases labelled N ("normal", at higher temperatures) and C ("commensurate", at lower temperatures). Since the vector **k** associated with the wave of atomic displacements, is temperature dependent, the incommensurate phase cannot be described by a simple order parameter but by a continuous set of order parameters, indexed by **k**. Nevertheless, one gets a simplification by introducing the order parameter η of the N-C phase transition, which is well defined if the symmetry group of phase C is a subgroup of that of N. Then the incommensurate phase can be described as a spatial modulation of phase C. For this reason, the term modulated phase is also used (for further details see e.g. ref /1/ and /2/).

An important class of materials presenting an incommensurate phase, is constituted by those whose order parameter η (of N-C structural change) does not fulfil the so called Lifshitz criterion . It implies the presence, in the free energy expansion, of an antisymmetric term which couples a component of η to the spatial derivative of another one. To be concrete, let us consider the case where η has two components p and q, the expression (p dq/dx - q dp/dx) (Lifshitz term) being invariant under the symmetry operations of phase N. The free energy density can be written as /2/

$$f(x) = f_o + \frac{\alpha}{2}(p^2 + q^2) + ... + \delta\ (p\frac{dq}{dx} - q\frac{dp}{dx}) + \frac{\varkappa}{2}(\ (\frac{dp}{dx})^2 + (\frac{dq}{dx})^2) + .. \quad (1)$$

Because of the Lifshitz invariant, the loss of stability when the coefficient α changes sign, does not lead to the C phase but to a state with inhomogeneous displacements. This can be more easily understood after introducing new variables ρ and θ :

$$p = \rho \sin \theta,\ q = \rho \cos \theta \quad (2)$$

$$f(x) = f_o + \frac{\alpha}{2}\rho^2 + ... - \delta\ \rho^2 \frac{d\theta}{dx} + \frac{\varkappa}{2}\{(\frac{d\rho}{dx})^2 + \rho^2(\frac{d\theta}{dx})^2\} + .. \quad (3)$$

The third term lowers the energy for inhomogeneous θ but spatial variations of θ are limited by the last positive term. To obtain the wave vector variation and the transition to the C phase, one has to add (at least) another term $\rho^n\cos(n\theta)$ which is called anisotropic because of its θ dependence. With e.g. a positive coefficient in front of it, it is minimal for $n\theta = (2l+1)\pi$. Close to T_i (transition temperature between phases N and I) the amplitude ρ is very small and the anisotropic term is negligible : the solution which minimizes the free energy is sinusoidal. On decreasing the temperature from T_i, there is a competition between the Lifshitz and the anisotropic terms, which results in a change of k and of the form of modulation. These solutions obey the Euler-Lagrange equations derived from the expression of f(x), which are non-linear and which in general do not have exact solutions. However, on approaching the transition temperature T_c (to C phase), in the major part of the period L of the modulation, the order parameter η is associated with values of θ which minimize the anisotropic term and correspond to the different domain states in the C phase. The structure of the incommensurate phase is then rather similar to a domain structure with domain walls regularly disposed in the x direction : far below T_i in the modulated phase, the sinusoidal structure changes into a periodic structure of domain walls of the C phase. One can attribute to them a negative energy in the I phase (because of Lifshitz invariant), and a positive energy in the C phase, where the anisotropic term is preponderant. At the I-C transition, their density theoretically goes to zero, but this is not the case in real crystals where walls are pinned by defects of crystal lattice.

b) case of quartz type phases

In the preceding paragraph, we presented incommensurate phases formed due to the violation of Lifshitz criterion. Modulated phases can, however, appear between N and C phases even if the order parameter satisfies the Lifshitz criterion /3/. Quartz belongs to this category, since its order parameter has a single component, which can be identified with the angle of rotation of SiO_4 tetrahedra around their twofold axes along x type directions /4/. This mechanism breaks the 6-fold symmetry of β phase (N phase) with the space group $P\ 6_2 2_x 2_y$. In the α phase (C phase) the twofold symmetry operations 2_y are lost and the hexagonal symmetry is reduced to trigonal $P\ 3_2 2_x$.

In the C phase, there are only two domain states, corresponding to positive and negative value of η. They are referred to as "Dauphiné twins". The loss of stability of β phase does not occur for displacements corresponding to the trigonal α phase, but for a modulated phase /5,6/. This is due to an invariant playing the role of the Lifshitz term, $(U_{xx} - U_{yy})\frac{\partial \eta}{\partial x} - 2\ U_{xy}\ \frac{\partial \eta}{\partial y}$, and which couples spatial derivatives of the order parameter η to the strain tensor U_{ij} . As a result, there occurs an instability with respect to modulated displacements, with three modulation wave vectors $\mathbf{k}_i$ (plus their opposites) lying at T_i along the directions of the lost twofold axes (y axes) /7/.

With such wave vectors at 120° from each other, there are several possibilities : the modulations are, e.g., spatially separated (forming so called stripe phases) or all modulations are superimposed. The latter case corresponds to the observations of the incommensurate phase by means of electron microscopy, where a regular pattern of equilateral triangles reflects the spatial modulation of the order parameter η /8,9/.

Both phases, stripe phase and equilateral triangles phase provide non trivial examples and will be used to illustrate our approach.

III . SYMMETRY OF SINGLE DOMAIN WALLS

Before investigating incommensurate structures, which we shall treat as a regular pattern of domain walls, we recall basic properties of single domain walls. A *domain wall* is a transient region between symmetrically equivalent C structures called domain states. The symbol

$S_i/\mathbf{n}/S_j$ is used for a planar coherent wall with domain state S_i on the negative side, and S_j on the positive side of the wall normal $\mathbf{n}$, which determines the orientation of the wall. Within this convention, the identity $S_i/\mathbf{n}/S_j = S_j/-\mathbf{n}/S_i$ always holds.

Symmetry group T_{ij} *of the domain wall* $S_i/\mathbf{n}/S_j$ contains operations that keep invariant orientation and position of the wall, i.e., T_{ij} is a layer group which consists of two parts /10/ :

$$T_{ij} = F_{ij} + \underline{t}'_{ij} F_{ij} \quad , \tag{4}$$

where F_{ij} comprises operations leaving invariant both domain states S_i, S_j and the normal $\mathbf{n}$, whereas $\underline{t}'_{ij}$ is an operation that exchanges domain states S_i, S_j and transforms $\mathbf{n}$ into $-\mathbf{n}$ (the former action is signified by a prime, the latter by underlining). Walls for which $\underline{t}'_{ij}$ exists are called *symmetric walls*.

Two walls are symmetrically equivalent, $S_i/\mathbf{n}/S_j \simeq S_k/\mathbf{n}'/S_l$, if there exists an operation from the symmetry group of the N phase, that transforms one wall into the other. Symmetrically equivalent walls have the same structure and the same energy. A wall $S_i/\mathbf{n}/S_j$ is called *reversible* if it is symmetrically equivalent with the reversed wall $S_j/\mathbf{n}/S_i$.

Symmetry group T_{ij} of a domain wall determines the number n_w of symmetrically equivalent domain walls according to a simple formula:

$$n_w = |G| : |T_{ij}|, \tag{5}$$

where $|G|, |T_{ij}|$ are orders (numbers of operations) of point groups G (symmetry of the N phase) and T_{ij}, resp.

The knowledge of wall symmetry T_{ij} is very useful in locating orientations with extremal wall energy and identifying the specific tensor properties of walls.

As an illustrative example, we shall investigate symmetry properties of domain walls in quartz. For simplicity, we use continuum description and consider walls parallel to z axis (6-fold axis of the β phase). Then, the orientation of a wall is specified by the angle Θ between the axis y_2 and the axis ξ attached to the wall (see fig.1c).

First examine the symmetry of a domain wall perpendicular to the coordinate axis x_2, $1/-\frac{\pi}{2}/2$, where $(-\frac{\pi}{2})$ specifies the angle Θ and 1 and 2 denote domain states $-\eta_o$ and $+\eta_o$, resp. (fig. 1a). This wall is transformed into itself by any 2-fold rotation axis 2_{x2} perpendicular to the wall, further by any 2-fold axis 2_{y2} and 2_z lying in the plane of the wall and parallel to the coordinate axis y_2 and z, resp. The former operations 2_{x2}, belonging to the symmetry group $3\,2_x$, preserve both domain states 1, 2 and also leave invariant the normal $\mathbf{n}$, whereas the latter rotations $\underline{2}'_{y2}$, $\underline{2}'_z$, which were lost at the phase transition,

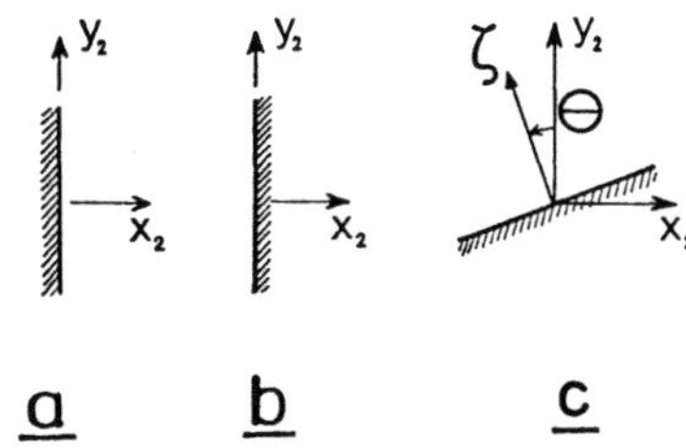

Fig.1 : Domain walls in quartz.
a) $1/(-\pi/2)/2$ b) $1/(+\pi/2)/2$
c) general orientation $1/\ \Theta\ /2$.

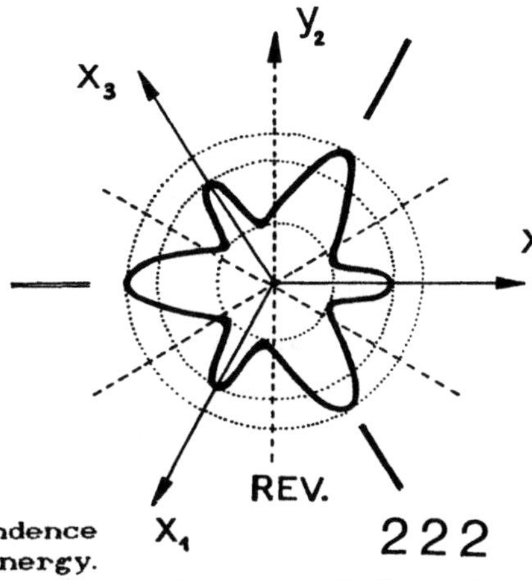

Fig.2 : Angular dependence $E(\Theta)$ of the wall energy. Reversible and prominent orientations are indicated.

exchange both domain states 1, 2 and, simultaneously, transform the wall normal **n** into antiparallel −**n**. All above mentioned operations transforming the wall into itself form the group $\ell\ 2_{x2}\underline{2}'_{y2}\underline{2}'_{z}$ where ℓ signifies a layer group.

The reversed domain wall $2/-\frac{\pi}{2}/1 = 1/\frac{\pi}{2}/2$ (see fig. 1b) has the same symmetry $\ell\ 2_{x2}\underline{2}'_{y2}\underline{2}'_{z}$ as the original wall $1/-\frac{\pi}{2}/2$. There is, however, no operation from the symmetry group $6\ 2_x 2_y$ of the β phase, which would transform the wall $1/-\frac{\pi}{2}/2$ into reversed wall $2/-\frac{\pi}{2}/1$. These two walls are, therefore, not symmetrically equivalent; i.e., they have different structure and different energy.

The wall 1/0/2, which is perpendicular to y_2 (fig.1c with Θ=0), has a symmetry expressed by the layer group $\ell\ \underline{2}'_{z}$. The operations $2'_{y2}$, that exchange domain states only, and operations $\underline{2}_{x2}$ reversing the wall normal only, are no more symmetry operations of the wall but they transform the wall 1/0/2 into reversed wall 2/0/1 and vice versa, i.e., these walls are reversible.

It is easy to show that a wall parallel to z, with a general orientation Θ (fig. 1c) has the symmetry $\ell\ \underline{2}'_{z}$ as well, but, contrary to walls with special orientation $\Theta = 0$, it is not reversible. From eq.(5) we find the number of symmetrically equivalent walls: for walls with $T_{12}=\ell\,2_{x2}\underline{2}'_{y2}\underline{2}'_{z}$ we obtain $n_w = 12 : 4 = 3$, for $T_{12}=\ell\,\underline{2}'_{z}$ we get $n_w = 6$. Symmetrically equivalent walls can be generated by applying operations of $6\ 2_x 2_y$ on one wall. Thus, e.g., from $1/-\frac{\pi}{2}/2$ one gets symmetrically equivalent walls $1/\frac{\pi}{6}/2$ and $1/\frac{5}{6}\pi/2$ with the symmetry groups $\ell\ 2_{x3}\underline{2}'_{y3}\underline{2}'_{z}$ and $\ell\ 2_{x1}\underline{2}'_{y1}\underline{2}'_{z}$, resp.

All walls parallel to z are symmetric walls, i.e., the order parameter η is an odd function of the distance ξ from the central plane of the wall. Symmetry of a wall does not, therefore, depend on the form of the function $\eta(\xi)$, unless the condition $\eta(-\xi) = -\eta(\xi)$ is violated. In particular, symmetry of a wall does not depend on the thickness of the wall.

Symmetry T_{12} of the wall is useful in locating the extrema of wall energy /11/. The angles $\Theta = (2k + 1)\frac{\pi}{6}$, $k = 0, \pm 1, \pm 2, \ldots$, represent symmetrically prominent orientations with $T_{12} = \ell 2_{xi} \underline{2}'_{yi} \underline{2}'_{z}$, $i = 1, 2, 3$, whereas for any other orientation, $T_{12} = \ell \underline{2}'_{z}$. Symmetrically prominent orientations present a sufficient condition for local extrema of wall energy. Taking into account experimental fact that the equilibrium orientation of Θ changes smoothly with temperature (and cannot therefore correspond to any symmetrically prominent orientations which are fixed), one can draw the quantitative form of the angular dependence of the wall energy $E(\Theta)$ (fig. 2). Orientations corresponding to minimal wall energy are rotated away from $\Theta = 2k\pi/3$ by an angle $\pm\varepsilon$. Orientation parallel to z, which we have been considering, is also a symmetrically prominent one, since any tilt of the wall decreases the wall symmetry to ℓ 1.

From the monoclinic symmetry $\ell \underline{2}'_{z}$ of an equilibrium wall, it directly follows that the wall carries spontaneous polarization P_z parallel to z; P_z has the same sign for three walls related by operations 3 and 3^2 and opposite sign for walls related by any 2-fold axis. The spontaneous deformation of the wall includes a volume dilatation and a shear. These spontaneous properties can also be deduced from a phenomenological theory /12/.

If n walls meet along a common line we call this line a *n-fold domain vertex*. For topological reasons, only 2-, 4-, and 6-fold vertices exist in quartz /8/. Their symmetry is described by rod groups, but we shall not discuss this point here.

IV . REGULAR LATTICE OF WALLS

1) Symmetry of a regular pattern of walls

As already mentioned in part II, an incommensurate phase close to the I-C transition can be approximated by a regular domain texture formed by domain walls of negative energy /13/. Symmetry of this domain texture is, within the continuum approach, described by a group $\mathcal{M}$ with translation subgroup $\mathcal{T}_m$. Domain texture of I phases modulated in one direction only (so called stripe phase) is formed by an array of parallel domain walls in which a certain basic sequence of walls is regularly repeated. The translation subgroup contains discrete translations along the modulation wave vector and translations parallel to the walls. Domain texture of I phases modulated in two or more directions is formed by a regular network of walls and wall vertices. Translational group $\mathcal{T}_m$ is then discrete in two or three directions.

The knowledge of the symmetry group $\mathcal{M}$ allows one to find the macroscopic symmetry M of the I phase as the factor group of $\mathcal{T}_m$ in $\mathcal{M}$: $M \simeq \mathcal{M}/\mathcal{T}_m$. The group M determines :

(i) macroscopic properties of the I phase, especially its spontaneous properties which are forbidden by symmetry in N and C phases and allowed in the I phase;

(ii) symmetry dictated extrema of energy. Similarly as for domain walls, symmetrically prominent orientations can be helpful in locating orientations with minimal energy of I phase. We shall see, however, that a texture can have prominent orientations different from that of single walls which form the texture;

(iii) number n_b of symmetrically equivalent variants of I phase, $n_b = |G|:|M|$. These variants, related by operations of G not contained in M, have the same energy and can coexist as textural *blocks*. Interfaces between blocks are called *block boundaries*.

2) Example of the stripe phase in quartz

To construct a stripe phase, in which walls are parallel (and under the assumption that they have the same energy), one needs reversible walls. From the preceding results it follows that this phase should be built up from walls 1/0/2 and 1/π/2 (=2/0/1) regularly

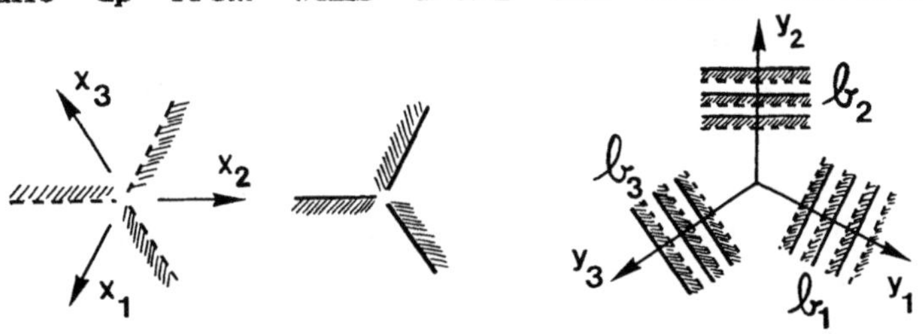

Fig.3 : Reversible domain walls with $-P_z$ and $+P_z$ polarizations, and symmetrically equivalent stripe phases.

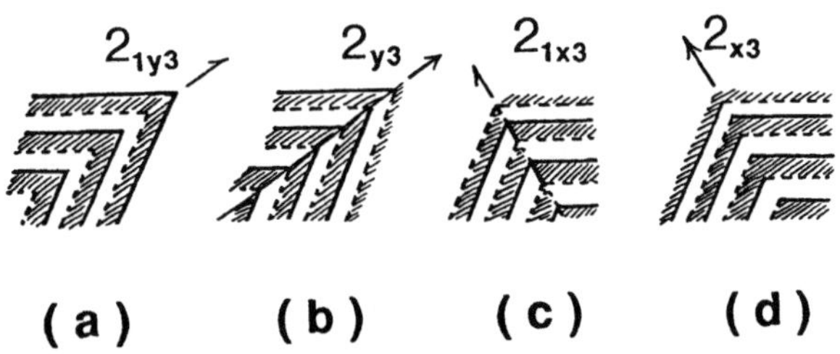

Fig.4 : Block boundaries between stripe phases with polar 2-fold symmetries.

disposed in the y_2 direction (fig. 3). Such a pattern has an orthorhombic symmetry $\mathcal{B} = 2_{x2}2'_{y2}2'_{z}$ with 2_{x2} axes in between two walls and $2'_z$ axes in the walls. There are $n_b = |G|:|\mathcal{B}| = 3$ symmetrically equivalent stripe phases which we call blocks. These blocks are built up from walls perpendicular to y_i (i= 1, 2, 3) and the corresponding macroscopic symmetries of blocks are $\mathcal{B}_i = 2_{xi}2_{yi}2_z$, resp. Each block can be identified with a ferroelastic domain since by symmetry one can predict that in block b_i $(U_{xi\,xi} - U_{yi\,yi}) \neq 0$, but the symmetry is non-polar, in agreement with the fact that consecutive walls carry opposite polarizations.

This macroscopic symmetry is important in considerations concerning block boundaries, which join two ferroelastic blocks. The orientation corresponding to minimal energy of the boundary is given by the condition of mechanical compatibility, as in usual ferroelastic materials. Let us consider, for instance, the pair of blocks b_1,b_2. Orientations of stress-free block boundaries are in this case perpendicular to x_3 and y_3. The symmetry of a boundary depends on the relative phase of modulations on opposite sides of the boundary. The highest symmetry cases are represented in figure 4 : since 2_{yi} axes exchange the stripe phases b_1 and b_2 , whereas 2_{xi} axes do not, the symmetries of block boundaries are expressed by following layer groups $\ell\ 2_{1y3}$ (fig. 4a), $\ell\ 2_{y3}$ (fig. 4b), $\ell\ 2_{1x3}$ (fig. 4c), or $\ell\ 2_{x3}$ (fig.4d). These higher symmetry configurations have extremal energy. It is interesting to note that in fig. 4 the boundaries represented in (a) and (d) do not involve other walls than 1/0/2 and 2/0/1 whereas this is not the case for (b) and (c). In (b), the supplementary walls are perpendicular to x_3 and have maximal energy (fig. 2 & 4b). In (c), walls are perpendicular to y_3 and then (for $\varepsilon = 0$) their energy is minimal (and negative). From this one can expect that the real structure of block boundaries corresponds to cases (a) and (c). Block boundaries are always formed by a regular array of two-fold or four-fold vertices.

Another interesting feature concerns ferroelectricity: whereas the blocks are ferroelastic but not ferroelectric, a spontaneous polarization either parallel to y_3 (fig. 4a) or to x_3 (fig. 4c) is allowed by symmetry.

Parallel block boundaries are reversible: $b_1/\Theta/b_2$ and $b_2/\Theta/b_1$ are related by the 2_z operation and have, therefore, opposite spontaneous polarizations ($\Theta = -\frac{\pi}{3}$ or $\frac{\pi}{6}$).

Evidence of stripe phase has been obtained by neutron diffraction in /14/ but observations in direct space are not available.

3) Analysis of the 3-q phase of quartz

The superposition of the three waves with equal amplitudes can lead to different patterns formed out of triangles or hexagons, depending on the relative phase of the modulations. Here, we shall consider the pattern of equilateral triangles since it corresponds to observations. To construct it, one has two possibilities, either from walls carrying a polarization P_z or from walls carrying $(-P_z)$. Let us now consider the symmetry of such a structure diperiodic in the plane (001). For any $\theta \neq \pm \frac{\pi}{2}$ (and symmetrically related orientations), the group leaving the figure invariant is $p6'_z$, whose symmetry axes are shown in fig.5. It can also be written as $p6'_z = p3 + 2'_z(p3)$, and one can note that this symmetry is polar (all walls carry the same polarization, and the macroscopic P_z is non zero). It is interesting to note that the modulated phase is ferroelectric, though neither α nor β phases is polar. For the special orientation $\theta = \pm \frac{\pi}{2}$, the structure has a higher symmetry since now 2_x and $2'_y$ leave the pattern invariant. For this special orientation, the polarization P_z therefore vanishes, and one has an extremum of energy, which is a maximum. However, as the structures with $\theta = + \frac{\pi}{2}$ and $\theta = - \frac{\pi}{2}$ cannot be related by any symmetry operation of $6'_z2_x2'_y$, there is no symmetry requirement that both maxima of energy coincide. From this one can conclude that the angular dependence of energy is similar to that of a single wall (fig. 2). In particular, the positions of minima are not determined by symmetry. It follows, that both blocks with $\pm P_z$, which are related, e.g., by $2'_y$, are rotated in opposite directions (fig. 6).

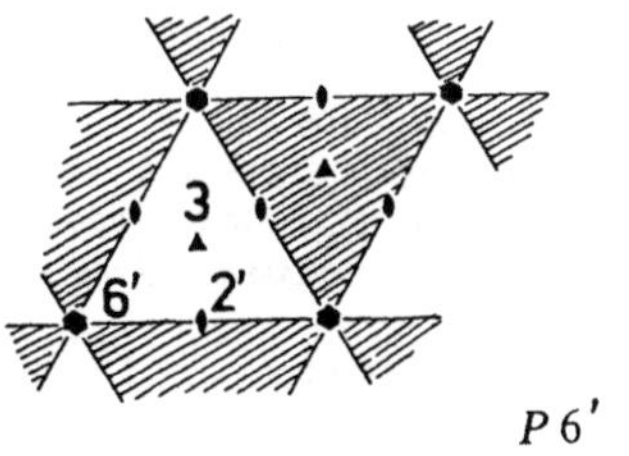

Fig.5 : Symmetry p6' of the 3-q phase.

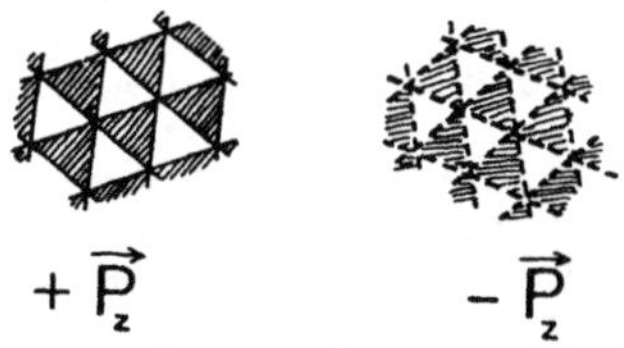

Fig.6 : 3-q phase: blocks with opposite polarization along z.

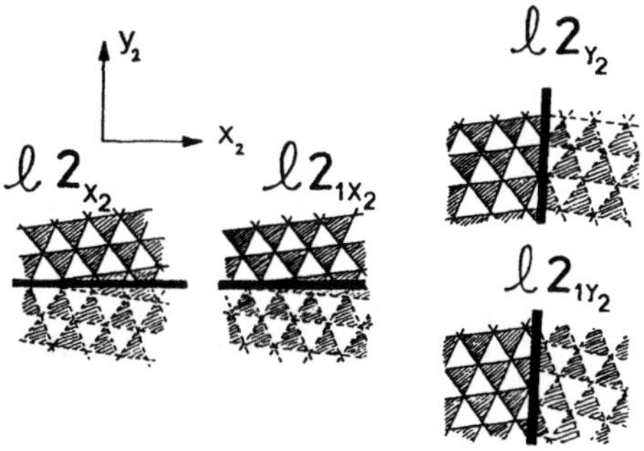

Fig.7 : 3-q phase : block boundaries with prominent orientations.

Since the blocks can be identified with ferroelectric non-ferroelastic domains, there is no strong requirement concerning block boundary orientation. However, special orientations normal to x and to y can have higher symmetries, depending on the difference of phase of modulations on opposite sides of the boundary (fig. 7). In all cases, a polarization normal to [001], lying within the boundary, is allowed. Within the boundary, the polarization rotates in an analogous manner as does magnetization in ferromagnetic Bloch walls.

V . CONCLUSION

On the example of quartz, we have shown that some important conclusions about properties of non-linear objects, like domain walls and incommensurate phases, can be obtained from symmetry analysis without solving non-linear equations. Though this approach has limitations (it cannot yield quantitative values, temperature dependences, etc), conclusions are rigorous, reflect specific features of complicated systems and disclose tiny effects not considered in simplified models. In crystal physics, symmetry analysis represents a useful first step in examining properties of real non-linear systems.

REFERENCES :

/1/ V.Dvořák, in *Modern Trends in the Theory of Condensed Matter*, Ed. A. Pekalski and J. Przystawa (Springer, Berlin),p447 (1980).

/2/ A.P. Levanyuk, in *Incommensurate Phases in Dielectrics*, Ed. R. Blinc and A.P. Levanyuk (North Holland, Amsterdam), Vol. 1, p1 (1986).

/3/ Y. Ishibashi, H. Shiba, J. Phys. Soc. Jpn 45, 409 (1978).

/4/ H. Grimm, B. Dorner, J. Phys. Chem. Solids 36, 407 (1975).

/5/ K. Gouhara, Y. Hao Li, N. Kato, J. Phys. Soc. Jpn 52, 3697 (1983) G. Dolino, J.P. Bachheimer, and C.M.E. Zeyen, Solid State Commun.45, 295 (1983).

/6/ G. Dolino, in *Incommensurate Phases in Dielectrics*, Ed. R. Blinc and A.P. Levanyuk (North Holland, Amsterdam), Vol. 2, p205 (1986).

/7/ T.A. Aslanyan, A.P. Levanyuk, M. Vallade, and J. Lajzerowicz, J. Phys. C 16, 6705 (1983).

/8/ J. Van Landuyt, G. Van Tendeloo, S. Amelinckx, M.B. Walker, Phys. Rev. B 31, 2986 (1985).

/9/ C. Roucau, E. Snoeck, P. Saint-Grégoire, J. Physique 47, 2041 (1986).

/10/ V. Janovec, Ferroelectrics 35, 105 (1981).

/11/ G. Kalonji, J. Physique, Colloque C4, sup.n°4, 46, 249 (1985).

/12/ M.B. Walker and R.J. Gooding, Phys. Rev. B 32, 7408 (1985).

/13/ V. Janovec and V. Dvorak, Ferroelectrics 66, 169 (1986).

/14/ P. Bastie, F. Mogeon, C.M.E. Zeyen, Phys. Rev. B 38, 786 (1988).

STRAIN SOLITARY WAVES IN FERROELASTIC-MARTENSITIC TRANSFORMATIONS. LATTICE MODEL AND CONTINUUM APPROXIMATION.

J. POUGET

Laboratoire de Modélisation en Mécanique, Associé au C.N.R.S.
Université Pierre et Marie Curie, Tour 66
4, Place Jussieu 75252 Paris FRANCE.

On the basis of a lattice model the dynamics and stability of nonlinear coherent structures made of elastic domains for ferroelastic-martensitic transformations are examined. Such transformations are usually accompanied by the formation of more or less complex domain patterns (deformed regions of the material). A two-dimensional lattice is first considered. The model involves competing and nonlinear interactions which allow for nonlinear excitations of soliton type. The continuum approximation leading, in fact, to the notion of quasi-continuum is presented next. The significant results are, for a one-dimensional reduced version, the softening and upward curvature of the transverse phonon branch, nonlinear periodic waves describing modulated structures, martensitic and austenitic solitons corresponding to the shearing motion of the atomic planes. The two-dimensional system is considered in order to investigate the instability of an elastic soliton with respect to the transverse direction.

1. -Introduction

We propose a lattice model in the view of studying **nonlinear dynamics** of **microstructure patterns** made of elastic domains involved in **martensitic-ferroelastic transformations**. More precisely, we are interested in the underlying micro-physics which induces special kinds of strain transformations associated with the nucleation of elastic domains. In the physical point of view, ferroelastic and related martensitic transformations have been defined as a subset of diffusionless phase transformations involving rather large lattice distorsions, that means the transformation is mostly dominated by strain energy [1]. Accordingly, the transition is of **first order** (or almost second order) where the spontaneous strain is the order parameter in the Landau theory of phase transition and the ferroelasticity is called proper [2,3]. For instance, In-Tl, Ni-Tl, Nb_3 Sn , Fe-Pd etc.... provide good canditates of such crystals. Crystals undergoing such transformations exhibit particularly interesting effects : elasticity; pseudo-elasticity, ferroelasticity with hystheresis as well as **shape memory effects** of which technological applications are especially promissing. All these effects are in fact intimately connected with phase transisition taking place in materials and they are only observed as the global behaviour of the crystal.

Nevertheless, our task will be limited to small region of a perfect material, that is low dimensional materials, where only few elastic domains coexist.The global behaviour is certainely monitored by the dynamics of twinned regions and the formation of coherent arrays of twins or ferroelastic domains is a consequence of the first order transition. These domain patterns are usually obversed experimentally by means of electron microscopy and the morphology of martensitic twins seems to be very rich and complex [4,5]. This is a motivation for examining the problem of the elastic domain patterns at the microscopic level.

In the present study we address our attention to martensitic-ferroelastic transformations and examine an intrinsic mechanism of the elastic twin formation as well as their dynamics. The nucleation process can be seen as pre-transitional modulated structure developing in the high temperature phase which forms thus a periodic array of parallel twin bands. The periodic structure can be **commensurate** or **incommensurate** with the parent phase lattice [5,6,7]. It follows therefore that the coherent motion of the twin boundaries is of great importance to understand martensitic transformations. In the mechanical point of view, we are concerned with the competition between the **nonlocal elasticity** (inhomogeneous state of deformation) and **local nonlinear** elastic energy (homogeneous state of deformation with stable, unstable or metastable regions), which plays a crutial role in motion of coherent elastic twin boundaries.It is, in fact, the combination of both nonlinear and competing interactions which generates the propagation of nonlinear excitations that we can describe as **solitons**. Our model is a good extention of the ideas underlying the nonlinear soliton models developed in other contexts [8] to the long wavelength elastic description of nonlinear structures in martensitic transformations. Based on nonlinear lattice dynamics we propose an alternative, since we have at hand the microscopic model and its continuum version at a time. This seems to more convenient because we can define the notion of quasi-continuum where the physical backgrounds of the lattice model have been kept.

The present work is particularly devoted to a two-dimensional lattice model which allows us to describe a cubic-tetragonal transformation. For instance In-Tl crystal undergoes a ferroelastic transition of the first order with f.c.c. (m3m) symmetry above the transition point and f.c.t. (4/mmm) below [9,10]. We consider the following types of interatomic interactions between particles : (i) by pair between the first neighbours and (ii) of three-body type (between three adjacent particles) or noncentral interactions. In addition, the lattice **anisotropy**, which derives from the dimensionality of the model, allows to place the predominance of the [110] direction in evidence. The cubic-tetragonal transformations is mostly characterized by the shearing motion of the close-packed atomic planes $(1\bar{1}0)$ stacked in the [110] direction. Because of the anisotropy and, of course, of the special kinds of interactions the material exhibits anormalously low anisotropy **shear modulus**

softening (transverse acoustic branch in [110] direction) which is strongly temperature dependent. The phonon mode softening usually occurs at non-zero wave vector in the phonon dispersion spectrum [10,11,12]. Structural modulations can then develop within the parent phase with periodic deformation patterns of which the wavelength corresponds to that of phonon anomalies [5,13,14]. The origin of such patterns emerges from competing and nonlinear interactions. Accordingly, a rather fine scale of the lattice description seems to be necessary.

2. -Construction of the microscopic model

a - Two-dimensional model

Let us consider an atomic plane extracted from a cubic lattice (austenitic or high temperature phase), for instance the f.c.c. symmetry of In-Tl. However, we assume homogeneous deformation of the lattice along a perpendicular direction to the atomic plane, say the [001] direction, that means we deals with plane deformation in the (001) plane. Then, our lattice, in its undeformed state, consists of squares parallel to the [100] and [010] crystallographic directions (see figure 1). A particle of the plane is located by (i,j) in the co-ordinate system {[100],[010]} or (**i**,**j**) or by (I,J) in {[110],[$1\bar{1}0$]} or (**I**,**J**) deduced from the former by a rotation 45° clockwise. After deformation of the lattice the particles perform displacements in the lattice plane defined by u(i,j) and v(i,j) which are the longitudinal and transverse displacements, respectively.

We define next the interatomic forces acting on the particles of the lattice. We assume that the particle at a site (i,j) interacts with the first nearest neighbours surrounding it. Two types of interactions are considered : (i) interaction by **pairs**

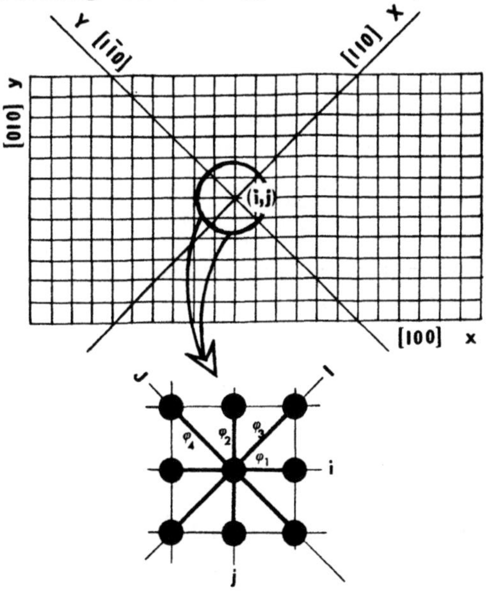

Fig.1 : Geometry of the lattice model, two-dimensional system made of squares. Interactions of the particle at (i,j) with the first neighbours.

between the first nearest particles in the four directions i, j, I, J and (ii) interactions of the **three-body** type in the same directions [15]. The latter interactions acts on particles as **non-central forces**, which is equivalent to bond-bending or torsonial forces [15,16] and involve three center forces due to the electronic covalent interatomic interactions. The lattice potential must then account for the change in angle between bond segments joining particle pairs, which represent, at the microscopic level, twisting and bending of the unit crystalline cell. This second kind of interaction may be equivalent, in some sense, to the interactions between first neighbouring "elastic dipoles". Moreover the lattice anisotropy is characterized by potentials of different strengths in the four directions. Since the lattice energy must be translationally and rotationally invariant, it depends only on the modulus of the relative particle positions. In fact, after some algebraic manipulations we are able to show that the lattice energy is a function of discrete Lagrangian deformation tensor including geometric nonlinearities and also of the first order finite differences of these discrete deformations. The lattice energy can be written as [17]

$$\mathcal{V} = \sum_{i,j} \{\varphi_1(\epsilon_{11}(i,j)) + \varphi_2(\epsilon_{22}(i,j)) + \varphi_3(\epsilon_{11}(i,j), \epsilon_{22}(i,j), \epsilon_{12}(i,j)) + \varphi_4(\epsilon_{11}(i,j), \epsilon_{22}(i,j), \epsilon_{12}(i,j)) + \psi(\Delta_L\epsilon_{11}(i,j), \Delta_T\epsilon_{11}(i,j), \Delta_L\epsilon_{22}(i,j), \Delta_T\epsilon_{22}(i,j), \Delta_L\epsilon_{12}(i,j), \Delta_T\epsilon_{12}(i,j))\} \tag{1}$$

where the discrete components of the Lagrangian strain tensor ϵ_{pq} are defined exactly as usually in continuum mechanics [18] and are functions of the first order finite differences of the discrete displacements Δ_L and Δ_T in the i and j directions, respectively. The potentials φ_α (α = 1,2,3,4) are functions of the particle pairs between the nearest neighbours in [100], [010], [110] and [1$\bar{1}$0], respectively and the potential ψ derives from the three-body interactions in the same directions.On introducing appropriate symmetrical strain components for the cubic symmetry

$$e_1 = (\epsilon_{11} + \epsilon_{22})/\sqrt{2}\ , \quad e_2 = (\epsilon_{11} - \epsilon_{22})/\sqrt{2}\ , \qquad e_3 = \epsilon_{12} \tag{2}$$

where the index (i,j) has been omitted, we can write the lattice energy as a function of the deviatoric and dilatational parts of the (discrete) strain tensor and their first order finite differences. A particular case of strain transformation of the two-dimensional system will be examined in a further section, but we now turn on a reduced one-dimensional version.

b - One-dimensional model

Since it is possible to work in the co-ordinate system {[110],[1$\bar{1}$0]}, the discrete quantities depend on the new index (I,J) with I=i+j and J=j-i. We assume next that the lattice displacements depend only on the I index. Important reductions of the model are then obtained and significant physical situations can be emphasized. After some algebra the Hamiltonian of the discrete system can be written as (in dimensionless notations)

$$\mathcal{H} = \sum_I \frac{1}{2} \dot{V}^2(I) + \sum_I \Big[\frac{1}{2}\alpha S^2(I) - \frac{1}{3}S^3(I) + \frac{1}{4}S^4(I) + \frac{1}{2}\beta(S(I+1) + S(I))^2$$

$$+\frac{1}{2}\delta(S(I) - S(I-1))^2 + \frac{1}{2}\eta(S(I+1) + S(I) - S(I-1) - S(I-2))^2\Big] \qquad (3)$$

where V(I) represents the transverse displacement in J direction (or $[1\bar{1}0]$) and S(I) is the only one strain component connected with V(I) through

$$S(I) = V(I+1) - V(I). \qquad (4)$$

The lattice parameters α, β, δ and η are linear combinations of the second order derivatives of the potentials φ_α and ψ with respect to the particle positions taken at equilibrium. The first part of the Hamiltonian (3) is the kinetic energy, the next three terms derive from the expansion of the lattice energy up to the fourth order with respect to the strain. the fifth term of (3) describes the interactions between second nearest neighbouring atomic planes and the last two terms emerge from the non-central interactions between first and second nearest neighbouring atomic planes. The combination $A=\alpha+4\beta$ corresponds to the elastic modulus $(C_{11}- C_{12})/2$ [19]. Along with the Landau theory of phase transitions for ferroelastic crystal the elastic modulus A depends on the temperature according to the usual Curie-Weiss law. We remark that the lattice deformation is now well described by S(I) which represents the **relative shear displacement** of the close-packed atomic planes $(1\bar{1}0)$ in the stacking direction [110].

3. -Equations of motion

The equations of motion for the transverse displacement V(n) are easely obtained by using the Hamiltonian principle and the definition of the discrete strain (4). This yields the following difference equations for the discrete shear deformation S(n)

$$\ddot{S}(n) = \alpha(S(n+1)-2S(n)+S(n-1)) - (S^2(n+1)-2S^2(n)+S^2(n-1))$$

$$+ (S^3(n+1)-2S^3(n)+S^3(n-1)) + \beta(S(n+2)-2S(n)+S(n-2))$$

$$-\delta(S(n+2)-4S(n+1)+6S(n)-4S(n-1)+S(n-2))$$

$$-\eta(S(n+4)-4S(n+2)+6S(n)-4S(n-2)+S(n-4)) \qquad (5)$$

This is a set of coupled nonlinear ordinary differential equations which governs the shear deformation travelling perpendicularly to the close-packed atomic planes $(1\bar{1}0)$. In the discrete case, these equations are not usually tractable except for the linearized situation. The strongly nonlinear problem can be however investigated by means of numerical simulations with appropriate initial and boundary conditons. An alternative situation happens in the case of the continuum approximation. This side of the problem will be studied further.

4. - Transverse phonon dispersion

Now, we consider linearized equations (5) about a uniform deformation S_0, where S_0 must satisfy one of the minima of the lattice potential. The state of deformation S_0 is either an austenitic or a martensitic phase. On looking for harmonic sinusoidal wave solutions of the linearized equations we obtain the dispersion relation connecting the angular frequency with the wave number. The dispersion relation curves are depicted in figure 2 for different values of the lattice parameters α, β, δ and η, which places the importance of competing interactions in evidence. Curve (a) differs slightly from the classical case (one-dimensional model of monoatomic chain). The dispersion branch (b) has a small bend and the curve (c) with a deeper bend undergoes a softening at **nonzero wave-number**, which can be interpreted as a **precursor of martensitic transformation** or premartensitic transition [14,20]. This softening is commonly observed by means of neutron inelastic scattering techniques [9,14,20]. The instability generated around the nonzero wave-number produces a modulated lattice distorsion, within the parent phase, with a q_0 pseudoperiod which is detected by electron microscopy and electron or x-ray diffraction [7].At last, the curve (d) exhibits an upward convexity. this situation is important for the existence of strain solitons.

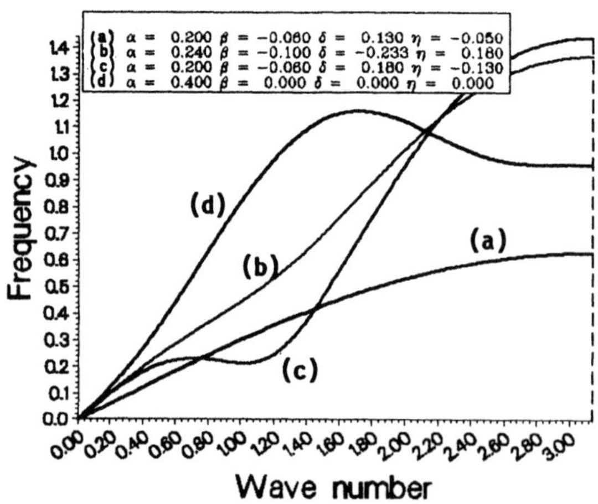

Fig.2 : Dispersion curves for the transverse phonons in [110] direction.

5. -Continuum model

We assume now that the discrete functions (displacements, strains) are slowly varying over a lattice spacing. The problem is then to relate the space of discrete functions of a discrete variable to a space of analytic functions. That means that it is possible to find a field S(x,t) which at every x=nb takes on the value S(t)=S(n,t) and the function S(x,t) can be considered as a **good interpolation** of the sampling (nb,S(n,t)) . After some classical algebras, the equation of motion takes on the new form

$$S_{,tt} = \Sigma_{,xx} \tag{5a}$$

where

$$\Sigma = \sigma - \chi_{,x} \tag{5b}$$

is the Cauchy stress, σ the maco-stress and χ the micro-stress. These stresses derive from an elastic potential as follows

$$\sigma = \frac{\partial\psi}{\partial S} \qquad \text{and} \qquad \chi = \frac{\partial\psi}{\partial(S_{,x})} . \tag{5c}$$

The elastic potential is given by

$$\psi(S,S_{,x},\theta) = \frac{1}{2}AS^2 - \frac{1}{3}\delta^3 + \frac{1}{4}\delta^4 + \frac{1}{2}\gamma(S_{,x})^2 \tag{6}$$

where we have set

$$\gamma = \delta + 16\eta - \beta - A/12. \tag{7}$$

Eq. (5) is a nonlinear dispersive partial derivative equation governing the continuous strain.The form of the elastic potential corresponds exactly to the Ginzburg-Landau free energy including a cubic term in strain which induces a first-order phase transition [21]. Note that the mico-stress is due to the non central interatomic interaction of the microscopic model. We can therefore compare the continuum approximation to the nonlinear elasticity exhibiting weakly nonlocal behaviour (strain gradient elasticity) [16].

Since we have introduced stresses we must define the corresponding boundary conditions. We will consider for numerical applications periodic boundary conditions which are compatible with the mechanical conditions on stresses and with the equation of motion. Moreover, because of special type of motion $U = U_0$ and $V = V(x)$ we have an incompressible motion. On the other hand the strain compatibility conditions are also satisfied [18].

6. -Nonlinear excitation solutions and strain solitons.

Since we have in mind to examine the possible existence of nonlinear excitations, we will solve the set of equations (5a-c) with appropriate boundary conditions. The simplest idea is to search for solutions of these equations subjet to the boundary conditions as functions only of the phase variable ξ=x-ct where c is a constant phase velocity. Then we transform the nonlinear partial derivative equation (5) into an ordinary derivative equation with respect to ξ which is easily to solve. Subsonic strain solitons can travel for $c^2 < A$, but such nonlinear excitations exist if we have $\gamma > 0$, this condition is just that of the upward curvature of the dispersion curve. In addition, we have obviously other conditions on the soliton amplitude.

Remark : On returning to the transverse displacement $V(x,t)$ related to the strain through $S = V_{,x}$, we can transform eq.(5). In the equation for V we introduce the right-running characteristic co-ordinate $\xi = x - C_T t$ ($C_T^2 = A$) and slow time $T = \epsilon t$ and by using multiple-scale technique we arrive at

$$Y_{,T} + 3(Y^2)_{,\xi} + 9\mu(Y^3)_{,\xi} - \gamma Y_{,\xi\xi\xi} = 0 \tag{8}$$

where $Y = -3S/\mu$, μ is the ratio of the nonlinear coefficients of the elastic potential and the small parameter ϵ is associated with the nonlinearity of the problem. It is clear now that eq.(8) is the combination of the Korteweg-de-Vries and Modified Korteweg-de-Vries equations.

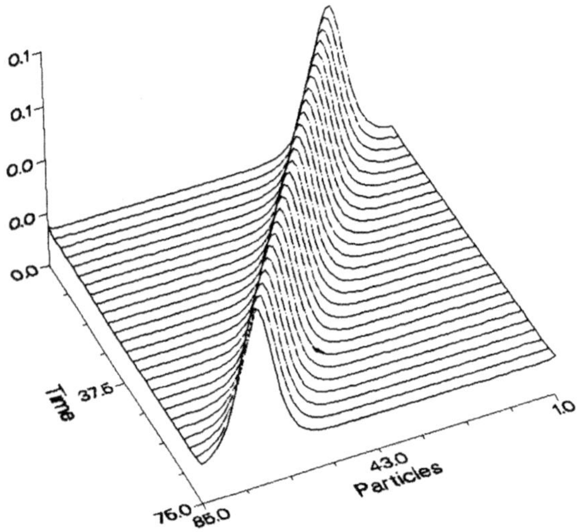

Fig.3 : Martensitic soliton moving in an austenitic matrix.

Numerical simulations: We consider now the fully dynamic processes, at this end the numerical scheme is provided by the set of discrete equations (5) with periodic boundary conditions at each end of the lattice. We begin with a martensitic soliton as initial conditions. The evolution of the wave is given in figure 3. This illustration represents a small layer of martensite (deformed lattice) moving in the parent phase (undeformed lattice). The corresponding deformed two-dimensional lattice is given in figure 4, the martensitic band is sweeping across the crystal in the stacking direction [110]. The complementary situation describing an austenitic soliton moving in a martensitic matrix is drawn in figure 5. Note that the amplitude of the strain state in the martensitic phase controls both the depth of the austenitic soliton and its

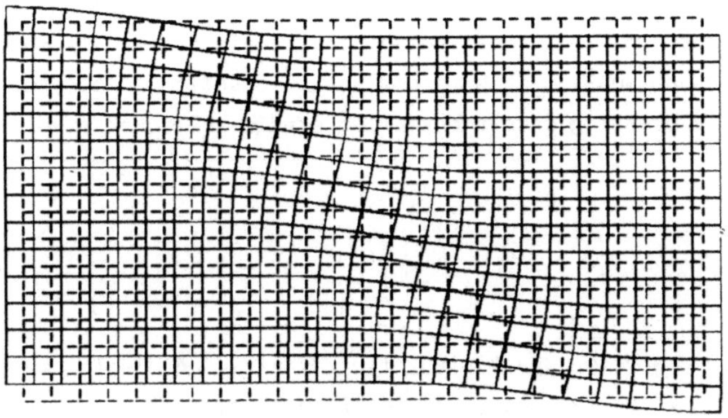

Fig.4 :The corresponding deformed lattice to the martensitic soliton sweeping across the crystal

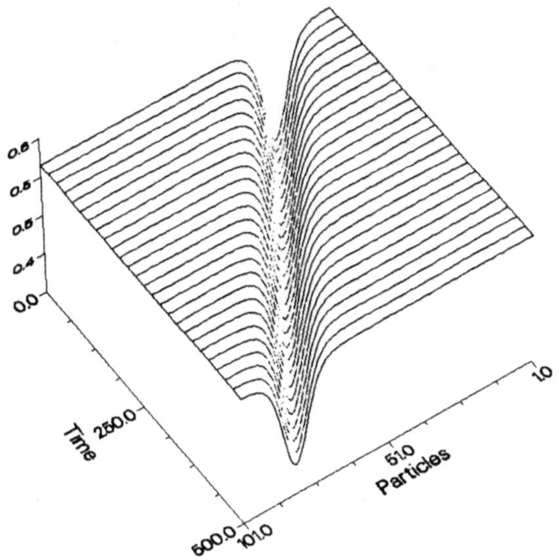

Fig.5 : Austenitic soliton moving in the martensitic phase.

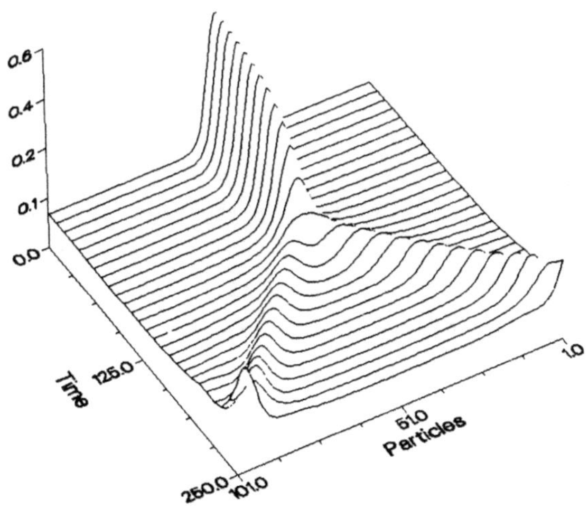

Fig.6 : Instability : the moving soliton (small velocity) is splitting into two strain solitons.

velocity. Finally, it is interesting to show an unstable strain soliton and this problem is represented in figure 6. The amplitude of the strain soliton is chosen such that its velocity is rather small, this is not a static soliton. However, we observe next that the soliton is breaking into two smaller strain solitons travelling in opposite directions.

7. -Two-dimensional problem

Now, we suppose that all the discrete quantities, i.e. strains and displacements, depend on i and j (or I and J if we use the (I,J) co-ordinate system). The lattice energy (1) is still valid, nevertheless we consider a simple case where the energy is a function of the discrete spherical and deviatoric parts of the strain tensor. In addition, we expand the lattice energy up to the fourth order in strain components and up to harmonic order for the first order finite differences of the deformation. The resulting expression must satisfy the symmetry of the cubic-tetragonal transformation. A further step in the simplification consists in considering a special case of transformation for which the lattice potential is function only of the deviatoric part, say $S(i,j)$ and keeping the spherical part constant. As for the one-dimensional problem we construct the Lagrangian of the discrete system and the equations of motion are deduced from the Lagrangian, this yields

$$\ddot{S}(i,j) = \Delta_L^2\Sigma(i,j) + \Delta_T^2\Sigma(i,j) \tag{9}$$

where we have introduced

$$\Sigma(i,j) = \sigma(i,j) - \Delta_L\chi_L(i,j) - \Delta_T\chi_T(i,j) \tag{10a}$$

and

$$\sigma(i,j) = \alpha S(i,j) - S^2(i,j) + S^3(i,j) \tag{10b}$$

$$\chi_L(i,j) = \delta(S(i,j) - S(i-1,j)) \tag{10c}$$

$$\chi_T(i,j) = \delta(S(i,j) - S(i,j-1)) \tag{10d}$$

where Δ_L and Δ_T are the first order finite differences in i and j directions, respectively, and Δ_L^2 and Δ_T^2 are the second order finite differences in the same directions. Eq.(10a) represents the total discrete stress with $\sigma(i,j)$ defined by eq.(10b) is due to the linear and nonlinear parts of the lattice energy, and $\chi_L(i,j)$ and $\chi_T(i,j)$ are the discrete micro-stresses deduced from the non-central interactions. Moreover we notice that the stress (10b) has the same form as in the one-dimensional problem (see eq.(5c) or (6)).

The problem is now to examine the existence of localized nonlinear coherent structures, for instance the formation of elastic domains. The idea is, in fact, to see into the stability of a strain soliton given by the one-dimensional model (see the preceding sections) with respect to the transverse direction, i.e. [010] direction.

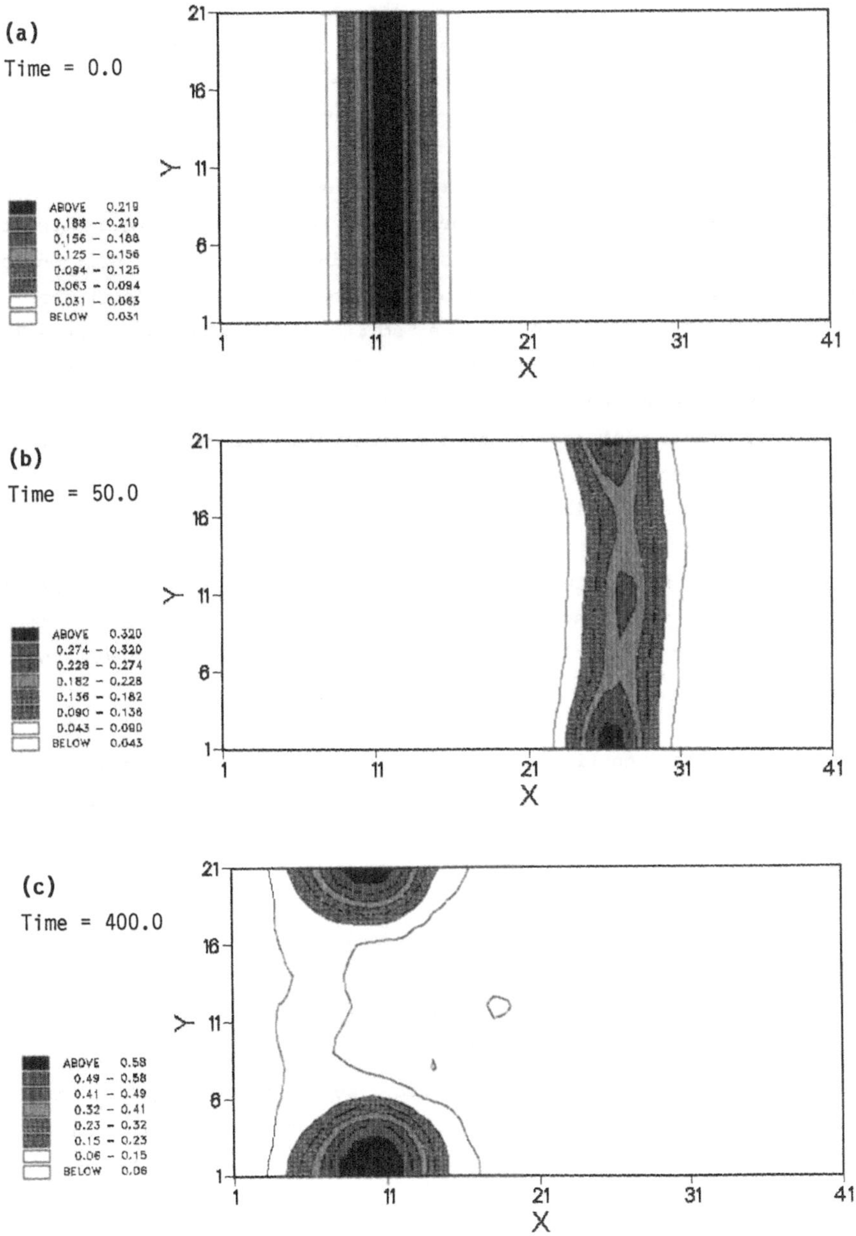

Fig.7 : Instability process for the two-dimensional system: (a) initial condition, the strain soliton, (b) small perturbations are developing in the transverse direction, (c) formation of a localized coherent structure (elastic domain).

We investigate the problem by means of numerical simulations. The numerical scheme is provided, once more, by the set of discrete equations (9,10a-d) with periodic boundary conditions in both directions. Moreover, these boundary conditions are compatible with the mechanical conditions on the macro-stress (eq.(10b)) and micro-stresses (eqs(10c,d)). Since the problem deals with a dynamical process we must consider initial conditions, the latter conditions are given by the solution of the one-dimensional system extended to the transverse direction. Figure 7 collects together the deformation state of the lattice for different times. The initial condition is presented in figure 7.a, this is a shaded-contour map of the strain in the lattice plane where each pattern corresponds to different strain strengths. We note that the initial condition which is a strain or martensitic soliton does not depend on the transverse variable. A lapse of time latter, small perturbations are developing in both directions while the soliton is moving (see figure 7.b). in particular, we observe that the strain amplitude is modulated with a period which seems to be the transverse length of the lattice. It turns out that the strain soliton is **no longer stable with respect to the transverse direction.** The instabilities are growing, this effect leads to a bifurcation of the initial solution and the strain soliton becomes a **localized elastic domain** in both directions (fig 7.c). Some small oscillations are still present, those are due to some discreteness effects of the lattice. However, the instability is not the result of the discrete problem but certainly the result of nonlinear instabilities monitored by the strain amplitude and lattice parameters. The analytical problem of such instability process, which is a very complex study, will be presented elsewhere.

Remarks : Insofar as the continuum approximation of the two-dimensional problem is concerned, we use exactly the same process as in the one dimensional situation. The partial derivative equation for the continuous function S(x,y,t) is then given by (for $\delta_L=\delta_T=\delta$ and $\Upsilon=\delta-A/12$)

$$S_{,tt} = \Delta\sigma - \Upsilon\Delta(\Delta S) \tag{11}$$

where Δ is the Laplacian operator in the lattice plane and σ is still defined by eq.(5c), but now S is a function of x and y. In order to explain the result of figure 7.c, i.e. the localized elastic structure, we transform eq.(11) into polar co-ordinates by setting $(x,y) \rightarrow (r\cos\theta,\ r\sin\theta)$. We assume next that the strain is independent of the polar angle. In addition, we consider that the radius r is large enough so that we can get rid of the terms of the order higher than 1/r and the approximate equation can be written as

$$S_{,tt} = \sigma_{,rr} - \Upsilon\chi_{,rrr}\ . \tag{12}$$

We notice that eq.(12) is exactly eqs(5a-b) by changing r into x. Moreover, the macro-stress σ and micro-stress χ are still given by eq.(5c) with the potential (6). Accordingly, the solution to eq.(12) has the same form as in the one-dimensional model, but it corresponds in fact to a localized strain soliton travelling in the lattice plane since the radius r depends both on x and y. Note however that this does not explain the instability process transforming figure 7.a into figure 7.c.

8. -Conclusions

This work has been devoted to the study of pattern formation related to elastic domains as met in martensitic-ferroelastic transformations exhibiting first order phase transition (the order parameter is a strain). With this in view, we have considered a lattice model on the basis of a two-dimensional lattice model possesses all the ingredients which allow for nonlinear excitation propagation. Moreover, the considered interactions by pairs and of the non-central type lead to an lattice energy which fulfils the physical requirements involved in the cubic-tetragonal transformation. The pertinent results are : (i) the partial softening of the transverse acoutic branch at nonzero wave-number, (ii) the upward convexity of the dispersion branch near the long wavelength limit, (iii) the propagation of nonlinear excitations describing the shearing motion of the atomic planes along the stacking direction (martensitic and austenitic solitons) and (iv) the instability process of an elastic soliton with respect to the transverse direction in the two-dimensional system leading to a localized martensitic domain. The patterns thus obtained correspond to the domain structures commonly observed by means of electron microscopy in various alloys such as Ni-Ti, Ti-Mn, In-Tl just to quote a few examples [4,5,7]. Nevertheless, further work can be investagated. We can study, on the basis of the lattice model the discreteness effects or the ground states of deformation which minimize the lattice energy in the case of rather tiny martensitic domains. It is also worthwhile examining the influence of an external force and damping on the dynamics of such structures in order to explain the hysteresis effect.

References

[1] ROITBURD A.L., 1978, in Solid State Physics vol.33, Advances in Research and Applications, eds H. Ehrenreich, F. Seitz and D. Turnbull (Academic Press, New York), p.317.

[2] LANDAU L.D., LIFSHITZ E., 1980, Statistical Physics (Pergamon Press, Oxford).

[3] TOLEDANO J-C., TOLEDANO P., 1987, The Landau Theory of Phase Transitions (World Scientific, Singapore).

[4] CHRISTIAN J.W., KNOWLES K.M., 1981, Proceedings of an Int. Conf. on Solid-Solid Phase Transformations (The Metall. Soc. of AIME), p.1185.

[5] KNOWLES K.M., CHRISTIAN J.M., SMITH D.A., 1982, J. de Physique, 43, C4.185.

[6] AUBRY S., 1978, in Solitons and Condensed Matter Physics, eds A.R. Bishop and T. Schneider (Springer Verlag, Berlin), p.264.

[7] TANNER L.E., PELTON A.R., GRONSKY R., 1982, J. de Physique 43, C4.169.

[8] BISHOP A.R., KRUMHANSL J.A., TRULLINGER S.E., 1980, Physica D 1, 1.

[9] FINLAYSON T.R., MOSTOLLER M., REICHARDT W., SMITH H.G., 1985 , Solid State Comm. 53, 461.

[10] MURAKAMI Y., 1975, J. Phys. Soc. Jpn. 38, 404.

[11] BRUINSMA R., 1982, Phys. Rev. B 25, 2951.

[12] MORI M., YAMADA Y., SHIRANE G., 1975, Solid State Comm. 17,127.

[13] BARSCH G.R., KRUMHANSL J.A., TANNER L.E., WUTTIG M., 1987, Scripta Metall. 21, p.1257.

[14] SHAPIRO S.M., LERESE J.Z., NODA Y., MOSS S.C., TANNER L.E., 1986, Phys. Rev. Letters 57, 3199.

[15] KRUMHANSL J.A., 1963, in Lattice Dynamics, ed. R.F. Wallis (Pergamon, Oxford, U.K.).

[16] MINDLIN R.D., ESHEL N.N., 1968, Int. J. Solid Struct. 4, 109.

[17] POUGET J., 1989, Phase Transitions 14, 251.

[18] TEODOSIU C., 1982, Elastic Models of Crystal Defects (Springer Verlag, Berlin).

[19] GUNTON D.J., SAUNDERS G.A., 1974, Solid State Comm. 14, 865.

[20] SHAPIRO S.M., 1988, in Competing and Microstructures : Statics and Dynamics, eds R. Lesar, A.R. Bishop and R. Heffner (Springer Verlag, Berlin), p. 84.

[21] ANDERSON R.W., BLOUNT E.I., 1965, Phys. Rev. Lett. 14, 217.

Critical fluctuations in the 2D XY antiferromagnet $BaNi_2(PO_4)_2$: NMR measurements.

Ph. Gaveau+, J.P. Boucher+, L.P. Regnault++ and Y. Henry++
DRF-G/Service de Physique/DSPE+ and MDN++ , Centre d'Etudes Nucléaires
de Grenoble, 85X - 38041 Grenoble cedex, France.

Abstract: Non-linear excitations are known to play a crucial role in low-dimensional magnetism. For the two-dimensional (2D) XY model these non-linear excitations are described in terms of vortices which govern the so-called Kosterlitz-Thouless transition occuring at a temperature $T_{KT}>0$. We present here results of ^{31}P nuclear magnetic resonance obtained on the compound $BaNi_2(PO_4)_2$ which is a good example of a quasi 2D XY type antiferromagnet. Above the 3D magnetic ordering occuring at a temperature $T_N \simeq 23.6$ K a diverging behaviour is observed for the corresponding nuclear spin-lattice relaxation (NSLR) rate $1/T_1$. We show that this behaviour can be interpreted in the recent model of Mertens et al. describing the dynamical behaviour of free vortices.

I) The $BaNi_2(PO_4)_2$ crystal.

Both the crystallographic structure of $BaNi_2(PO_4)_2$, elucidated by Eymond and Durif [1], and its magnetic properties, described by Regnault et al.[2], have established that this compound is a good two-dimensional(2D) XY type antiferromagnet.

The structure consists of 2D layers of magnetic Ni ions located on a honeycomb lattice of parameter a = 2.8 Å. As a first approximation, one can consider that the magnetic interactions within the XY plane result from an antiferromagnetic superexchange coupling $\tilde{J} = -11$ K between neighbouring spins. The effective coupling $\tilde{J}'$ between planes is small and the ratio $\tilde{J}'/\tilde{J} \simeq 10^{-3}- 10^{-4}$ ensures a good two-dimensional character for this compound. Moreover, due to a strong single ion anisotropy (D = 7.3 K), the spin S=1 is essentially confined in the XY plane at low temperature.

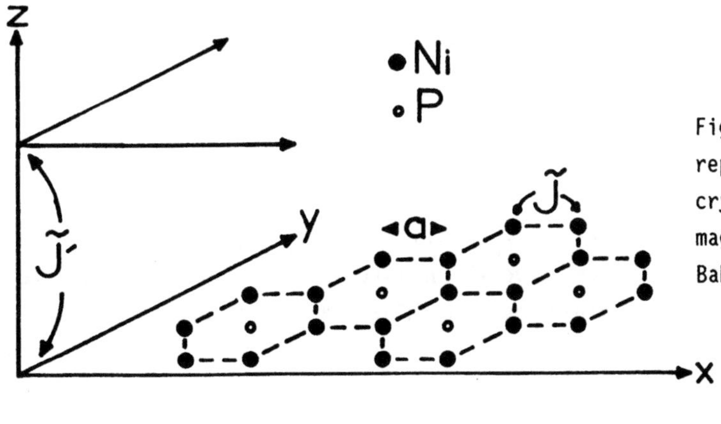

Figure 1:schematic representation of the crystallographic and magnetic stucture of $BaNi_2(PO_4)_2$.

II) The 2D XY model

1) Static properties

This model corresponds to the following hamiltonian :

$$\mathcal{H} = \sum_{i,j} J_{ij} \left(S_i^x S_j^x + S_i^y S_j^y \right)$$

where J_{ij} is the exchange coupling between neighbouring spins in the XY plane.

Kosterlitz and Thouless [3] have shown that, in such a model, there exists a topological phase transition at a temperature $T_{KT} \simeq 1.8\ |J|S^2$, governed by localised non-linear excitations which can be described in terms of vortices. For temperature $T < T_{KT}$, the system goes to a state of vortex-antivortex bound pairs. For temperature $T > T_{KT}$, the pairs begin to unbind, leading to a short range order defined by the correlation length $\xi(\tau)$:

$$\xi(\tau) = \xi_0 \exp\left(\frac{b}{\sqrt{\tau}}\right) \qquad \text{with} \quad \begin{cases} \xi_0 \simeq a \quad , & b = \dfrac{\pi}{2} \\ \tau = \dfrac{T-T_{KT}}{T_{KT}} & \end{cases} \tag{1}$$

This correlation length can be interpreted as the half mean distance between free vortices [4] so that the density of free vortices n_v is given by :

$$n_v = \frac{1}{(2\xi)^2} = \frac{1}{4\xi_0^2} \exp\left(\frac{-2b}{\sqrt{\tau}}\right)$$

2) Dynamical properties.

For $T > T_{KT}$ the vortices are expected to move freely in the XY plane. The expression of the root-mean-square velocity $\bar{u}$ of the vortices has been evaluated by Huber [5] :

$$\bar{u} = \sqrt{\frac{\pi}{2}\left(\frac{JS^2 a^2}{\hbar}\right) n_v \, \mathrm{Ln}\left(\frac{k_B T_{KT}}{JS^2 n_v a^2}\right)}$$

In the critical regime ($\tau \to 0$) , all these quantities $\xi(\tau)$, $n_v(\tau)$ and $\bar{u}(\tau)$ vary rapidly with τ, exponentially :

$$\xi(\tau) \to \infty \quad , \quad n_v(\tau) \to 0 \quad , \quad \bar{u}(\tau) \to 0$$

In this temperature range, several contributions are expected in the magnetic fluctuation spectrum. The spin-waves can give an appreciable contribution which should not depend strongly on temperature. The contributions of the moving free vortices have recently been investigated by Mertens et al. [6]. In their model, the fluctuations due to the vortices are proportional to n_v and, hence, they decrease rapidly with τ. However, there exists also an important in plane (IP) contribution, which is associated with the spin flipping induced by the motion of the vortices. It results, for the local fluctuations, in a "central peak" :

$$S_{IP}^{loc}(\omega) = \sum_q S_{IP}(q,\omega) \simeq \frac{\gamma}{\gamma^2 + \omega^2}$$

where ω is the frequency and γ provides a direct measure of the flipping rate. In a ballistic model, this flipping rate is given by :

$$\gamma = \sqrt{\pi n_v}\ \bar{u}$$

At very low frequency ($\omega \to 0$), the intensity of this contribution $[S_{IP}^{loc}(\omega \simeq 0) \simeq 1/\gamma]$ is diverging as $\tau \to 0$ since $\gamma \to 0$. The observation of such an effect could be the signature of moving free vortices.

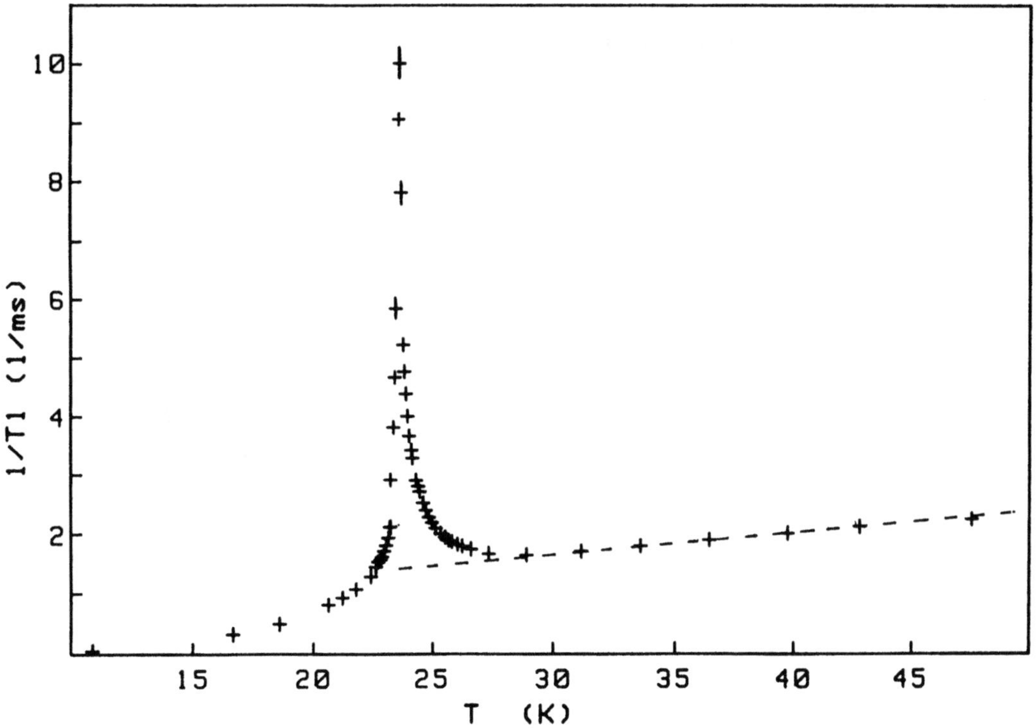

Figure 2: experimental values of the ^{31}P nuclear spin-lattice relaxation rate $1/T_1$ as a function of temperature. The dashed line is an evaluation of the spin-wave contribution (see text).

III) Experimental results.

By nuclear magnetic resonance, one can probe the low frequency fluctuation spectrum of magnetic systems. In $BaNi_2(PO_4)_2$, we have measured the nuclear spin-lattice relaxation (NSLR) rate $1/T_1$ of ^{31}P as a function of temperature. The phosphorus is located in the centre of the honeycomb stucture of the magnetic Ni ions and $1/T_1$ is proportional to the local fluctuation spectrum $S^{loc}(\omega)$ probed at the nuclear Larmor frequency ω_N :

$$\frac{1}{T_1} = A\, S^{loc}(\omega_N)$$

where A is a geometrical coefficient related to the hyperfine coupling. In our measurements ω_N = 34 MHz for an applied field $H_0 \sim 2$ Teslas.

For the temperature range 10 K < T < 50 K, our results are shown on figure 2. A rapid variation of $1/T_1$ is observed between T = 23.6 K and 27 K. The peak maximum, which is defined accurately, can be attributed to the 3D ordering temperature $T_N = 23.6 \pm 0.05$ K . This value is in good agreement with other evaluations [2]. We would like to show that the rapid variation <u>above</u> T_N can be well analysed in terms of 2D fluctuations. In real magnetic systems, the Kosterlitz-Thouless transition can never be observed due to the interplane exchange coupling $\tilde{J}'$. However, as shown by Regnault et al.[2], T_N should be very close to T_{KT} for this compound :

$$T_{KT} \simeq 0.95 - 0.98\ T_N$$

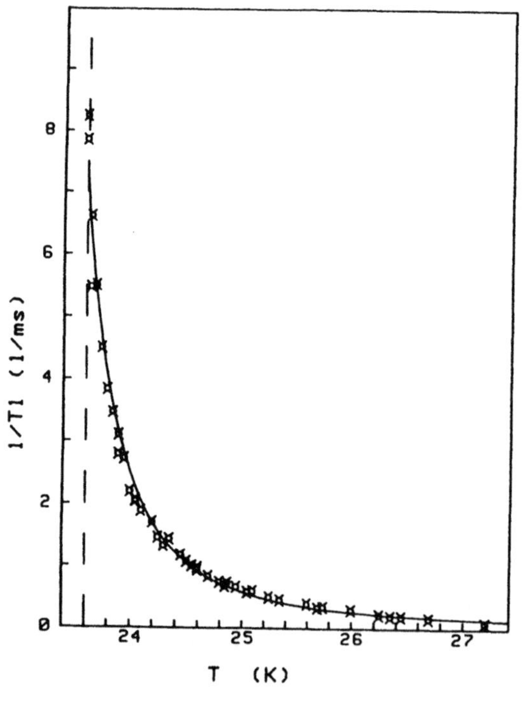

Figure 3:expected contribution of the "central peak" to the nuclear spin-lattice relaxation rate as a function of temperature. The full line is a theoretical curve obtained from eq.(2). The dashed line corresponds to the 3D magnetic ordering at T_N = 23.6 K

The slow temperature dependance observed above $T \simeq 27$ K is extrapolated down to T_N and attributed to the spin-wave contribution which, as discussed above, should not depend drastically on T. The remaining part -i.e. the part above the dashed line on figure 2- is tentatively attributed to the central peak resulting from the flipping process occuring in the model of moving free vortices. The corresponding data and the theoretical curve $1/T_1$:

$$\frac{1}{T_1} = A\, S_{IP}^{loc}(\omega_N) \propto \frac{1}{\gamma} \qquad \text{for } \omega_N \ll \gamma \qquad (2)$$

are shown on figure 3. A very good agreement (see the full line on this figure) is obtained for the values $T_{KT} = 0.95\ T_N$, $\xi_0 = a$ and $b = 0.95$. Although the latter value differs from $\pi/2$ given in eq.(1), it is in reasonable agreement with more recent theoretical evaluations [6,7]. In this comparison the corresponding values for γ which range from 0.5 GHz for T = 23.8 K to 10 GHz for T = 27 K, are much larger than ω_N and are consistent with the approximation of eq.(2).

IV) Conclusion

Our measurements on the compound $BaNi_2(PO_4)_2$ have clearly established that low frequency fluctuations develop rapidly when approaching T_N. According to recent elastic neutron scattering measurements, the 3D critical regime does not exceed 0.5 K above T_N [2]. Therefore, it seems reasonable to attribute the large increase of $1/T_1$ observed below 27 K and down to 23.6 K as resulting essentially from 2D fluctuations. Our analysis, relying on the model of moving free vortices as discussed by Mertens et al.[6] is shown to provide a good agreement with the data.

The results discussed in the present work were obtained with the external field H_0 applied perpendicular to the XY plane. In that case, the rapid variation of $1/T_1$ does not depend on the field value : the same results are actually obtained for H_0 = 1 and 2 Teslas . However, when the field is applied within the XY plane, the diverging behaviour observed for $1/T_1$ is strongly field dependant : the variation increases with H_0. This interesting effect is now under investigation.

References :

[1] S. Eymond and A. Durif, Mat. Res. Bull. **4**, 595 (1969)

[2] L.P. Regnault and J. Rossat-Mignod , in *Magnetic properties of layered transition metal compounds*, edited by L.J. de Jongh and R.D. Willet (Reidel, Dordrecht, 1989)

[3] J.M. Kosterlitz and D.J. Thouless, J. Phys. C **6**, 1181 (1973)
J.M. Kosterlitz, J. Phys. C **7**, 1047 (1974)

[4] D.J. Bishop and J.D. Reppy, Phys. Rev. Lett. 40, 1727 (1978)

[5] D.L. Huber, Phys. Rev. B **26**, 3758 (1982)

[6] F.G. Mertens,A.R. Bishop,G.M. Wysin and C.Kawabata, Phys.Rev. B **39**, 591 (1989)

[7] S.W. Heinekamp and R.A. Pelcovitz,Phys. Rev B **32**, 4528 (1985)

1/S QUANTUM CORRECTIONS FOR THE MAGNETIC SOLITONS IN VARIOUS HEISENBERG MODELS

D.V. Kapor, M.J. Škrinjar and S.D. Stojanović
Institute of Physics, Faculty of Sciences, 21000 Novi Sad, Yugoslavia

Introduction

From the early days of quantum theory of magnetism, i.e. from the times of Heisenberg ferromagnet, two approaches appeared: one treating spins (magnetic moments) as the operators of the angular momentum and the other one trying to represent them by Bose-operators. Both approaches have their pro's and contra's, but today one can read in most of the textbooks that boson representations "work" well at low temperatures, while for higher temperatures one has to take into account proper nature of spins.

When the soliton theory was first applied to the Heisenberg ferromagnet, the spins were treated as classical vectors [1] . Only later, various quantum corrections were looked for. The question that arose naturally was: if boson approach can describe spins for some other purposes, can it work here? The most suitable representation was Holstein-Primakoff (HP) representation [2] for at least two reasons: it is not limited to low-temperatures in the most general form and its validity becomes better for higher spins ($S \gg 1$) which agrees well with the classical limit ($S \to \infty$, $\hbar \to 0$, $S\hbar = S_c$).

For the system with the ground state of all spins pointing "up", HP representation takes the form:

$$\hat{S}^z_n = S - \hat{B}^+_n \hat{B}_n \tag{1.a}$$

$$\hat{S}^+_n = \sqrt{2S}\sqrt{1 - \frac{\hat{B}^+_n \hat{B}_n}{2S}}\;\hat{B}_n \qquad \hat{S}^-_n = (\hat{S}^+_n)^+ \tag{1.b}$$

where $\{\hat{B}^+_n, \hat{B}_n\}$ are Bose-operators sastisfying commutation relations

$$[\hat{B}_n, \hat{B}^+_m] = \delta_{n,m} \qquad [\hat{B}^+_n, \hat{B}^+_m] = [\hat{B}_n, \hat{B}_m] = 0 \tag{2}$$

(We must notice that this reperesentation is valid for the system $\hbar=1$, so in order to perform the classical transition correctly, we shall introduce factors of $\hbar$ into the Hamiltonian.)

The general procedure for the application of boson representations in the soliton theory is rather simple and consists in the following steps:

a) expressing spin operators in terms of Bose-operators ("bosonisation");

b) averaging the Hamiltonian over Glauber's coherent states (CS) [3]:

$$\hat{B}_n|\alpha_n\rangle = \alpha_n|\alpha_n\rangle \qquad (2)$$

c) treating the averaged Hamiltonian as the classical Hamilton's function of the system with α and $i\hbar\alpha^*$ as conjugated variables [4];

d) expressing the results in terms of physical relevant quantities (energy, momentum, magnetization).

How it occured that such simple procedure failed to give reasonable results in the earliest attempts can be explained by the fact that the application of CS demands the ordered product of operators, which is not easy to obtain working with the square root is HP representation (1).

Quantum corrections for spin operators

We shall now present our approach which produces correct results in the classical limit and also allows the calculation of quantum corrections up to any order of 1/S. This will be ilustrated with the correction of order 1/S. Since we are interested in the classical limit, we suppose S to be large enough that we can perform the expansion of the form [5]:

$$\left(1 - \frac{\hat{B}_j^+\hat{B}_j}{2S}\right)^{1/2} = \sum_{k=0}^{\infty}\binom{1/2}{k}\left(-\frac{1}{2}\right)^k\left(\frac{\hat{B}_j^+\hat{B}_j}{S}\right)^k \qquad (3)$$

Putting $\left(\hat{B}_j^+\hat{B}_j\right)^k$ into the normal ordered form, it can be written generally as

$$\left(\hat{B}_j^+\hat{B}_j\right)^k = \hat{B}_j^{+k}\hat{B}_j^{\,k} + a_{k-1}^k\hat{B}_j^{+k-1}\hat{B}_j^{\,k-1} + a_{k-2}^k\hat{B}_j^{+k-2}\hat{B}_j^{\,k-2} + a_2^k\hat{B}_j^{+2}\hat{B}_j^{\,2} + \hat{B}_j^+\hat{B}_j \qquad (4)$$

It can be seen by inspection that $a_k^k = a_1^k = 1$. This expression shows that for any order $\hat{B}_j^{+l}\hat{B}_j^{\,l}$ there appear also the contributions arising from $\left(\hat{B}_j^+\hat{B}_j\right)^p$ for any $p > l$. Yet, expression (3) shows that such contributions are devided with S^p $(p > l)$ and these terms are lost in the classical limit.

We can determine these terms, for example, by calculating $\left(\hat{B}_j^+\hat{B}_j\right)^{k+1}$

in two ways:

$$\left(\hat{B}_j^+\hat{B}_j\right)^{k+1} = \hat{B}_j^{+\,k+1}\hat{B}_j^{\,k+1} + a_k^{k+1}\hat{B}_j^{+\,k}\hat{B}_j^{\,k} + \ldots =$$

$$= \hat{B}_j^+\hat{B}_j\left(\hat{B}_j^+\hat{B}_j\right)^k = \hat{B}_j^+\hat{B}_j\left(\hat{B}_j^{+\,k}\hat{B}_j^{\,k} + a_{k-1}^k\hat{B}_j^{+\,k-1}\hat{B}_j^{\,k-1} + \ldots\right)$$

This leads to the imporatant recurrent relation:

$$a_l^{k+1} = l\, a_l^k + a_{l-1}^k \tag{5}$$

In order to obtain the corrections of order 1/S, we have to evaluate a_{k-1}^k:

$$a_k^{k+1} = a_{k-1}^k + k$$

leading to

$$a_k^{k+1} = 1 + 2 + \ldots + k = \frac{k(k+1)}{2}$$

Using these two corrections, we have

$$\sqrt{1 - \frac{\hat{B}_j^+\hat{B}_j}{2S}} \simeq \sum_{k=0}^{\infty}\binom{1/2}{k}\left(-\frac{1}{2}\right)^k\left[\frac{\hat{B}_j^{+\,k}\hat{B}_j^{\,k}}{S^k} + \frac{1}{S}\frac{k(k-1)}{2}\frac{\hat{B}_j^{+\,k-1}\hat{B}_j^{\,k-1}}{S^{k-1}} + O\left(\frac{1}{S^2}\right)\right] \tag{6}$$

We shall treat only one-dimensional chains, so we use the simple product of coherent states

$$|\alpha\rangle = \prod_{j=1}^{N}|\alpha_j\rangle \tag{7}$$

to obtain

$$\langle\alpha|\hat{S}_j^+|\alpha\rangle = \sqrt{2S}\langle\alpha_j|\sqrt{1 - \frac{\hat{B}_j^+\hat{B}_j}{2S}}\,\hat{B}_j|\alpha_j\rangle \approx \sqrt{2S}\left[\sum_{k=0}^{\infty}\binom{1/2}{k}\left(-\frac{1}{2}\right)^k\langle\alpha_j|\frac{\hat{B}_j^{+\,k}\hat{B}_j^{\,k}}{S^k}\hat{B}_j|\alpha_j\rangle\right.$$

$$\left. + \frac{1}{S}\sum_{n=0}^{\infty}\binom{1/2}{k}\left(-\frac{1}{2}\right)^k\frac{k(k-1)}{2}\langle\alpha_j|\frac{\hat{B}_j^{+\,k-1}\hat{B}_j^{\,k-1}}{S^{k-1}}\hat{B}_j|\alpha_j\rangle + O\left(\frac{1}{S^2}\right)\right] \approx \tag{8}$$

$$\approx \sqrt{2S}\left[\sum_{k=0}^{\infty}\binom{1/2}{k}\left(-\frac{1}{2}\right)^k\left(\frac{|\alpha_j|^2}{S}\right)^k\alpha_j + \frac{1}{S}\sum_{k=0}^{\infty}\binom{1/2}{k}\left(-\frac{1}{2}\right)^k\frac{k(k-1)}{2}\left(\frac{|\alpha_j|^2}{S}\right)^{k-1}\alpha_j + O\left(\frac{1}{S^2}\right)\right]$$

The main purpose of the averaging over coherent states was to obtain c-numbers,so we can easily manipulate with them. It is more natural to go over to the notation

$$\tilde{\alpha}_j = \alpha_j/\sqrt{S} \tag{9}$$

because, the proper orders of 1/S appear. The first term on the right -

hand side is easily summed back to the square-root, while the second one demands some algebra. The final result is

$$\langle\alpha|\hat{S}_j^{+}|\alpha\rangle = \sqrt{2}\, S \sqrt{1 - \frac{|\tilde{\alpha}_j|^2}{2}}\; \tilde{\alpha}_j \left[1 - \frac{1}{32S}\, \frac{|\tilde{\alpha}_j|^2}{\left(1 - \frac{|\tilde{\alpha}_j|^2}{2}\right)^2}\right] + O\left(\frac{1}{S^2}\right) \tag{10}$$

and we have also

$$\langle\alpha|\hat{S}_j^{z}|\alpha\rangle = S(1 - |\tilde{\alpha}_j|^2) \tag{11}$$

It is also convenient to express the result using trigonometrical parametrization. If we denote with (α_c) CS corresponding to the classical limit, we have:

$$\langle\alpha_c|\hat{S}_j^{z}|\alpha_c\rangle = S\cos\Theta \tag{12.a}$$

$$\langle\alpha_c|\hat{S}_j^{\pm}|\alpha_c\rangle = S\sin\Theta\, e^{\pm i\phi} \tag{12.b}$$

giving:

$$\langle\alpha|\hat{S}_j^{z}|\alpha\rangle = S\cos\Theta \tag{13.a}$$

$$\langle\alpha|\hat{S}_j^{\pm}|\alpha\rangle = S\sin\Theta\, e^{\pm i\phi}\left[1 - \frac{1}{8S}\, \frac{1-\cos\Theta}{(1+\cos\Theta)^2}\right] \tag{13.b}$$

These expressions allow us to perform the analysis of various Heisenberg models.

Isotropic Heisenberg chain

Let us start with isotropic Heisenberg model of the form

$$H - H_0 = -\mu f \hbar \sum_j (\hat{S}_j^z - S) - J\hbar^2 \sum_j (\vec{S}_j \vec{S}_{j+1} - S^2) \qquad (\mu = g\mu_B) \tag{14}$$

Here f is an external field directed along z-axis, $\mu\hbar$ is the magnetic moment of the ion and, $J\hbar^2$ is the exchange integral between nearest neighbours. There exist classical [1] and semiclassical [6] solutions of the problem. So it came as a surprise that first attempts [7,8] failed to reproduce them. First successful result was obtained by de Azevedo *et al.*[9] who managed to reproduce classical results by working with the "truncated" HP representation, using the expansion of square root up to four Bose-operators. We generalized their approach to the anisotropic case [10] including the terms up to six Bose-operators. It was this success that initiated the present work with the complete series.

Performing above mentioned procedure, we obtain

$$\langle\alpha|H - H_o|\alpha\rangle = H_c + H_{quant} \tag{15}$$

It is usual to perform the classical limit $S \to \infty$, $\hbar \to 0$ with $S\hbar \to S_c$, $JS_c^2 \to J_c$, $\mu S_c \to \mu_c$. Next step is also the continuum limit, which implies expansion up to terms of order a^2 (a is the lattice constant):

$$\sum_j \to \frac{1}{a}\int dx \qquad \alpha_{j\pm 1} \to \alpha(x \pm a) \simeq \alpha \pm a\alpha_x + \frac{a^2}{2}\alpha_{xx}$$

The Hamiltonian density is given as

$$\mathcal{H}_c = \mu_c f|\tilde{\alpha}|^2 + \frac{1}{4}J_c a^2(\tilde{\alpha}^2\tilde{\alpha}_x^{*2} + \tilde{\alpha}^{*2}\tilde{\alpha}_x^2) + \frac{1}{16}J_c a^2 \frac{|\tilde{\alpha}|^2}{1 - \frac{|\tilde{\alpha}|^2}{2}}(|\tilde{\alpha}|_x^2)^2 \tag{16}$$

$$\mathcal{H}_{quant} = \frac{J_c}{32S}\left[\frac{4|\tilde{\alpha}|^4}{1 - \frac{|\tilde{\alpha}|^2}{2}} + a^2\frac{\left(-1+|\tilde{\alpha}|^2 + \frac{|\tilde{\alpha}|^4}{8}\right)\left(\tilde{\alpha}_x^{*2}\tilde{\alpha}^2 + \tilde{\alpha}_x^2\tilde{\alpha}^{*2} + 4|\tilde{\alpha}|^2|\tilde{\alpha}_x|^2\right)}{\left(1 - \frac{|\tilde{\alpha}|^2}{2}\right)^3}\right] \tag{17}$$

This expression was derived under the assumption of cyclic boundary conditions and that all derivatives of $|\tilde{\alpha}|^2$ vanish at infinity.

The equation of motion for the system is

$$iS_c\frac{\partial\tilde{\alpha}}{\partial\hbar} = \frac{\partial\mathcal{H}_c}{\partial\tilde{\alpha}^*} - \frac{\partial}{\partial x}\left(\frac{\partial\mathcal{H}_c}{\partial\tilde{\alpha}_x^*}\right) + \frac{\partial\mathcal{H}_{quant}}{\partial\tilde{\alpha}^*} - \frac{\partial}{\partial x}\left(\frac{\partial\mathcal{H}_{quant}}{\partial\tilde{\alpha}_x^*}\right) \tag{18}$$

and it obtains the following form:

$$iS_c\tilde{\alpha}_t = \mu_c f\tilde{\alpha} - J_c a^2\tilde{\alpha}_{xx} - \frac{1}{2}J_c a^2(2\tilde{\alpha}|\tilde{\alpha}_x|^2 + \tilde{\alpha}^2\tilde{\alpha}_{xx}^* - \tilde{\alpha}^*\tilde{\alpha}_x^2) -$$

$$- \frac{1}{16}J_c a^2\left[\frac{\tilde{\alpha}(|\tilde{\alpha}|_x^2)^2}{(1 - \frac{1}{2}|\tilde{\alpha}|^2)^2} + \frac{2\tilde{\alpha}|\tilde{\alpha}|^2|\tilde{\alpha}|_{xx}^2}{1 - \frac{1}{2}|\tilde{\alpha}|^2}\right] + \frac{1}{S}f(\tilde{\alpha}) \tag{19.a}$$

where $f(\tilde{\alpha})$ follows from $\mathcal{H}_{quant}$:

$$f(\tilde{\alpha}) = \frac{J_c}{16}\tilde{\alpha}|\tilde{\alpha}|^2\frac{4 - |\tilde{\alpha}|^2}{\left(1 - \frac{|\tilde{\alpha}|^2}{2}\right)^2} + \frac{J_c}{32}a^2\left[(2|\tilde{\alpha}|^2\tilde{\alpha}_{xx} + \tilde{\alpha}^2\tilde{\alpha}_{xx}^*) \times\right.$$

$$\times\frac{2-2|\tilde{\alpha}|^2 - \frac{|\tilde{\alpha}|^4}{4}}{\left(1 - \frac{|\tilde{\alpha}|^2}{2}\right)^3} + \frac{1}{2}\tilde{\alpha}^3\tilde{\alpha}_x^{*2}\frac{1 - \frac{5}{2}|\tilde{\alpha}|^2 - \frac{|\tilde{\alpha}|^4}{8}}{\left(1 - \frac{|\tilde{\alpha}|^2}{2}\right)^4} + \tilde{\alpha}|\tilde{\alpha}_x|^2\frac{4-5|\tilde{\alpha}|^2 - |\tilde{\alpha}|^4 + \frac{|\tilde{\alpha}|^6}{8}}{\left(1 - \frac{|\tilde{\alpha}|^2}{2}\right)^4} +$$

$$\tilde{\alpha}^{*}\tilde{\alpha}_x^{2}\,\frac{2-\frac{3}{2}|\tilde{\alpha}|^2+4|\tilde{\alpha}|^4-\frac{|\tilde{\alpha}|^6}{16}}{\left(1-\frac{|\tilde{\alpha}|^2}{2}\right)^4}\Bigg] \qquad (19.b)$$

This system of equations represents the initial point of our calculations.

Classical solutions α_c are obtained if the terms $\mathcal{H}_{quant}$ and $f(\tilde{\alpha})$ are neglected. The equation (19) is then solved by the substitution

$$\tilde{\alpha}_c(x,t) = A(\xi)\, e^{i[\varphi(\xi)+\Omega t]} \qquad \xi = x - vt \qquad (20)$$

v is the soliton velocity. The imaginary part of (19) gives

$$\varphi_\xi = \frac{V}{2}\,\frac{1}{1-\frac{A^2}{2}} \qquad V = \frac{S_c v}{J_c a^2} \qquad (21)$$

and the real part gives

$$A_\xi^2 = \gamma_o A^2\left[1-\frac{V^2}{4\gamma_o}-\frac{A^2}{2}\right] \qquad \gamma_o = \frac{S_c\Omega+\mu f}{J_c a^2} \qquad (22)$$

The solution is:

$$A^2 = \frac{2\varphi_o^2}{\mathrm{ch}^2\frac{2}{\Gamma}\xi} \qquad \varphi_o^2 = \sin^2\frac{Pa}{4S_c} \qquad \Gamma = \frac{Ma}{2S_c}\,\frac{1}{\varphi_o^2} \qquad (23)$$

where P and M are momentum and magnetization respectively. We did not quote all the equations in detail, but it can be seen from (22) that working with the complete boson series leads to the equation in which only terms up to four Bose-operators contribute, explaining the success of "truncated" series calculations [9].

Let us now look for the parameters of the system with qunatum correction included. It can be done easily, because in the calculations of order 1/S, it is sufficient to calculate terms of order 1/S with classical solutions. (In this part, spin components will mean their values averaged over coherent states).

Magnetization of the system is given by

$$M = \frac{\hbar}{a}\int (S - S^z)dx = S_c\,\frac{1}{a}\int_{-\infty}^{\infty}|\tilde{\alpha}|^2 dx = S_c\frac{1}{a}\int_{-\infty}^{\infty}|\tilde{\alpha}_c|^2 dx \qquad (24)$$

In HP representation, there is no quantum correction for M!

As for the momentum [6]:

$$P = \frac{\hbar}{a}\int dx\,\frac{1}{S_c + S^z}\left(S^x\,\frac{\partial S^y}{\partial x} - S^y\,\frac{\partial S^x}{\partial x}\right) = \frac{\hbar}{2i}\,\frac{1}{a}\int dx\,\frac{1}{S_c + S^z}\left(S^-\,\frac{\partial S^x}{\partial x} - S^+\frac{\partial S^-}{\partial x}\right) \tag{25}$$

We obtain

$$P_c = \frac{S_c}{2i}\,\frac{1}{a}\int_{-\infty}^{\infty} dx\,(\tilde{\alpha}^*_c\tilde{\alpha}_{cx} - \tilde{\alpha}_c\tilde{\alpha}^*_{cx}) = \frac{S_c}{a}\int_{-\infty}^{\infty} dx\,A^2\varphi_x \tag{26}$$

Quantum correction becomes:

$$P_{quant} = -\frac{1}{16S}\,\frac{S_c}{2i}\,\frac{1}{a}\int_{-\infty}^{\infty} dx\,(\tilde{\alpha}^x_c\tilde{\alpha}_{cx} - \tilde{\alpha}_c\tilde{\alpha}^x_{cx})\,\frac{|\tilde{\alpha}_c|^2}{\left(1 - \frac{|\tilde{\alpha}_c|^2}{2}\right)^2} =$$

$$= -\frac{1}{32S}\,VS_c\,\frac{1}{a}\int_{-\infty}^{\infty} dx\,\frac{A^4}{\left(1 - \frac{A^2}{2}\right)^3} \tag{27}$$

Finally, the energy of the system can be expressed as:

$$E_c = \frac{1}{a}\int \mathcal{H}_c\,dx = \mu f M + \frac{16S_cJ_c}{M}\sin^2\frac{P_c a}{4S_c} \tag{28}$$

Integrating (17) leads to a complicated function of P_c, P_{quant}, M_c, V and S_c with interesting property $E_{quant} \sim M$. The problem with this expression is that it is divergent at zone boundary for $S=1/2$, which is a typical drawback of the application of HP representation.

Let us now comment on the effect of the introduction of the anisotropy of the following form:

$$H_{an} = -\sigma J \hbar^2 \sum_j (\hat{S}^z_j\,\hat{S}^z_{j+1} - S^2) \tag{29}$$

where σ is the anisotropy parameter. The advantage of this type of anisotropy is that it has no 1/S corrections, yet the solutions of the classical part are changed. Without further discussion, we only note that the equation for φ_ξ is not changed, while the equation for A has a contribution of the term with six Bose-operators [10].

Unfortunately, it is vather difficult fo find some other calculations of the same type for the sake of comparison, so, in order to see the effect of our approach, we shall study a more "popular" case: sine-Gordon problem.

Quantum corrections to sine-Gordon equation

We shall study the easy-axis ferromagnetic (J > 0) and anti-ferromagnetic (J < 0) chain with the Hamiltonian of the form [11]:

$$\hat{H} = -J \sum_j \hat{\vec{S}}_j \hat{\vec{S}}_{j+1} + A \sum_j (\hat{S}^z_j)^2 - \mu f \sum_j \hat{S}^x_j \qquad (30)$$

(In order to follow the common notations, we accept $\hbar=1$ and $S_c=S$.) In the parametrisation described by eq. (12), we have the correction to $\hat{S}^{\pm}_j$ of the form (13). In order to obtain correct results, we must calculate $(\hat{S}^z_j)^2$ so, to keep the contribution of four Bose- operators:

$$(\hat{S}^z_j)^2 = S^2 - (2S-1)\, \hat{B}^+_j \hat{B}_j + \hat{B}^{+2}_j \hat{B}_j^{\,2} \qquad (31)$$

giving

$$\langle\alpha|(\hat{S}^z_j)^2|\alpha\rangle = S^2\left[\cos^2\Theta_j + \frac{1}{S}(1 - \cos\Theta_j)\right] \qquad (32)$$

Using these expressions, we obtain:

$$H_c = -JS^2 \sum_j \left[\sin\Theta_j \sin\Theta_{j+1} \cos(\phi_j - \phi_{j+1}) + \cos\Theta_j \cos\Theta_{j+1}\right] + AS^2 \sum_j \cos^2\Theta_j -$$

$$- \mu Sf \sum_j \sin\Theta_j \cos\phi_j \qquad (33)$$

$$H_{quant} = \frac{JS}{4} \sum_j \sin\Theta_j \sin\Theta_{j+1} \cos(\phi_j - \phi_{j+1}) \left[\frac{1-\cos\Theta_j}{(1+\cos\Theta_j)^2} + \frac{1-\cos\Theta_{j+1}}{(1+\cos\Theta_{j+1})^2}\right]$$

$$+ \frac{\mu f}{8} \sum_j \frac{1-\cos\Theta_j}{(1+\cos\Theta_j)^2} \sin\Theta_j \cos\Theta_j + AS \sum_j (1 - \cos\Theta_j) \qquad (34)$$

Going over to the continuum we have

$$\mathcal{H} = \frac{J\,S^2}{2}\,\Theta_x^2 + \frac{J\,S^2 a^2}{2}\left(1 - \frac{1}{2S}\right)\phi_x^2 \sin^2\Theta + AS^2\cos^2\Theta - \mu Sf\left(1 - \frac{1}{8S}\right)\sin\Theta\cos\phi$$

$$(35)$$

where $\mathcal{H}_q$ was calculated up to $\cos\Theta$.

In order to work with more familiar expressions, we change the angle $\Theta \to \frac{\pi}{2} - \Theta$, so SG - approximation is $\cos\Theta \approx 1$, $\sin\Theta \approx \Theta$. Performing the usual calculations, we obtain the SG -Hamiltonian and equations of motion [12]

$$H_{SG} = JS^2 a^2 \left(1 - \frac{1}{2S}\right) \frac{1}{a} \int \left\{\frac{1}{2}\phi_x^2 + \frac{1}{2c^2}\phi_t^2 + m^2(1 - \cos\phi)\right\} dx \qquad (36)$$

$$\phi_{xx} - c^2\phi_{xx} = -\omega^2\sin\phi \tag{37}$$

with

$$c^2 = 2AJSa^2\left(1 - \frac{1}{2S}\right) \qquad \omega^2 = 2AS\mu f\left(1 - \frac{1}{8S}\right) \qquad m^2 = \omega^2/c^2$$

For one-soliton (kink) solution, the energy of the system linear in $1/S$ is

$$E = E_s^o\left(1 - \frac{5}{16S}\right) \qquad E_s^o = \frac{8aS}{\sqrt{1-\frac{v^2}{c^2}}}\sqrt{\mu f JS} \tag{38}$$

This correction is surely larger than the one obtained by Mikeska [13], and for $S=\frac{1}{2}$ gives weak agreement with the experiment [14] which is typicall for HP-representation.

On the other hand, for antiferromagnet, we can use the result of Wysin *et al* [15] by making the substitution $J \to J\,(1-\frac{1}{2S})$, $\mu f \to \mu f(1-\frac{1}{8S})$ to obtain

$$E_s = E_s^o\ (1 - \frac{1}{8S}) \qquad E_s^o = \frac{1}{\sqrt{1-\frac{v^2}{c^2}}}\ \mu f S \tag{39}$$

and for TMCC ($S=\frac{5}{2}$) this is the correction of order 5 %, in good agreement with the experiment [16].

Are there any more quantum corrections?

All that we have presented so for, are the corrections calculated using α_c. In fact, the appearance of the term $\frac{1}{S}\ f(\alpha_c)$ in the equations of motion introduced also some corrections in α, and one should look for the solution of the form

$$\alpha = \alpha_c + \frac{1}{S}\ \alpha_1 \tag{40}$$

Linearizing the equations of motion for α, one obtains a system of linear non-homogeneous differential equations for real and imaginary part of α_1. We know that the contributions of α_1 to the parameters of system, vanish in most cases due to the symmetry, yet the complete influence of this part can be evaluated only after the numerical solution of equations for α_1.

Conclusion

Our aim was to indicate that a consistent application of well-known Boson techniques can produce, a better insight into quantum corrections to classical solitons, but limited of course by the limitations of the Holstein-Primakoff representation.

References

[1] J. Tjon and J. Wright: Phys. Rev. **B15**, 3470 (1977).

[2] T. Holstein and H. Primakoff: Phys. Rev. **58**, 1098 (1940).

[3] R.J. Glauber: Phys. Rev. **131**, 2766 (1963).

[4] L.R. Mead and N. Papanicolaou: Phys. Rev. B28, 1633 (1973).

[5] M.J. Škrinjar, D.V. Kapor and S. Stojanović: J. Phys. Cond. Matt. **1**, 725 (1989).

[6] D.V. Kapor, S.D. Stojanović and M.J. Škrinjar: J. Phys. **C19**, 2963 (1986).

[7] D.I. Pushkarov and Kh.I. Pushkarov: Phys. Lett. **61A**, 339 (1977).

[8] V.G. Makhankov, R. Myrzakulov and A.V. Makhankov: Phys. Scr. 35, 233, (1987)

[9] L.G. de Azevedo, M.A. de Moura, C. Cordeiro and B.F. Žekš: J. Phys. **C15**, 739, (1982).

[10] M.J. Škrinjar, D.V. Kapor and S.D. Stojanović: J. Phys. **C20** 2243 (1987).

[11] A.R. Bishop, J.A. Krumhansl and S.F. Trullinger: Physica D1, 1 (1980).

[12] R.K. Dodd, J.C. Eilbeck, J.D. Gibbon and H.C. Morris: Solitons and Nonlinear Wave Equations, Acad. Press. London (1982).

[13] H.J. Mikeska: Physica 120B, 235, (1983).

[14] K. Kopinga, A.M.C. Tinus and W.J.M. de Jonge: Phys Rev. B29. 2868 (1984)

[15] G. Wysin, A. Bishop and J Oitmaa: J. Phys. **C19**, 221 (1986)

[16] Y. Endoh, Y. Ajiro, H. Shiba and H. Yoshizawa: Phys. Rev. **B30**, 4074 (1984).

PART III

NONLINEAR WAVE PROPAGATION, NUMERICAL AND THEORETICAL STUDIES

DEFECTS IN NON-LINEAR WAVES IN CONVECTION

A. Joets and R. Ribotta
Laboratoire de Physique des Solides, Bât.510
Université de Paris–Sud
91405 Orsay Cedex France

ASTRACT :Time–dependent structures in convection such as traveling patterns are examples of nonlinear waves and represent space–time ordered structures. They can be described by an amplitude equation (the complex Landau–Ginzburg–Newell model). Defects occur as the result of localized modulations of the phase and play a major role in the mechanisms of progressive disorganization of the coherent structures. Most of our experimental findings concerning the nucleation and the stability of the defects are numerically simulated by the appropriate Landau–Ginzburg–Newell model.

1. INTRODUCTION

The propagating convective structures of rolls that can be developed in fluid layers are simple examples of nonlinear waves [1]. These waves are ordered states which can become locally unstable against phase perturbations [2]. It has been shown that singularities arise as a result, in form of defects in the space–time ordering [3]. The progressive disordering of the initial wave is shown to be mainly due to the erratic nucleation and motion of those defects.

For 1–D systems which break at onset simultaneously the invariance in space and time, a non–linear analysis of the problem usually leads to an amplitude evolution equation of the Landau–Ginzburg (L.G.) type[4–7]. The physical state is described by a local field $w \sim Re[A(X,T)e^{i(\omega t+kx)}+B(X,T)e^{i(\omega t-kx)}]$, where k is the wavevector (along x) and ω is the pulsation of the basic wave (solution of the linearized problem). The complex amplitudes A and B include small and slow variations of the local field and follow the two coupled equations [8,9]:

$$\frac{\partial A}{\partial T} = \mu A - c\frac{\partial A}{\partial X} + (1+i\alpha)\frac{\partial^2 A}{\partial X^2} - (1+i\beta)\,|A|^2A - (\gamma+i\delta)\,|B|^2A$$

$$\frac{\partial B}{\partial T} = \mu B + c\frac{\partial B}{\partial X} + (1+i\alpha)\frac{\partial^2 B}{\partial X^2} - (1+i\beta)\,|B|^2B - (\gamma+i\delta)\,|A|^2B$$

The slow scales X,T are such that $X = \sqrt{\epsilon}\,x$, $T = \epsilon t$, and ϵ represents the relative stress parameter. A and B are associated with respectively a left and a right traveling wave. The coefficient α measures the wave dispersion, β couples the amplitude and the frequency of the wave, and $\mu = \epsilon^2$. Basic solutions to this equation correspond to :

–single traveling wave (TW): $A = \sqrt{\mu}\,\exp(-i\beta\mu t)$, $B = 0$ (left–going wave), or $B = \sqrt{\mu}\,\exp(-i\beta\mu t)$, $A = 0$ (right–going wave). They are stable for $\gamma > 1$, $1+\alpha\beta > 0$.

–standing waves (SW): $|A| = |B|$, (they are stable for $|\gamma| < 1$).

We shall show experimentally and numerically, that slow wavetrain modulations may become locally unstable and trigger shocks. As a consequence of a shock, spatio–temporal defects are created in the wave pattern.

We use for the numerical simulations a split–step integration method involving the calculation of Fourier transforms over 256 collocation points. Periodical boundary conditions are imposed. The length of the box is L=100.

2. TRAVELING WAVE CONVECTION IN A NEMATIC LIQUID CRYSTAL

A nematic liquid crystal is an anisotropic fluid which can develop convective instabilities when driven by an AC electric field [10]. A layer of nematic of thickness d becomes unstable to convection when the applied voltage V exceeds some well defined threshold V_{th} (the "conduction regime"). Here the relative stress parameter is $\epsilon = (V^2 - V_{th}^2)/V_{th}^2$. The anisotropy raises the degeneracy of the direction of the wavevector **k** inside the plane of the layer. Therefore, the ordered homogeneous pattern consists of equally spaced rolls which are aligned along the perpendicular (hereafter **y**) direction to the anisotropy axis (**x**). Inside a well–defined range of the parameters the convective pattern propagates along **x** with a uniform velocity u [11] and a typical value of the order of 0.15 λ/sec, where $\lambda = 2\pi/k$. The sample is prepared so as to develop quasi 1–D structures (direction **x**). The extension L_x along x is typically 100 L_y ($L_y \simeq 1$ to 5 d).

The evolution of the pattern is represented in a space–time diagram. In a nematic liquid crystal, which is strongly birefringent, the periodic convective flow produces an optical pattern of parallel focal lines for light transmitted accross the layer, on a plane parallel to the layer. The position of the focal lines is related to that of the upwards or downwards motions, i.e. to the position of the rolls. An optical intensity profile recorded along **x** at some position y_0 consists then of a series of peaks separated by the wavelength λ, on a flat baseline. The successive profiles recorded at equal time intervals (here $\delta t \simeq 0.04$ sec.) for the same line $y = y_0$ are plotted equally spaced on top of each other along the t axis. One obtains a space-time diagram on which the peaks are aligned on oblique lignes, the slope of which is $dt/dx = u^{-1}$. For a uniform motion (progressive wave), the space–time diagram is a perfectly ordered 2–D structure of oblique lines.

3. LOCALIZED MODULATIONS OF THE PHASE AND NUCLEATION OF DEFECTS

An initially uniform progressive wave i.e. a fully ordered wave with a unique wavevector k may become unstable against long wavelength disturbances. Experimentally, the conditions for such perturbations are not yet well understood although they are found to be effective relatively close to the threshold for the basic wave ($\epsilon = 2.10^{-2}$). Localized modulations of the phase φ develop, apparently at random in space and in time. They are evidenced by a sudden change Δu in the phase velocity u over some wavelengths λ and might have a large relative amplitude $\Delta u/u \simeq 0.2$. Typically, two types of phase perturbations are found [12]:

a) the velocity amplitude jumps suddenly at some point and its profile along **k** is kink-like;

b) the profile of the velocity jump is symmetrical and has a lump shape.

The phase variation $\Delta\varphi = \varphi - (\omega t \pm kx)$ has a kink–shape in the first case and a lump–shape in the latter one.

4. NUCLEATION OF SPACE TIME–DISLOCATIONS

In the first case (**a**) hereabove mentioned, the wavetrain is separated into two halves with a different velocity and a shock occurs at the separation region. The effect of the shock is to change the magnitude of the local wavevector k so as to drive it beyond its stability limits. Restabilization is obtained by expelling one spatial period λ, within a time of the order the period of the wave $T = 2\pi/\omega$. When represented in the $\{x,t\}$ diagram (Fig.1a), this process corresponds to a phase jump of 2π, and we name it a space–time dislocation by analogy with the dislocations in solids (Fig.1a).

Isolated dislocations can be numerically simulated by taking as an initial condition a phase winding of 2π in a uniform (left–going) wave A but localized over a small length $l \simeq 2\lambda$ (Fig.1b). Then the total phase variation of A inside the box is equal to 2π. The amplitude of the wave B is taken equal to 0. In order to insure stability of the progressive wave, the coefficients of the L.G. equation are here taken such that: $\alpha=1.3$, $\beta=-0.7$, $\gamma=1.2$, $\delta=0$, $\mu=1$, $c=0$. Figures 1.c–f show the evolution of the profile of $|A(x)|$. The annihilation of one wave-period occurs when A passes through zero. Simultaneously, the total phase variation of A jumps from 2π to zero. After the appearance of the defect, A slowly relaxes to a stable TW solution. Note that, in this process, B does not depart from its zero initial value (the numerical noise has not a sufficient amplitude here).

5. SOURCES, SINKS AND SPACE–TIME GRAIN–BOUNDARIES

The case (**b**) is a more typical one, since it corresponds to a localized perturbation of the velocity in a structure perfectly ordered (i.e. with constant velocity and constant k) over infinitely large distances . Then, the acceleration and the deceleration parts induce two shocks that follow respectively a compression and a dilation in local wavelength. These shocks trigger a wave–breaking which results in the appearance of two sigularities: a sink S_i (periodic annihilation of periods λ) and a source S_0 (periodic creation of an additional period λ) (Fig.2a). As a consequence the local wave velocity is reversed inside the space separating the two defects. These defects are then the topological boundaries that limit the new counterpropagating wavetrain (right-going wave B) inside the initial one (left-going wave A). In the $\{x,t\}$ diagram the defects separate two different domains of oblique lines and we name them grain boundaries by analogy with the case of solids.

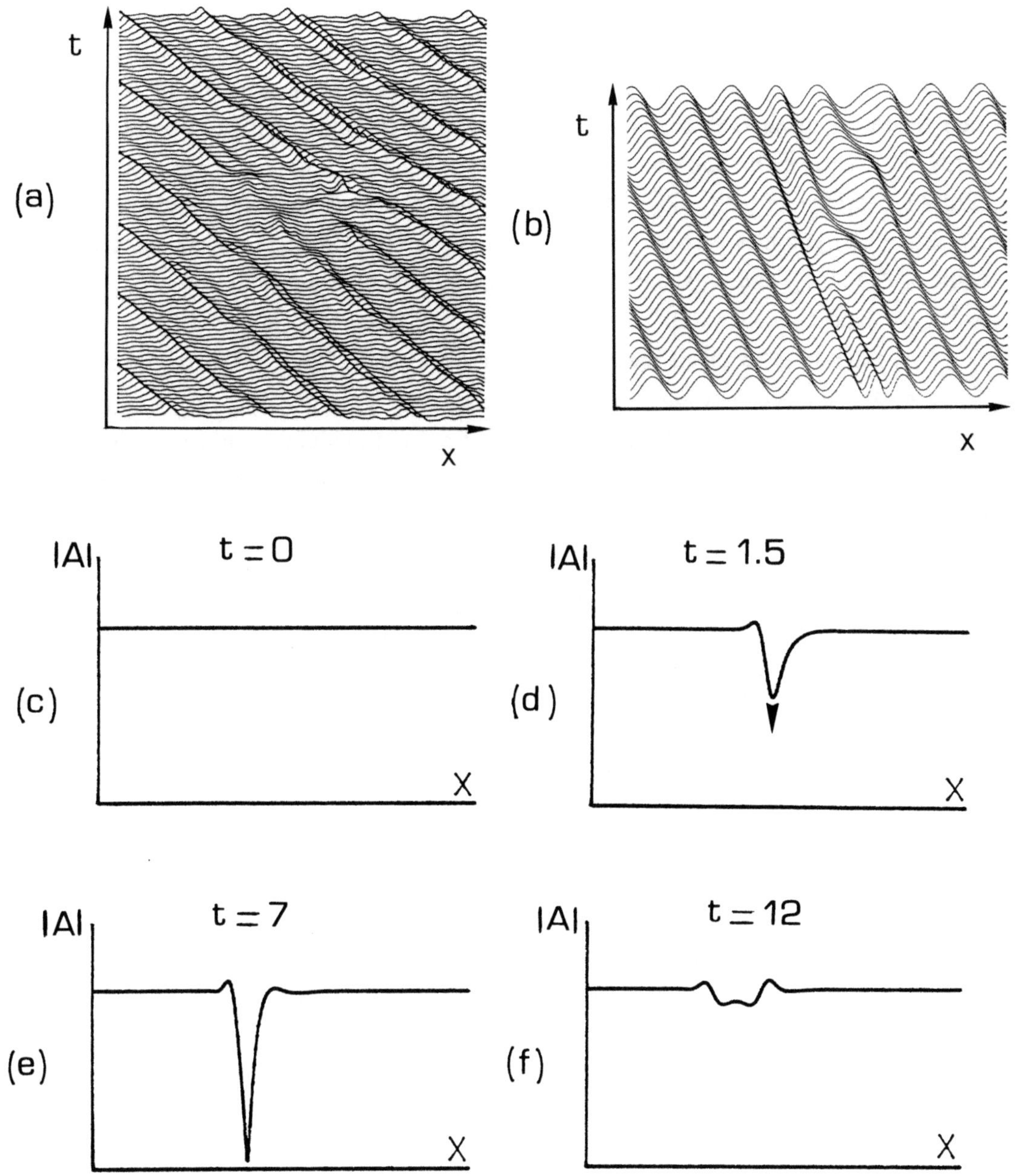

Figure 1 : Point defect (dislocation) in a traveling wave represented in the space time x–t. **a)** Experiment. **b)** Simulation of the local variable w(x,t) and **c–e)** of the evolution of the envelope profile $|A(x)|$. The defect occurs at t= 7, when A=0. Here, α=1.3, β=–0.7, γ=1.2, δ=0, μ=1, c=0.

It is possible to simulate numerically the creation of a pair of such defects by taking as initial conditions for A the same as in the previous case, and for B a hump of small amplitude (0.2) localized at the place of the local phase winding of A, with the same width in space (Fig.2b). Here the parameters of the equation are $\alpha=1$, $\beta=0.6$, $\gamma=1.2$, $\delta=1$, $\mu=1$, c=0. The process is shown on Fig. 2c–f. The first step is, as before, a localized decrease of $|A|$. The new fact is that the decrease of $|A|$ allows $|B|$ to increase. The dynamics of this process are complex and depending both on the initial height of the hump and on the relative rates of growth of B and of decay of A, different results are expected:

i) the initial height of the hump is relatively small. Then the correlative decay in $|A|$ is weak ($|A|$ does not vanish) and $|B|$ relaxes back to zero. The result appears as a simple weak perturbation of the phase in form of a rapid modulation of the wave A.

ii) the height of $|B|$ allows a large decay of $|A|$ (not necessarily to zero) but only for a time short compared to one time–period. Then, the two amplitudes cross each other for a while and a pair of dislocations is created. Next, $|B|$ relaxes to zero. At further times the initial wave structure A is recovered.

iii) $|A|$ decays to zero and simultaneously $|B|$ increases until saturation. At some time (here at t=10), $|A|$ and $|B|$ become locally equal. As the time goes $|A|$ continues to decrease down to zero and $|B|$ to increases up to the limit $\sqrt{\mu}$ (Fig 2f). This situation is locally stable. Then, the result is that over some distance the initial wave A is now replaced by the opposite wave B. The regions where one amplitude dominates, correspond to simple traveling waves, whereas the intersection points of the two profiles $|A|$ and $|B|$ correspond to the location of the defects (sink and source). At these points the two waves have an equal amplitude. Thus, they have the characteristics of local standing wave. Since the standing wave is unstable for these parameters values, the kink–like profiles of $|A|$ and $|B|$ act as a topological constraint which forces the two waves to be simultaneously present at some well–defined place.

Experimentally, the case ii) has never been observed up to now, whereas the case iii) is the most commonly observed. Case i) is rarely found in ordinary conditions.

6. STABILITY AND MULTIPLICATION OF THE DEFECTS

A source or a sink may become unstable in space when the external stress increases slightly (e.g. $\epsilon \simeq .05$). The defects can move in either direction by steps. Here again, the stability of the new complex state seems to depend on the existence of localized phase modulations [13].

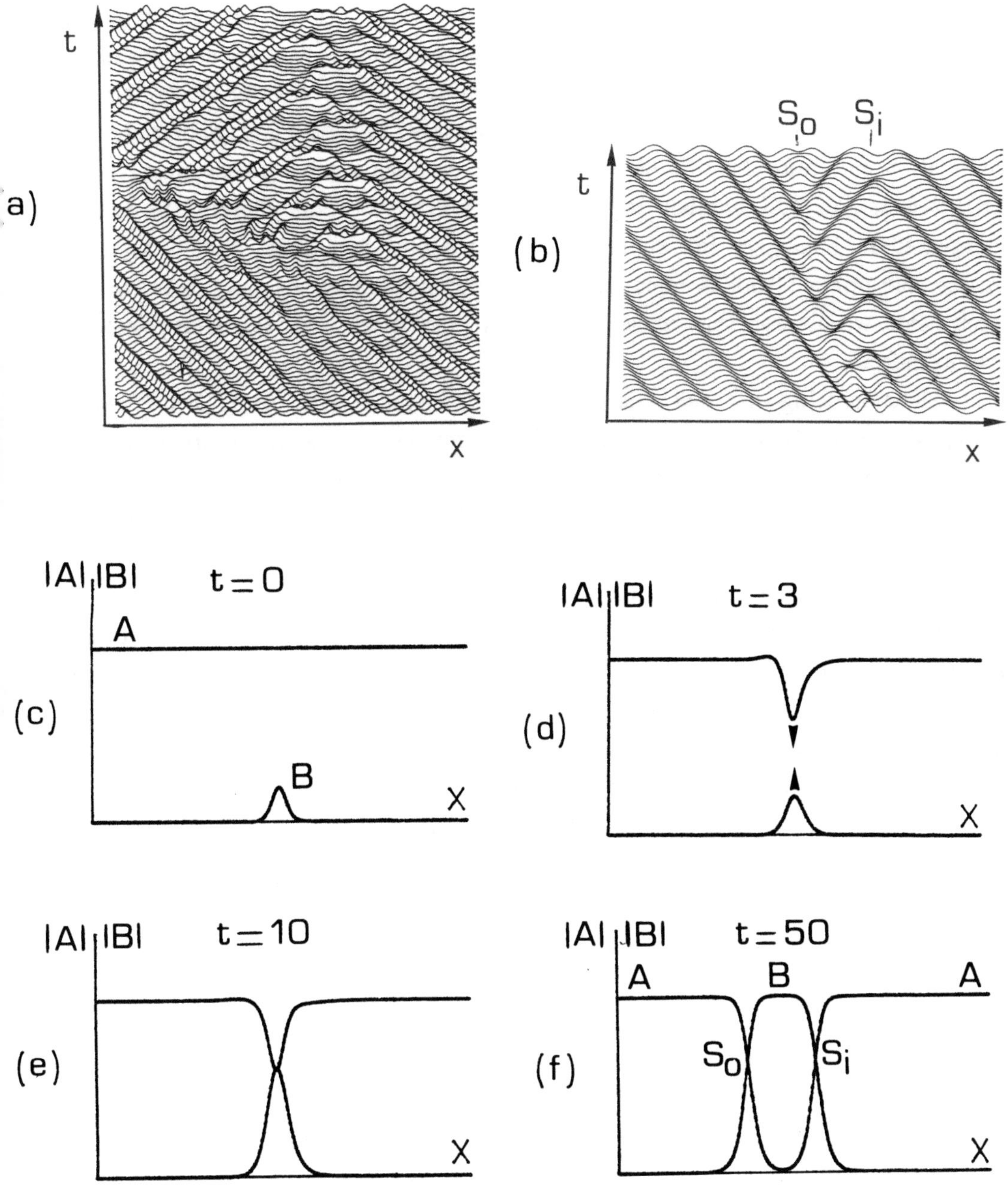

Figure 2 : Nucleation of a sink S_i and of a source S_o. **a)** Experiment. **b)** Simulation of the local variable w(x,t) and **c–e)** of the evolution of the profiles |A| and |B|. The two defects appear at t= 10, when |A|=|B|. In the space between the two defects B (opposite wave) is stable. Here α=1, β=0.6, γ=1.2, δ=1, μ=1, c=0.

They may also become unstable in time. Sinks and sources are topological singularities which can be associated with a topological charge (e. g. +1 for the sinks and –1 for the sources). In any process of destabilization the neutrality of the system must be conserved. One may think of two elementary processes: recombination with annihilation and multiplication by splitting. For instance a given defect may split into three defects (an odd number in order to conserve to total charge). In the case represented in Fig.3a, a sink splits into a source plus two sinks. Note that in that process, the original sink is replaced by a source surrounded by two sinks. In usual experiments we observe that nucleation occurs more often than the multiplication of existing defects.

The splitting of a defect is simulated for instance with: α=1, β=0.4, γ=1.02, δ=0.7, μ=1, c=0 (Fig. 3b). The initial conditions for the wave A and B are a sine and a cosine profile. The two profiles of $|A|$ and $|B|$ intersect at some point which corresponds to the location of the initial sink S_i (Fig 3c). The splitting process is represented on Fig 3c–f. It consists of a progressive inflexion of the two kink–profiles, until two local extrema (a maximum and a minimum) are created on each profile. Then two new intersection points appear, between which two new counterpropagating waves are created now. In this process two new defects are created as in the nucleation process of grain boundaries, but here two additional wavetrains with opposite sign are created whereas only one is created in the nucleation. In this respect, the splitting of existing defects appears as more efficient in the disorganisation process of the initial wave. However it is less frequently observed than the nucleation of two grain boundaries.

7. DISCUSSION

The traveling wave is an ordered spatio–temporal structure, in which the defects result from localized phase perturbations. Increasing ϵ further above the threshold leads to an increase of the number of defects and bifurcations to secondary homogeneous state are difficult to encounter before the fully disordered state is reached (ϵ>0.5). This situation is reminiscent of the rapid transition to the chaotic state found for the stationary structures where the defects play a major role [14]. There the transition to

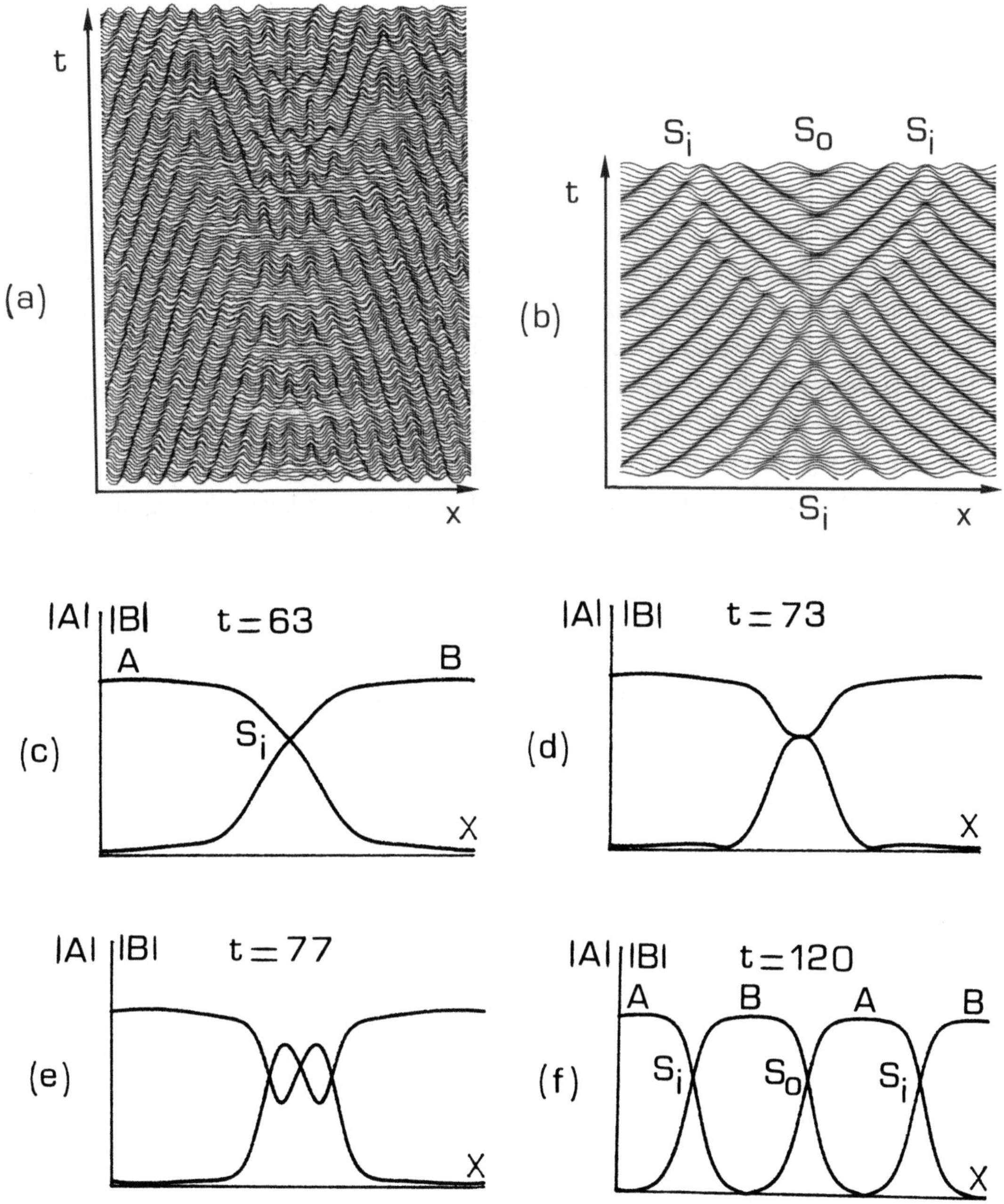

Figure 3 : Destabilization of a sink S_i by splitting. **a)** Experiment. **b)** Simulation of the local variable w(x,t) and **c–e)** of the evolution of the profiles $|A|$ and $|B|$. Two new defects (a source and a sink) appear at t=73. Here $\alpha=1$, $\beta=0.4$, $\gamma=1.02$, $\delta=0.7$, $\mu=1$, c=0.

chaos can be obtained as well, after a series of bifurcations to homogeneous states when the rate of increase of ϵ is rather small [14]. Here, on the contrary, the main mechanism for the transition comes from the localization of phase modulations which trigger shocks and nucleate defects.

The L–G model allowed us to simulate the creation of defects starting with ad_hoc initial conditions. We remark that the L–G model, which was elaborated after the hypothesis of small and slow variations of the envelope amplitude, describes quite satisfactorily the defects which correspond to rather large variations of the amplitude over small space scales. Spontaneous nucleations of dislocations can also be simulated for $1+\alpha\beta < 0$ (Newell–Benjamin–Feir criterion for phase instability [15]). But in that case, the disorganization (loss of coherence) comes not only from defects but is also associated with phase disorder. The main question which arises now, would concern the origin of the phase (or velocity) modulations and the mechanism of their localization. Another open problem is the range of existence of eventual secondary homogeneous (fully ordered) spatio–temporal states. It appears now necessary to determine from the experiment, the values of the coefficients in the model equation. The alternative way would be to derive the analogous of the amplitude equation from the constitutive (hydrodynamic) equations. Unfortunately, from the actual microscopic equations one is still unable to account even for the traveling–wave state. Experiments are under way in order to give values of some of the coefficients of the Landau–Ginzburg equation.

5 CONCLUSION

The traveling–wave state of convection inside a layer of an anisotropic fluid offers quite rich examples of singularities in 1–D non–linear waves. Even right above the threshold for the basic state, localized modulations of the envelope occur on scales comparable to the spatial period of the wave and give rise to defects in the space–time ordering. They appear as the key mechanism in a transition to the disorder by local destabilization of the initial wave. It is quite possible to reproduce numerically most of the spatio–temporal effects experimentally observed by using a set of coupled envelope equations of the Landau–Ginzburg type for the two counterpropagative waves.

This work was done under the DRET contract n° 87/142. We thank P. Coullet, J. Friedel, L. Gil, J. Lega, J. Lajzerowicsz and A.C. Newell for useful discussions.

References

[1] Whitham, G.B. Linear and Nonlinear Waves, John Wiley & Sons, New York, 1974.

[2] Bekki, N. and Nozaki, K. Phys. Lett. A110, 133 (1985).

[3] Joets, A. and Ribotta, R. in P. Coullet and P. Huerre (eds.), New trends in nonlinear dynamics and pattern forming phenomena: the geometry of nonequilibrium, Nato Asi series B, Plenum Press (Cargese Workshop Aug. 1988) (1989); also in the Proceedings of "Nonlinear Coherent Structures in Physics, Mechanics and Biological Systems" (Paris, June 1988), J. de Phys. C3, 171 (1989).

[4] Newell, A.C. and Whitehead, J.A., J. Fluid. Mech., 38, 279, (1969).
Segel, L.A., J. Fluid Mech., 38, 203, (1969).
Newell, A.C., Lect. Appl. Math. 15, 157, (1974).

[5] Kogelman, S. and DiPrima, R.C., Phys. Fluids 13, 1 (1970).

[6] Blennerhassett, P.J., Phil. Trans. Roy. Soc. Lond. A298, 451, (1980).

[7] Stewartson, K., and Stuart, J.T., J. Fluid Mech. 48, 529, (1971).

[8] Brand, H.R., Lomdahl, P. and Newell, A., Phys. Let. A118, 67, (1986).

[9] Coullet, P., Elphick, C., Gil, L. and Lega, J., Phys. Rev. Lett. 59, 884 (1987); Coullet, P. and Lega, J., Europhys. Lett. 7, 511 (1988).

[10] de Gennes, P. G., The Physics of Liquid Crystals, Clarendon, Oxford, 1974.

[11] Joets, A. and Ribotta, R., Phys. Rev. Lett. 60, 2164 (1988).

[12] Ribotta R. and Joets A., to appear in the Proceedings of "Partially Integrable Nonlinear Evolution Equations and their Physical Applications", NATO ASI series (Les Houches, march 1989).

[13] Ribotta R. and Joets A., in preparation.

[14] Joets, A. and Ribotta, R., J. Phys. (Paris) 47, 595 (1986).
Yang, X. D., Joets, A. and Ribotta, R., Physica 23D, 235 (1986).
Ribotta, R., in L. Kubin, G. Martin, (eds.), Non–linear phenomena in materials science, Trans Tech Publications, Switzerland (1988).

[15] Stuart, J.T. and DiPrima, R.C., Proc. R. Soc. Lond. A 362, 27 (1978).

GAP SOLITONS IN NONLINEAR ELECTRICAL TRANSMISSION LINES

J.M. Bilbault and M.Remoissenet
Laboratoire O.R.C , Université de Bourgogne
6,Bd Gabriel,21100 Dijon,France

Abstract.We study theoretically and numerically the properties of monochromatic waves in a nonlinear electrical transmission line,whose capacitance has a periodic spatial variation.In the continuum limit and weak amplitude limit ,we reduce the characteristic equations of this system to NLS equation.We find analytical solutions for the voltage envelope, which propagate with frequency in the gap induced by the capacitance periodicity.Our numerical experiments show that, when the input voltage increases, the transmissivity in the gap increases and the voltage envelope approaches the stationnary shape predicted by theory.

I. Introduction

The nonlinear optical response of periodic matter has been the subject of considerable interest[1,2,3,4].In the linear approximation, it is known that the periodical spatial variation of the dielectric optical constant induces gaps in the dispersion relation for the propagation of electromagnetic waves.Waves with frequency in a gap do not propagate,but decay exponentially along the propagation direction.However ,if the intensity of the wave is such that nonlinear effects can not be neglected , Chen and Mills [1] have shown, by numerical studies, that transmissivity of wave through the periodic nonlinear stack can attain unity at frequency in the gap.The shape of its envelope presents all characteristics of a soliton like wave, which was first investigated by Mills and Trullinger [4]. Using some approximations and appropriate boundary conditions, Sipe and Winful [2] reduced the equations for this slow envelope to the Nonlinear Schrödinger Equation, with soliton solutions.Dispersion,which balances nonlinearity to create solitons, is provided by the curvature of the branches of the dispersion relation.

The purpose of the present paper is to show that gap solitons can also exist in nonlinear electrical transmission lines with a periodic structure due to the capacitance.Specifically, we consider a simple electrical line where the evolution of the slow voltage envelope can be modeled by a Nonlinear Schrödinger Equation as shown in Section II .In Section III ,we develop the whole solution for a finite,nonlinear line.In Section IV ,our preliminary numerical results are presented and compared with the analytical predictions.

II.Derivation of the envelope-function equation

We consider the circuit illustrated in figure 1.In the following analysis to a first approximation,we neglect the weak losses due to the resistance of the inductor,as well as the conductance of the capacitances.

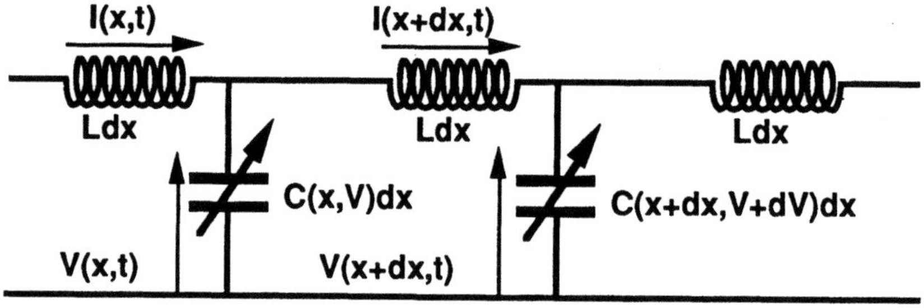

Figure1.Elementary cell model for the nonlinear transmission line in the continuum limit.

The capacitance C(x,V) is given by

$$C(x,V)=C_0(x) - \alpha V^2 \qquad (2.1)$$

$C_0(x)$,the linear part of the capacitance ,is a periodic function of x (figure 2a).Namely,we have $C_0(x)$= C1 for $0<x<d/2$ (mod. d) and $C_0(x)$=C2 ,for $d/2<x<d$ (mod. d).The line length l is chosen to be a multiple of d in order to have an integer number of periods in the line. The nonlinear part of C(x,V) depends on the square of the voltage V, a being the nonlinear coefficient (figure 2b).

A set of the fundamental equations for the line is

$$L\frac{\partial}{\partial t} I(x,t) = -\frac{\partial}{\partial x} V(x,t) \tag{2.2}$$

$$\frac{\partial}{\partial t} Q(x,t) = -\frac{\partial}{\partial x} I(x,t) \tag{2.3}$$

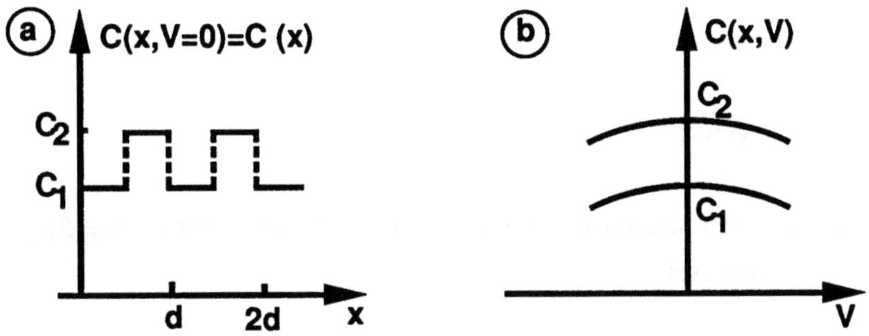

Figure 2.nonlinear periodic capacitance C(x,V) a- versus x,with V=0
b- versus V at fixed value of x

Here, Q(x,t) denotes the charge per unit length stored in the capacitor.From (2.1) one gets

$$Q(x,t) = C_o(x)\, V(x,t) - \frac{\alpha}{3}[V(x,t)]^3 \tag{2.4}$$

Eqs (2.2), (2.3) and (2.4) yield the wave equation

$$-\frac{\partial^2}{\partial x^2} V(x,t) + LC_o(x)\frac{\partial^2}{\partial t^2} V(x,t) - \frac{\alpha L}{3}\frac{\partial^2}{\partial t^2}[V(x,t)]^3 = 0 \tag{2.5}$$

The equation (2.5) is analogous to eq (2.3) of ref. [3]. As a formal simplification, we consider new parameters, c, a, β defined as :

$$L\frac{(C_1 + C_2)}{2} = \frac{1}{c^2} \tag{2.6}$$

$$\frac{C_2 - C_1}{C_2 + C_1} = a \tag{2.7}$$

$$\alpha L = \beta \tag{2.8}$$

Inserting (2.6) to (2.8) in (2.5) gives:

$$-\frac{\partial^2}{\partial x^2} V(x,t) + \frac{1 \pm a}{c^2} \frac{\partial^2}{\partial t^2} V(x,t) - \frac{\beta}{3} \frac{\partial^2}{\partial t^2} [V(x,t)]^3 = 0 \tag{2.9}$$

We now look for the nonlinear dispersion relation for monochromatic waves. We set:

$$V(x,t) = e^{-j\omega t} f(x) + c.c. \tag{2.10}$$

where c.c. denotes the complex conjugate.We find that f(x) must satisfy

$$\frac{d^2 f}{dx^2}(x) + \frac{\omega^2}{c^2} (1 \pm a) f(x) - \beta \, \omega^2 |f(x)|^2 f(x) = 0 \tag{2.11}$$

This problem has been studied for a number of years in the linear case[5]($\beta = 0$).For the nonlinear case, we must set :

$$f(x) = V e^{jkx} \tag{2.12}$$

where V is a constant amplitude.Setting

$$\eta_1 = \frac{\omega}{c} (1-a - \beta V^2)^{1/2} \tag{2.13}$$

$$\eta_2 = \frac{\omega}{c} (1+a - \beta V^2)^{1/2} \tag{2.14}$$

Writing continuity relations for $x = 0$ and $x = \frac{d}{2}$ for V(x,t) and I(x,t) leads to the relation :

$$\cos kd = \cos \eta_1 \frac{d}{2} \cos \eta_2 \frac{d}{2} - \frac{1}{2} [\frac{\eta_1}{\eta_2} + \frac{\eta_2}{\eta_1}] \sin \eta_1 \frac{d}{2} \sin \eta_2 \frac{d}{2} \tag{2.15}$$

The main result of this implicit equation is represented on figure 3. Dispersion curve presents gap frequencies for wave numbers multiple of π/d .The gap width increases with a; ω_- and ω_+ are the lower and upper band edges of the gap occuring at $k=\pi/d$:

$$\omega_- \approx \frac{\pi . c}{d} [\frac{1}{(1-\beta V^2)^{1/2}} - \frac{a}{\pi(1-\beta V^2)^{3/2}}]$$

(2.16)

$$\omega_+ \approx \frac{\pi . c}{d} [\frac{1}{(1-\beta V^2)^{1/2}} + \frac{a}{\pi(1-\beta V^2)^{3/2}}]$$

As the voltage V increases, ω_- and ω_+ are shifted to the high frequencies, whereas the gap width increases

$$\Delta\omega_{gap} = \omega_+ - \omega_- = \frac{2ac}{d}(1 - \beta V^2)^{-3/2} \quad (2.17)$$

In the following,we shall denote $\omega_1=\omega_-$ (for V=0) and $\omega_2=\omega_+$(for V=0).

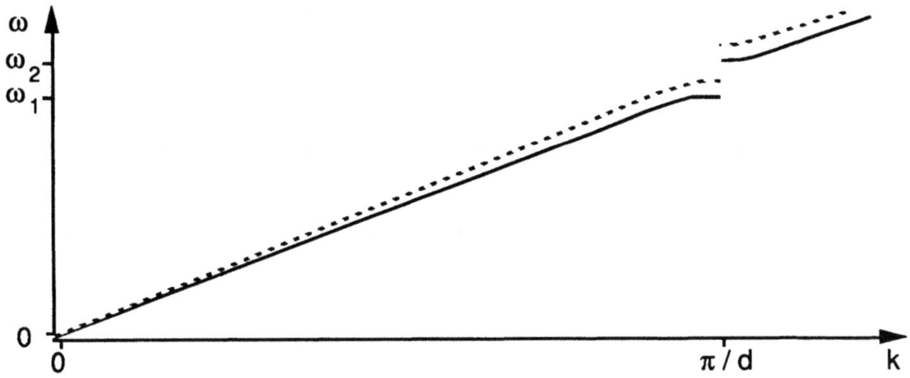

Figure 3.dispersion curve of a periodic nonlinear line α=0.1,V=0 (solid),V=0.6(dashed)

So,it is possible to propagate waves with frequency in the gap, if the amplitude of the wave is sufficient for the nonlinearity to shift this frequency out of the gap.

Starting from eq.(2.9),we now look for the evolution of the slow voltage envelope V(x,t).Using the reductive perturbation method[3], we obtain

$$V(x,t) = l^{1/2} \sum_{m=1} \varepsilon^m V_m(\zeta, \tau) f_m(x) e^{-j\omega_1 t} + c.c. \quad (2.18)$$

where $l^{1/2}$ is a normalizing coefficient.

$$\zeta = \varepsilon(x - V g t), \quad \tau = \varepsilon^2 t \quad (2.19)$$

and where the functions $f_m(x) e^{-j\omega_m t}$ are the solutions of the linear part of eq. (2.11). Inserting eqs (2.18) and (2.19) in eq. (2.9), we obtain a series of equations for the different orders of ε [3]. To order ε^3 ,one gets the Nonlinear Schrödinger Equation (NLS) for the component $V_1(\zeta, \tau)$.

$$j\frac{\partial}{\partial\tau} V_1(\zeta,\tau) + P\frac{\partial^2}{\partial\zeta^2} V_1(\zeta,\tau) + Q|V_1(\zeta,\tau)|^2 V_1(\zeta,\tau) = 0 \quad (2.20)$$

where P is the group velocity dispersion and Q is a new nonlinear coefficient

$$P = \omega''/2 = \frac{1}{2}\left(\frac{\partial^2\omega}{\partial k^2}\right)_{\omega 1} \quad \text{and} \quad Q = \frac{\omega\, l}{2}\, \beta \int_0^l |\, f_1(x) \,|^4\, dx \qquad (2.21)$$

To order ε^2, the term $V_2(\zeta, \tau)$ (see eq .(2.18)) is related to $V_1(\zeta, \tau)$ by

$$V_2(\zeta, \tau) = \frac{2}{\omega_2{}^2 - \omega_1{}^2} \frac{\partial V_1}{\partial \zeta}(\zeta, \tau) \int_0^l f_1{}^*(x) \frac{df_2(x)}{dx}\, dx \qquad (2.22)$$

In this equation, we consider only ω_1 and ω_2(edges of the lowest gap), which is valid [3,6] if $a << 1$.

We now consider a particular point of the linear dispersion curve, which corresponds to $k = \frac{\pi}{d}$, $V_g = 0$, $P < 0$, $\varepsilon \text{-->} 1$, $\omega_1 = \frac{\pi.c}{d}\left[1 - \frac{a}{\pi}\right]$, $\omega_2 = \frac{\pi.c}{d}\left[1 + \frac{a}{\pi}\right]$ (see eq.(2.16))

Then $\zeta = x, \tau = t$ and the functions $f_1(x)$ and $f_2(x)$ become :

$$f_1(x) = \frac{c}{c_o} \frac{1}{l^{1/2}} \left(\frac{2\pi}{\pi+2a}\right)^{1/2} \cos\left(\frac{\pi x}{d} + \frac{\pi}{4}\right)$$

$$(2.23)$$

$$f_2(x) = \frac{c}{c_o} \frac{1}{l^{1/2}} \left(\frac{2\pi}{\pi-2a}\right)^{1/2} \sin\left(\frac{\pi x}{d} + \frac{\pi}{4}\right)$$

In these equations, $c_o = (L'\, C')^{-1/2}$ is the velocity of waves propagating in a linear line which is connected to the nonlinear one, as represented on figure 4.

Seeking a voltage function of the form :

$$V_1(\zeta, \tau) = \Psi(x)\, e^{-j\delta t} \qquad (2.24)$$

gives

$$V(x,t) = \left(l^{1/2}\ \Psi(x)\, f_1(x)\, e^{-j(\omega_1 + \delta)t} + c.c.\right) + \ldots \qquad (2.25)$$

$$V(x,t) = V_0(x)\, e^{-j\omega t} + c.c. \qquad (2.26)$$

This wave propagate with the angular frequency $\omega_1 + \delta$, which is on the bottom of the gap, δ being a small positive detuning term. With eq (2.24), (eq. 2.20) becomes :

$$\frac{\partial^2}{\partial x^2}\Psi(x) + \frac{\delta\,\Psi(x)}{P} + \frac{Q}{P}\,|\,\Psi(x)\,|^2\,\Psi(x) = 0 \qquad (2.27)$$

In practice,the length l of the transmission line is finite,and one must consider the solutions of (2.27) for finite boundary conditions as we shall see in the following.

III. Solution for a finite nonlinear line

Following Chen and Mills [1], we write :

$$\Psi(x) = \gamma(x)\, e^{i\phi\,(x)} \qquad (3.1)$$

$$A(x) = \gamma^2(x) \qquad (3.2)$$

$\phi(x)$ and $A(x)$ are related by :

$$\phi(x) = W \int^{x} \frac{dx'}{A(x')} \qquad (3.3)$$

$$\text{with } W = \mathrm{Im}(\Psi^*.d\Psi/dx) \qquad (3.4)$$

W is related to $\mathrm{Im}(V_0{}^*.dV_0/dx)$[3] by

$$W = \frac{c_0{}^2}{\omega\omega''}.\mathrm{Im}(V_0{}^*\cdot dV_0/dx) \qquad (3.5)$$

Using eq.(2.2),one gets

$$W = \frac{c_0{}^2\,L}{2\omega''}.< V(x,t).I(x,t) > \qquad (3.6)$$

where the brackets indicate a time-average.So,W depends directly on the electrical power through the whole line.Denoting by A_m the minimum of A(x), and choosing the origin at the maximum of A(x),one gets [3] the solution of eq. (2.27)

$$A(x) = A_+ - (A_+ - A_m)\,\mathrm{sn}^2\,[\,(A_+ - A_-)^{1/2}\,\frac{Qx}{2P}\,|\;\kappa\;] \qquad (3.7)$$

where $\mathrm{sn}\,(x|\,\kappa\,)$ is Jacobi elliptic function [7] , whose modulus is

$$\kappa = \left[\frac{A_+ - A_m}{A_+ - A_-}\right]^{1/2} \qquad (3.8)$$

$$\text{with } A_\pm = -\frac{\delta}{Q} - \frac{A_m}{2} \pm \left[\left(\frac{\delta}{Q} + \frac{A_m}{2}\right)^2 + \frac{2W^2 P}{A_m Q}\right]^{1/2} \qquad (3.9)$$

Solution (3.7) corresponds to a nonlinear wave train in the gap,which reduces for an infinite line(i.e. for W=0 and κ=1) to a stationnary gap soliton on the form:

$$A(x)=(-2\delta/Q)^{1/2}\,\mathrm{sech}((-\delta/P)^{1/2}\,x) \qquad (3.10)$$

We now limit our study to a particular experimental case.Namely, the input of the nonlinear line is connected (figure 4) to a linear line of characteristic impedance $Z_c=(L'/C')^{1/2}$, where L' and C' are assumed to be unity.This linear line is connected to an external signal generator of impedance $Z_i=1$.The output of the nonlinear line is matched by a charge $Z_l=1$.

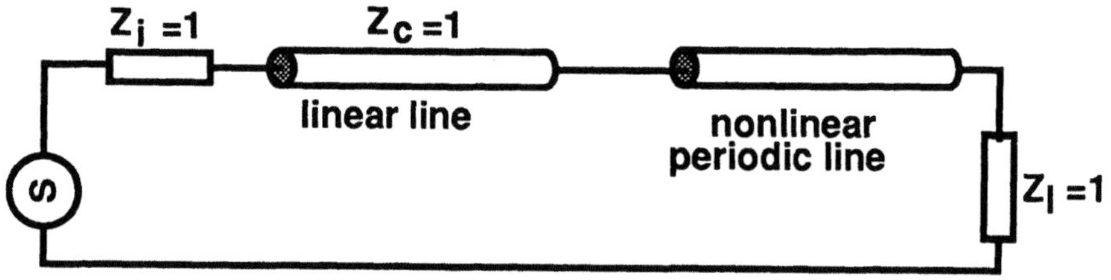

Figure 4. electrical device used to study the voltage envelope.The nonlinear periodic line is identical to this of figure 1,while in the linear line, C'=1 and L'=1.

In the general case,the voltage in the linear line is composed by an incident wave(V_i) and a reflected one(V_r).We observed pseudo-stationnary waves with maxima and minima gived by

$$V_{max}=V_i+V_r$$
$$V_{min}=V_i-V_r \qquad (3.11)$$

And so are the electrical currents in the linear line ; the power transmissivity is

$$|T|^2=\frac{4\,V_S^2}{(V_{max}+V_{min})^2} \qquad (3.12)$$

When $|T|^2$ reaches unity, no reflected wave exists and along the whole linear line, $V = V_i$, so does the voltage at the input and at the output of the nonlinear line, and its length l is related to the period $2K(\kappa)$ of the Jacobi elliptic function [7].We get

$$(A_+ - A_-)^{1/2} \frac{Q\,l}{2P} = 2K(\kappa) \tag{3.13}$$

W and A_m depend on voltage $V_s = V_i$: By eq. (2.23), (2.25), (3.1), (3.2) and (3.6), we obtain

$$A_m = \frac{\pi+a}{8\pi} \cdot \frac{c_0^2}{c^2} \cdot V_s^2 \tag{3.14}$$

and

$$W = \frac{c_0^2\, L}{4\omega''} . V_s^2 \tag{3.15}$$

IV.Numerical results

In the numerical experiments, we simulate exactly the experimental arrangment described on figure 4.Specifically, the source generates a sinusoidal plane wave which propagates in the linear line and enter in the nonlinear line of finite length *l*, where we observe the voltage envelope evolution.The numerical method basically consists in solving with a fourth-order Runge-Kutta method the equations of motion (2.2) and (2.3).The total number of unit cells in the nonlinear line is 900,with a capacitance periodicity of 20,which allows to use the continuum limit in the calculations.The time step Δt is chosen so that we have enough points during an oscillation (typically 160 pts./osc.)

Our preliminary results show that, if the voltage at the input of the nonlinear line is too weak for eq.(3.13) to be satisfied,the voltage envelope is evanescent (fig. 5)in good agreement with the linear approximation.On the other hand,if the input voltage increases and becomes close to solutions of eq. (3.13), we observe (fig.6) a quite different shape of the voltage envelope:it approaches the shape described by eq.(3.7) and the transmissivity increases.Complete evolution of $|T|^2$ with incident power will be presented elsewhere.

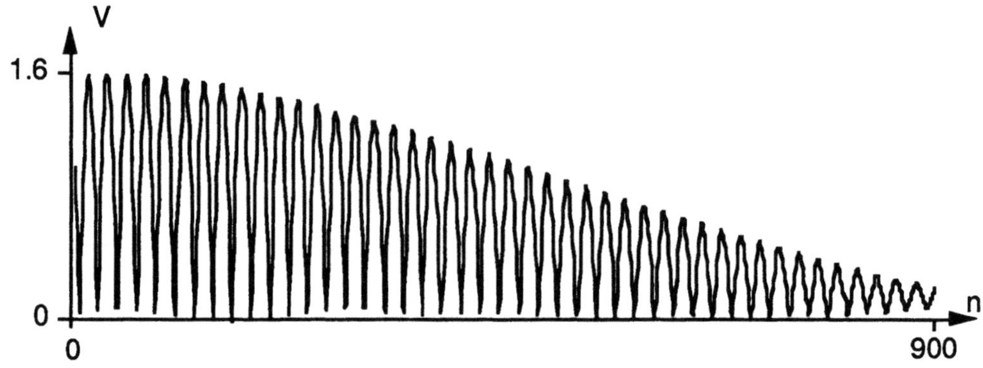

Figure 5. Voltage envelope in the nonlinear line of figure 4 vs. cell number .
L=1 , C_1=3.6 , C_2=4.4 , α=0.1 , input voltage amplitude is 1.335

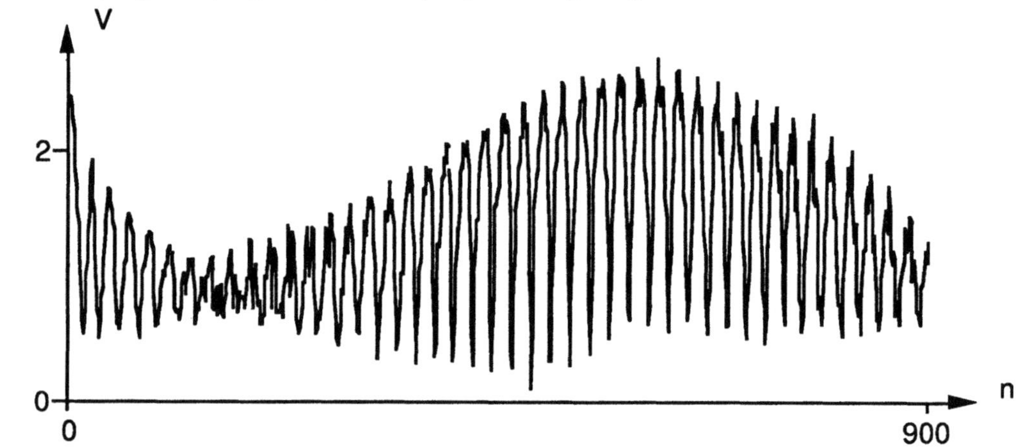

Figure 6.Same as in figure 5,but input voltage amplitude is 1.336

References

[1] W.Chen and D.L.Mills,Phys.Rev.B **36**,6269(1987)

[2] J.E.Sipe and H.G.Winful,Opt.Lett.**13**,132(1988)

[3] C.Martijn de Sterke and J.E.Sipe,Phys.Rev. A **38**,5149(1988)

[4] D.L.Mills and S.E.Trullinger,Phys.Rev.B **36**,947(1987)

[5] L.Brillouin and M.Parodi ,"Propagation des ondes dans les milieux périodiques", (Masson,Dunod,1956)

[6] H.Kogelnik,Bell Syst.Tech.J.**48**,2909(1969)

[7] P.F.Byrd and M.D.Friedman,Handbook of Elliptic integrals(Springer,Berlin,1954)

CAN A NONTRIVIAL SOLITONIC MODE BE OBSERVED IN A SQUARE JOSEPHSON JUNCTION?

E. Turlot, D. Esteve, C. Urbina and M. Devoret
Orme des Merisiers – DPHG-SPSRM –
CEN-SACLAY, 91191 GIF sur YVETTE Cedex France
and
J.C. Fernandez, R. Grauer*, H. Politano and G. Reinisch
Observatoire de la Côte d'Azur
Boite Postale No 139
F-06003 NICE Cedex France

We show by use of energy considerations and numerical simulations the existence of a stable 2-d isoperimeter solitonic mode propagating simultaneously along the two diagonals of a square Josephson junction of intermediate length.

I INTRODUCTION

The zeroth-order — i.e. without any perturbation — modelization of the dynamics of nonlinear modes which may propagate in a long Josephson junction (LJJ) is known to obey the following 1-dimensional (1-d) sine-Gordon (SG) partial differential equation (PDE), written here in reduced units [1-4] :

$$\Phi_{tt} - \Phi_{xx} + sin\Phi = 0 \qquad , \tag{1}$$

whose boundary conditions (b.c.'s) are "open-ended", and correspond to a zero external magnetic field :

$$\begin{aligned} \Phi_x(x = \pm l/2) = 0 \qquad , \\ l \gg 1 \qquad . \end{aligned} \tag{2}$$

Equations (1-2) describes the space-and-time evolution of the quantum mechanical phase Φ in a LJJ of length l much greater than the Josephson penetration depth.

The most stable nonlinear mode — which also has the highest amplitude, i.e. 2π — is the so-called "(anti) kink" soliton mode propagating at velocity v :

$$\Phi(x,t) = 4\,tan^{-1}\,exp\Big[\pm\frac{x - vt}{\sqrt{1 - v^2}}\Big] \qquad , \tag{3}$$

(+ : kink ; - : antikink) . It corresponds to a 2π-jump of the phase ϕ when the space variable x describes the length of the system.

The problem to decide when a "real" LJJ behaves according to the simple above-sketched 1-d soliton dynamics is not obvious. In other words, for which ratio values γ of the LJJ width w over the LJJ length l the fluxon physics is correctly described by the above 1-d PDE (1) in the lowest order of approximation ?

The idealistic approximation of a 2-d rectangle area by a 1-d line obviously requires $\gamma \ll 1$. Surprisingly, previous numerical simulations of the soliton-like nonlinear modes propagating in a 2-d LJJ, where γ is of order unity, showed the existence — and, which is more unusual, the

---

* Present address: Dept. of Mathematics, University of California of Santa Barbara, Santa Barbara, CA 93106 USA.

relative stability — of trivial 2-d extended kink-like wave fronts, which basically mean that the analytical description of the 2-d wave profile is degenerated with respect to the second y-spatial dimension [5]. Actually, the very existence of such modes is indeed trivial, since the dynamics of the phase Φ of a square LJJ is known to be described by the obvious 2-d extension of the SG PDE, in which the second space derivative Φ_{xx} is replaced by the Laplacian operator:

$$\Phi_{tt} - \Phi_{xx} - \Phi_{yy} + sin\Phi = 0 \quad . \tag{4}$$

On the other hand, their stability is related to the b.c.'s which are assumed at both limit lines bounding the y direction. In the case of reference [5], an overlap geometry over the rectangle area corresponding to $\gamma = 0.3$ was assumed. As a consequence, the y-b.c.'s were :

$$\Phi_y(y = \pm w/2) = \pm\eta \quad , \tag{5}$$

(or reverse, depending on the direction of the d.c. bias current) , where η is assumed small :

$$\eta \ll 1 \quad , \tag{6}$$

(actually, ref. [5] considered a rather large value of η :$\eta = 0.4$).

Therefore, an initial condition basically consisting in the superimposition of a trivial y-degenerated (say) kink and a x-degenerated parabola fitting the above b.c.'s (5), according to :

$$\begin{aligned} \Phi(x,y,t=0) &= 4\,tan^{-1}\,exp\Big[\frac{x}{\sqrt{1-v^2}}\Big] + (\eta/w)y^2 \quad , \\ \Phi_t(x,y,t=0) &= -v\Phi_x(x,y,t=0) \quad , \end{aligned} \tag{7}$$

was acceptable in the numerical simulations of the PDE (4), in order to lead to 2π wave fronts propagating in the x-direction (including reflections at $x = \pm l/2$ over a time interval of order 75 in reduced units [6]).

The reason of this relative stability is twofold : i) as emphasized above, the parameter η scaling the fit of the y-quasi-degenerated wave profile to both LJJ boundaries in the y-direction is small. ii) a small damping term $-\alpha\Phi_t$ is added to the r.h.s. of the 2-d PDE (4). This means that, if the corresponding attractor is close to a y-degenerated kink profile — and such is obviously the case for the system (4-7) — ,any initial condition which does not lie too far from it will develop, in the appropriate configuration space, a trajectory converging to it.

If the following approximation is assumed for all time values:

$$\begin{aligned} \Phi(x,y,t) &= 4\,tan^{-1}\,exp\Big[\frac{x-vt}{\sqrt{1-v^2}}\Big] + (\eta/w)y^2 \quad , \\ \Phi_x(x &= \pm l/2) = 0 \quad , \end{aligned} \tag{8}$$

it is worth noting that the above coarse approximation of a simple superimposition of two profiles as described in formula (7) leads indeed to the so-called "d.c. driven" 1-d SG PDE :

$$\Phi_{tt} - \Phi_{xx} + sin\Phi = I \quad , \tag{9}$$

where :

$$I = 2\eta/w \quad . \tag{10}$$

The question now arises to know if a "non-trivial" 2-d solitary wave front could propagate in a, say for sake of simplicity, square LJJ ($l = w$). By "non-trivial", we mean that the y-dependence of the solitary wave profile could not be reduced to a simple parabolic one like in equation (8). The present paper addresses this question for the particular case of an *intermediate-length Josephson junctions (IJJ)*:

$$l = 4 \quad . \tag{11}$$

The choice of the b.c.'s will be of crucial importance, as possible symetries in an infinite geometry, say, rotations, will no longer survive in a particular bound square system. We shall adopt in the present paper the following b.c.'s :

$$\Phi_x(x = -l/2) = \Phi_y(y = -l/2) = \eta \quad ,$$
$$\Phi_x(x = +l/2) = \Phi_y(y = +l/2) = 0 \quad . \tag{12}$$

These b.c.'s describe the physics of a square cross-type IJJ which the bias current enters through the $x = -l/2$ boundary and exits through the $y = -l/2$ boundary (see figure 1). For experimental reasons, we believe that such b.c.'s are most appropriate for the detection of the solitary mode described in this paper.

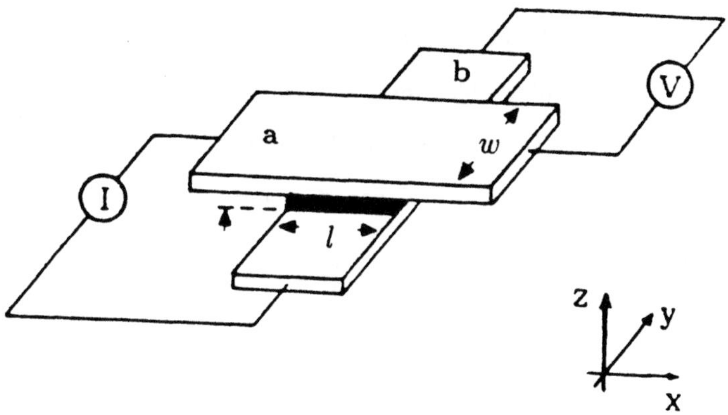

FIGURE 1

Josephson junction of cross-type geometry. The dimensions are l and w; a and b are the two superconducting films. After A. Barone.

In the next section, we shall perform a rapid survey of the situation in which the system is assumed fully linear, in the sense that the $sin\Phi$ term is simply dropped in the PDE (4). This cursory glance will help one's -physical feeling of how nonlinear solitary modes could be obtained in a square IJJ of the type (11-12). We shall show how the so-called (1,1) stationary cavity mode — whose existence in a square LJJ was already pointed out in [7] — can suggest the structure of the 2-d closed isoperimeter solitonic mode,the dynamics of which is the subject of the present paper.

Section III will provide a numerical description of this mode : it appears as a couple of kink-antikink wave front pairs propagating along, say, the first diagonal and extending according to a quasi-degenerated topology along the second one . At each time value, there exist two such kink-antikink wave front pairs propagating along both diagonals, in such a way that the 2π wave fronts build in the $[x,y]$ plane a variable geometrical pattern — actually a rectangle of constant perimeter (or equivalently of constant energy) — inside the original square IJJ. Indeed, the energy of a wave front whose profile is spatially degenerated along a given direction is obviously proportional to its length along this direction, since this energy is the result of the integration over the whole area of the SG hamiltonian density.

Section IV reports a short discussion about the possible experimental detection of such non-trivial solitonic modes.

II LINEAR MODES IN 2-D SQUARE GEOMETRIES

Consider the following linear wave equation :

$$\Phi_{tt} - \Phi_{xx} - \Phi_{yy} = 0 \quad , \tag{13}$$

corresponding to the nonlinear PDE (4), and assume, for sake of simplicity, open-end b.c.'s at $x = \pm l/2$ and $y = \pm l/2$:

$$\Phi_x(x = \pm l/2) = 0 ; \; \Phi_y(y = \pm l/2) = 0 \tag{14}$$

As is well-known, the stationary modes which may resonate according to b.c.'s (14) are :

$$\Psi_{m,n}(x.y.t) = [f_m(m\pi x/l)]\,[f_n(n\pi y/l)]\; exp[2i\pi\nu_{m,n}t] \quad , \tag{15}$$

where the function f is a *sine* for odd values of the index, and a *cosine* for even values of it. The eigenfrequencies $\nu_{m,n}$ are :

$$\nu_{m,n} = (1/2l)\sqrt{m^2 + n^2} \quad . \tag{16}$$

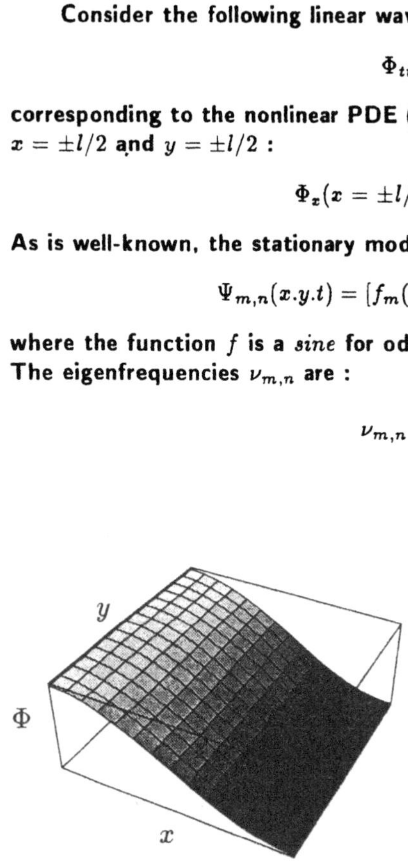

FIGURE 2a

The linear mode (1,0) of amplitude $\varepsilon \ll 1$.

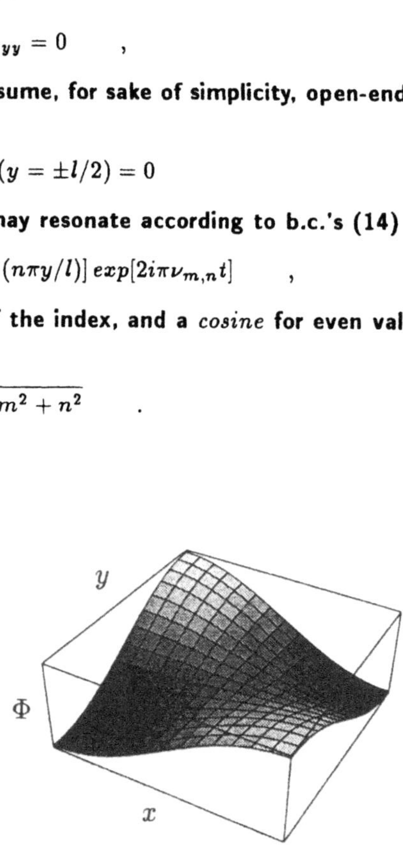

FIGURE 2b

The linear mode (1,1) of amplitude $\varepsilon \ll 1$.

Figures (2a,b) respectively displays the topology of the (1,0) and the (1,1) mode. Note that the former (1,0) mode may be regarded to as the superimposition of both $e^{\pm i[(\pi/l)x \pm 2\pi\nu_{1,0}t]}$ linear propagative modes (running waves), while the latter (1,1) mode may be written in terms of $e^{+i[(\pi/l)(x\pm y)+2\pi\nu_{1,1}t]}$ and $e^{-i[(\pi/l)(x\pm y)-2\pi\nu_{1,1}t]}$ linear modes respectively propagating along the diagonals $x \pm y = constant$. This means in particular that the (1,0) mode displayed on figure (2a) is clearly the linear "equivalent " of the "trivial" nonlinear mode described by equation (8).

Now the question is : what is the nonlinear propagating mode corresponding to figure (2b), i.e. to the (1,1) standing wave ? A glance to figure (3) immediately shows that it seems very close to the superimposition of a kink-antikink pair propagating in $x+y$ direction and an antikink-kink pair propagating in $x-y$ direction, according to the following profile, illustrated on figure (3), which is deduced from the analytical expressions corresponding to bions (kink-antikink pairs)[8] :

$$\begin{aligned}\Phi(x,y,t) = 4tan^{-1}\Big\{ (1/v)sinh\Big[\frac{(l/2\sqrt{2})-vt}{\sqrt{1-v^2}}\Big]\, sech\Big[\frac{x+y}{\sqrt{2(1-v^2)}}\Big]\Big\} \\ - 4tan^{-1}\Big\{ (1/v)sinh\Big[\frac{(l/2\sqrt{2})+vt}{\sqrt{1-v^2}}\Big]\, sech\Big[\frac{x-y}{\sqrt{2(1-v^2)}}\Big]\Big\} \quad .\end{aligned} \tag{17}$$

We show in the next section that this very simple heuristic argument indeed leads to the existence of such a non-trivial nonlinear solitary mode running along both diagonals. Its stability will be pointed out by use of reasonably long-time numerical simulations up to 160 periods) of the 2-d PDE (4) — including the small damping term $-\alpha\Phi_t$ — with the b.c.'s (12).

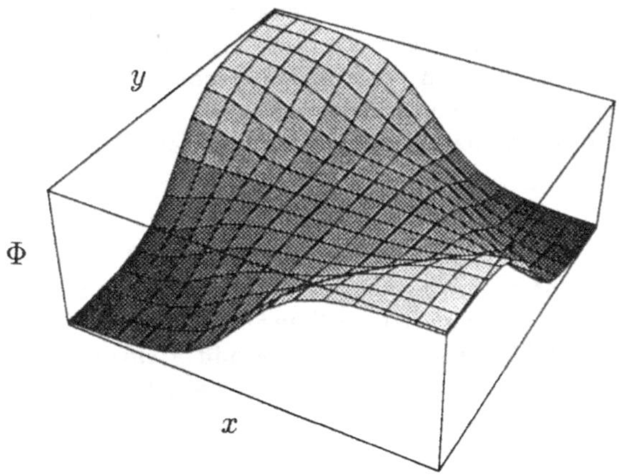

FIGURE 3

The nonlinear mode constituted by a kink-antikink pair along each diagonal (the amplitude of the wave is 2π.

III : THE 2-d ISOPERIMETER SOLITONIC MODE

We look for a nonlinear solitary mode propagating in the square area defined by $-l/2 < x < l/2$ and $-l/2 < y < l/2$, according to the 2-d PDE (4) (including damping) and the b.c.'s (12).

1)-The first obvious physical requirement to be fulfilled is to recover, in our 2-d geometry, the translational invariance that defines the steady state dynamics of the 1-d (anti)kink. This means that all possible dynamical states of the wave must have the same energy.

As emphasized in the introduction of this paper, this means that the *total length of the $\pm 2\pi$ wave front must be conserved, i.e. the wave front must be an isoperimeter one.*

2)-The second constraint that we demand is an obvious extension to 2-d geometries of the monotonic decrease (or increase, depending on the sign of η adopted in (12)) of the phase Φ during shuttling regimes of 1-d LJJ's at a given point. Actually, *a full cycle must be a net increase of the phase Φ by a given integer multiple value of $\pm 2\pi$.* Assume, say, -2π (i.e. a monotonic decrease of the phase Φ). Hence, each part of the wave front will move in such a way that, locally, the phase decreases. Assuming, in addition, a *closed geometry of the wave front*, in order to fulfill the first above requirement, we see that the motion of this (small) part of the front actually induces the *global wave front dynamics*, in a way which is displayed in the series of figures (4a-h).

The first figure (4a) corresponds to the initial condition (17) — displayed on figure 3 —, i.e. to two bions located along the two diagonals with an initial velocity *very close to the limit velocity* equal to 1, in our dimensionless units. Actually, we adopt, for the numerical simulations :

$$v = 0.99 \quad . \tag{18}$$

Then, the seven remaining pictures (figures 4b-h) display the expected dynamics of the *2-d solitonic wave front pattern*, according to both above constraints 1) and 2). Note the *8π jump* of the phase, which occurs at the corners when the two diagonal fronts collide in figure (4c) and (4g). Finally, compare figures (4a) and (4h), and it becomes clear the the inferred 2-d cycle has indeed a *periodic decrease of the phase Φ equal to 8π.*

In order to check this idea by numerical simulations, we consider the following weakly damped 2-d SG PDE:

$$\Phi_{tt} - \Phi_{xx} - \Phi_{yy} + sin\Phi = -\alpha\Phi_t \quad , \tag{19}$$

together with the b.c.'s (12), which model the effect of the induced magnetic field generated by the d.c. bias current entering the square Josephson junction through the $x = -l/2$ boundary and exiting it through the $y = -l/2$ boundary (see figure 1). Moreover, we adopt the initial condition (17) and, in order to fulfill the b.c.'s (12), we add a small-amplitude parabola in x and y, which also produces from the very beginning the right uniform force (given by the Laplacian operator in the PDE (19)) to accelerate the wave fronts:

$$\begin{aligned}\Phi(x,y,t) = {} & 4tan^{-1}\Big\{ (1/v)sinh\Big[\frac{(l/2\sqrt{2}) - vt}{\sqrt{1-v^2}}\Big] sech\Big[\frac{x+y}{\sqrt{2(1-v^2)}}\Big]\Big\} \\ & -4tan^{-1}\Big\{ (1/v)sinh\Big[\frac{(l/2\sqrt{2}) + vt}{\sqrt{1-v^2}}\Big] sech\Big[\frac{x-y}{\sqrt{2(1-v^2)}}\Big]\Big\} \\ & - (\eta/2l)\Big[(x-l/2)^2 + (y-l/2)^2\Big] \quad .\end{aligned} \tag{20}$$

In this simulation the numerical scheme is a modification of the one used by Eilbeck et. al. (1984). It is a stabilized (using the Lax method) leapfrog discretization second order in space and time.

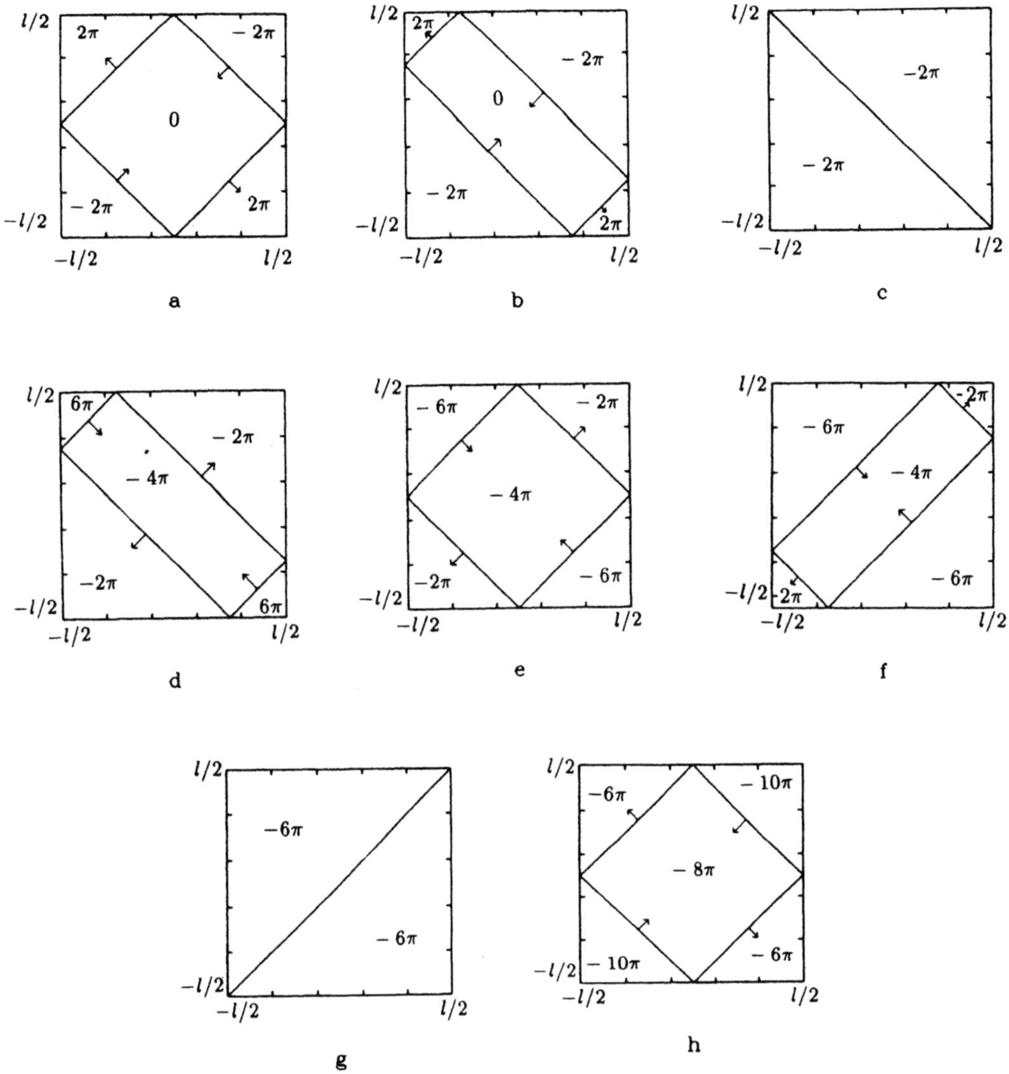

FIGURE 4

These figures display schematically the geometrical pattern of the wave fronts in the $\{x, y\}$ plans over one period.

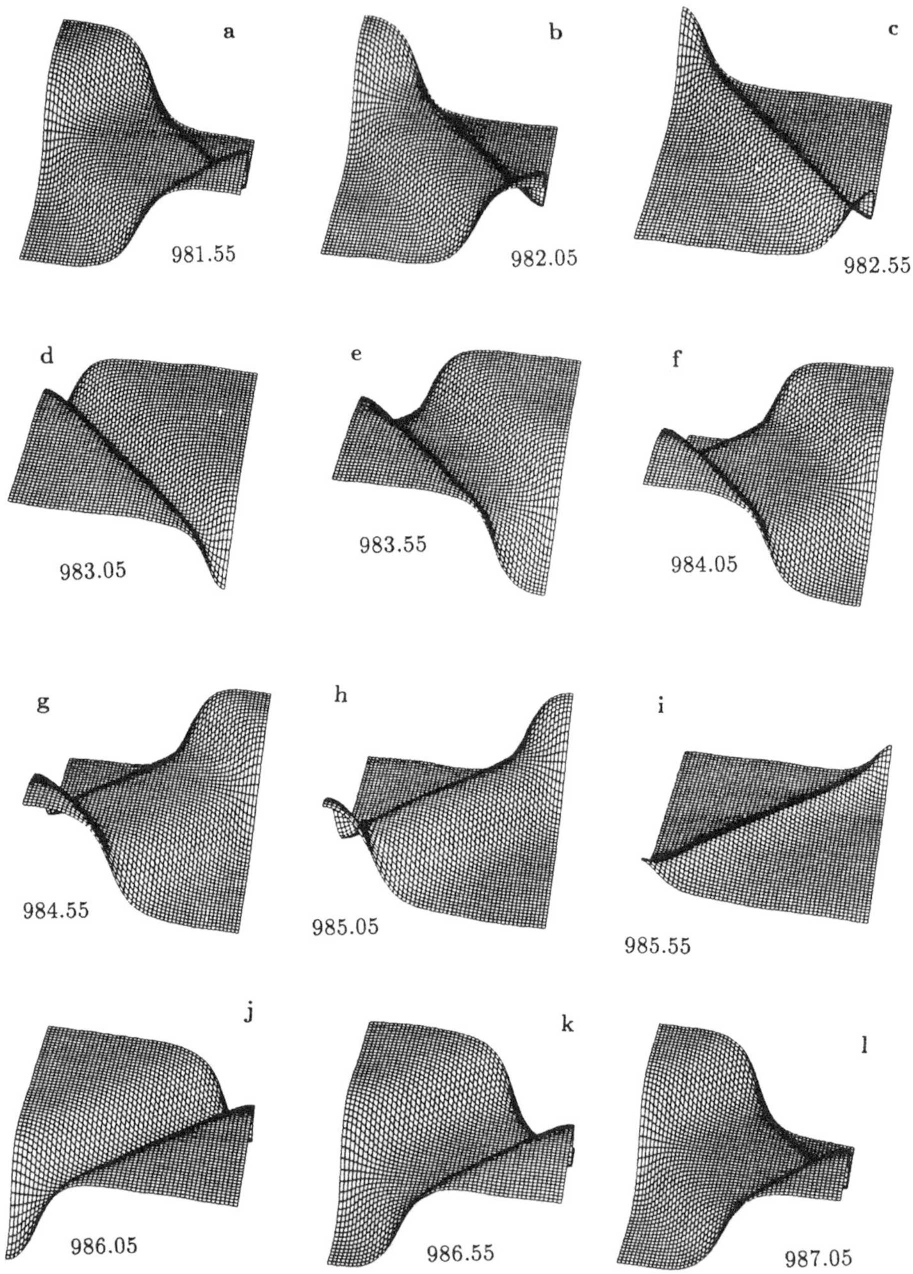

FIGURE 5

This set of pictures show the result of the numerical integration of equation (19) with b.c.'s (12) and starting at $t = 0$ with initial conditions (20). This set display the time evolution of the phase Φ over a full period between $t = 981.55$ and $t = 987.05$.

Using the discretization

$$\Phi(i\Delta, j\Delta, n\Delta t) = \Phi^n_{i,j} \ , \quad 1 \le i,j \le N \tag{21}$$

where Δ represents the square mesh and Δt the time step, the numerical scheme reads

$$\begin{aligned}\Big(\frac{\alpha\Delta t}{2} + 1\Big)\Phi^{n+1}_{i,j} &= \Big(\frac{\alpha\Delta t}{2} - 1\Big)\Phi^{n-1}_{i,j} + \Big[2 - 4\Big(\frac{\Delta t}{\Delta}\Big)^2\Big]\Phi^n_{i,j} \\ &+ \Big(\frac{\Delta t}{\Delta}\Big)^2 \left[\Phi^n_{i+1,j} + \Phi^n_{i-1,j} + \Phi^n_{i,j+1} + \Phi^n_{i,j-1}\right] \\ &- (\Delta t)^2 \sin\left[\frac{1}{4}(\Phi^n_{i+1,j} + \Phi^n_{i-1,j} + \Phi^n_{i,j+1} + \Phi^n_{i,j-1})\right] \ .\end{aligned} \tag{22}$$

The boundary conditions are introduced by using an extra set of points outside the mesh

$$\begin{aligned}\Phi^n_{0,j} &= \Phi^n_{2,j} - 2\eta\Delta \qquad & \Phi^n_{N+1,j} &= \Phi^n_{N-1,j} \\ \Phi^n_{i,0} &= \Phi^n_{i,2} + 2\eta\Delta \qquad & \Phi^n_{i,N+1} &= \Phi^n_{i,N-1}\end{aligned} \tag{23}$$

The linear stability analysis gives the so called Courant-Friedrichs-Lewy condition: $c = \frac{\sqrt{2}\Delta t}{\Delta} \le 1$. Due to the fact that this is a linear condition, for stability we use a constant c lower than 1. In order not to introduce too much numerical viscosity we choose $c = 0.95$.

The result of the numerical simulation of the 2-d PDE problem defined by (12,18-20) is displayed in the series of figures (5a-l), ranging from $t = 981.55$ to $t = 987.05$, i.e. over a full period of a 8π jump of the phase of the system in the already well-established asymptotic dynamical regime. Figure (6) shows the regular decrease of the phase, considered at the center $x = y = 0$, about the same time interval. Due to the b.c.'s (12), and to the choice of the initial condition (18)(20), which lie very close to the attractor, the transients do not

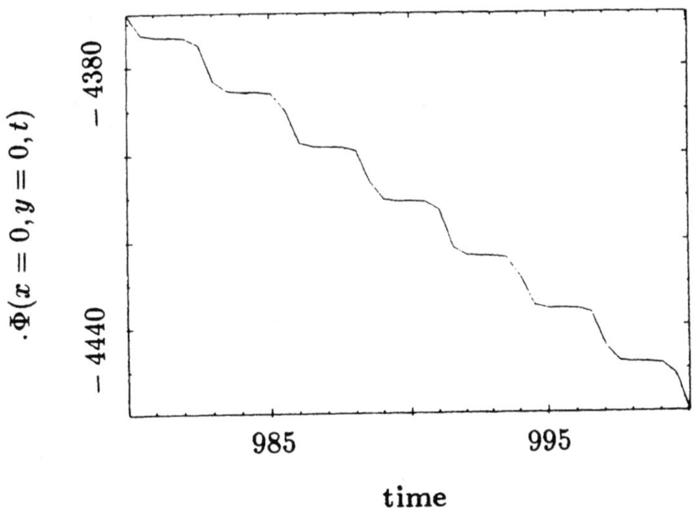

FIGURE 6

The time evolution of the phase Φ at the point $x = 0$; $y = 0$ over few periods.

exceed a few tens of periods T, where:

$$T = 4\sqrt{2} \quad , \tag{24}$$

with an accuracy of 1 %, because of the choice of the parameters $l = 4$ and $v = 0.99$ (cf. formulas (11) and (18)).

We conclude that the numerical simulation provides an excellent confirmation of the non-trivial 2-d solitonic wave front dynamics inferred in the series of schemes (4a-h).

Many questions still remain open:

a)-are other kinds of such non-trivial solitary wave front patterns possible, in particular a more "elementary" one, consisting in the nonlinear superimposition of two (anti)kinks running perpendicular to each other ? Then the above-described isoperimeter solitonic wave front loop would appear as a combination of four such "elementary" 2-d structures, being located each at the four corners of the loop. The fact that the b.c.'s of the square system considered in the present paper are very close to — or exactly equal to perfectly reflective ones (cf. formulas (6) and (12)) is a strong argument favouring this conjecture, as these boundaries actually define virtual structures outside the square area equivalent to those enclosed inside.

b)-what are the requirements concerning the velocity of the moving wave fronts ? In the numerical simulations we did, the greatest stability corresponded to the highest velocity, reached as the result of balance between damping and "edge-feeding" through the parameter η.

c)-what is the influence of the b.c.'s ? In particular, could such b.c.'s as in (2) and (5) — which experimentally just mean a different bias input into the system — lead to similar solitonic wave fronts as those described in the present paper ?

d)-what is the influence of the length l of the system ? The choice (11) appeared in our numerical simulations as suitable for the rapid onset of a stable asymptotic dynamical state, and it seems that IJJ's do more likely allow such isoperimeter moving wave front structures than LJJ's ($l \gg 1$).

e)-which initial conditions lie in the basin of attraction of the 2-d isoperimeter solitonic attractor ? To answer this question, a possible theoretical approach could concern the search of a few collective-coordinates which would determine most of the wave dynamics, and then use for the resulting ODE system the powerfull "cell-mapping techniques" [9]. The proportion of the phase space of the system occupied by the basin of attraction of the 2-d isoperimeter soliton attractor would then provide a good estimation of the probability for a real experiment — i.e. which has no reliable control of the initial conditions — to obtain such a state [10].

A detailed study of the whole 2-d dynamical problem addressing the above questions is out of the scope of this paper. At the present stage, we only wish to emphasize the richness of the field which the existence of non-trivial solitonic modes opens in strongly nonlinear 2-d wave dynamics.

IV A POSSIBLE EXPERIMENTAL DETECTION OF THE 2-D ISOPERIMETER SOLITONIC WAVE FRONT LOOP.

The two macroscopic quantities which determine the dynamics of an IJJ and which can easily be measured are the bias current, basically determined by the value of η (see for instance formula (8)), and the voltage across the junction, which is given, in units of the flux quantum $\hbar/2e$, by [1-4]:

$$V = \Phi_t = \frac{\Delta\Phi}{\Delta t} \quad . \tag{25}$$

The plot of the former as a function of the latter is the so-called "current-voltage characteristics". In the steady-state regime described by figures (4a-h), the voltage V can be estimated as the ratio of the $\Delta\Phi = 8\pi$ jump of the IJJ phase over the period $\Delta t = T$ given by equation (24) :

$$V_{2-d} = \frac{8\pi}{(4\sqrt{2})} = \sqrt{2}\pi \quad . \tag{26}$$

This voltage has to be compared to the so-called 1-d "Zero-Field-Step" (ZFS) voltage of the trivial y-degenerated shuttling kink front (7), which is clearly :

$$V_{1-d} = \frac{(\Delta\Phi = 4\pi)}{(\Delta t = 8)} = \pi/2 \quad , \tag{27}$$

since the shuttling period of a relativistic kink is 8 (cf. equation (11)), while the net phase change over this period is 4π. Therefore :

$$\frac{V_{2-d}}{V_{1-d}} = 2\sqrt{2} \quad . \tag{28}$$

As a consequence, an experiment which would provide both these 2-d and 1-d ZFS's by changing the bias feeding into a square cross-type Josephson junction of intermediate length l (of the order of a few Josephson penetration depths) from the symetrical configuration described by equations (5) into the asymetrical one, described by equations (12), could presumably display the existence of the 2-d isoperimeter solitonic mode pattern which is predicted by the present paper.

Note that such an experiment will have to distinguish between the two above-mentionned solitonic modes and the *linear modes of the type displayed on figures [2a,b]*, which are described by formulas (15) and (16). Assuming that, in one oscillation of the standing wave, the net change of the phase is equal to $\Delta\Phi = 2\pi$, we obtain indeed the so-called "Fiske steps" in the current-voltage characteristics (see equation (16)):

$$V_{m,n} = 2\pi\nu_{m,n} \quad . \tag{29}$$

Therefore, the first steps (up to the value $3\pi/2$) in the whole voltage scale are :

$$\begin{aligned} &V_{1,0} = \frac{\pi}{4}\,;\; V_{1,1} = \frac{\pi}{2\sqrt{2}}\,;\; V_{1-d} = V_{1.0}(2) = \frac{\pi}{2}\,;\; V_{1,1}(2) = \frac{\pi}{\sqrt{2}}\,;\; V_{1,0}(3) = \frac{3\pi}{4}\,;\\ &V_{1-d}(2) = V_{1,0}(4) = \pi\,;\, V_{1,1}(3) = \frac{3\pi}{2\sqrt{2}}\,;\; V_{1,0}(5) = \frac{5\pi}{4}\,;\\ &V_{2-d} = V_{1,1}(4) = \pi\sqrt{2}\,;\; V_{1-d}(3) = V_{1,0}(6) = \frac{3\pi}{2} \quad , \end{aligned} \tag{30}$$

where $V_{1-d}(j)$ is the second ($j = 2$) and the third ($j = 3$) 1-d ZFS, while the $V_{m,n}(i)'s$ stand for the ith harmonic of the (m,n) Fiske step.

In order to distinguish the V_{2-d} step from the $V_{1,1}(4)$ one requires an external applied magnetic field along appropriate directions. This important aspect of the experiment is stiil being investigated.

This work was performed in the frame of the EEC "Stimulation" contract ST2-0267-J-C. Rainer Grauer was supported by this contract. The authors thank N. F. Pedersen and G. Costabile for interesting discussions.

REFERENCES

[1] A. Barone and G. Paterno, "Physics and applications of the Josephson effect", John Wiley, New York, (1982) and references therein.

[2] Pedersen N.F., "Solitons in long Josephson junctions", In Advances in Superconductivity, Ed. B. Deaver and J. Ruvalds,Plenum, NY, 149, (1982).

[3] Pedersen N.F., "Solitons in Josephson transmission lines", In Solitons,Ed. by S.E. Trullinger, Zakharov and V.L. Pokrovsky,Elsevier Sc. Pub. 469, (1986).

[4] Pagano S., "Nonlinear dynamics in long Josephson junctions", Phd Thesis, Lyngby, (1987) and references therein.

[5] Eilbeck J.C., Lomdahl P.S., Olsen O.H. and Samuelsen M.R., "Comparison between one-dimensional and two-dimensional models for Josephson junctions of overlap type", J. Appl. Phys. 57, 861, (1985).

[6] Lomdahl P.S., Olsen O.H., Eilbeck J.C. and Samuelsen M.R., "How good are one-dimensional Josephson junction models?", J. Appl. Phys. 57, 997, (1985).

[7] Mygind J. and Pedersen N.F., "On the behaviour of a two-dimensional Josephson tunnel junction in the weakly nonlinear regime", IEEE Trans. Magn. MAG-21, 632, (1985).

[8] G. Eilenberger, "Solitons", Mathematical Methods for Physicists, Springer series in Solid-State Sciences 19, Springer Verlag, (1989).

[9] F. Varosi, C. Grebogi and J.A. Yorke, Phys. Lett. A 124, 59, (1987).

[10] J.C. Fernandez, R. Grauer and G. Reinisch, "Phase-locked dynamical regimes to an external microwave field in a long, unbiased Josephson junction", in the same proceedings.

Radiative Damping of Driven Sine-Gordon Breathers

R. Döttling, J. Eßlinger, W. Lay, and A. Seeger
Max-Planck-Institut für Metallforschung, Institut für Physik, and
Universität Stuttgart, Institut für Theoretische und Angewandte Physik
Postfach 80 06 65, D-7000 Stuttgart 80, Fed. Rep. Germany

Abstract: The paper studies the behaviour of breather solutions of the sine-Gordon (Enneper) equation under the action of a constant external force s. As a first step the modification of the breather solution by the external force is calculated in the so-called adiabatic approximation. In a second step the coupling of the breather motion to the "heavy phonons" of the system is obtained by a perturbation treatment based on the inverse scattering transform. The results are used to calculate to the order s^2 the radiative damping of breathers in the limits of small or large breather amplitudes.

1. Introduction

The non-linear hyperbolic differential equation

$$\frac{\partial^2 \Phi}{\partial x^2} - \frac{\partial^2 \Phi}{\partial t^2} = \sin \Phi \tag{1.1}$$

is the only Lorentz-invariant partial differential equation with soliton properties and possesses therefore particular importance among the soliton equations. It was also the first non-linear partial differential equation whose soliton properties were recognized (see below).

We denote (1.1) as Enneper equation since it was originally considered by A.Enneper [1] in its characteristic form

$$\frac{\partial^2 \Phi}{\partial \xi \partial \eta} = \sin \Phi \tag{1.2}$$

in the differential geometry of surfaces of constant negative Gaussian curvature. A wide-spread alternative name is "sine-Gordon equation". The study of (1.2) in the context of differential geometry revealed very interesting properties, including the possibility to generate from known solutions of (1.1) new solutions by means of the Bäcklund transformation [2,3].

In physics, (1.1) found its first applications in crystal physics in the study of one-dimensional dislocation models [4,5], with x and t having the meaning of spatial and temporal coordinates. It was in this context that solitonic solutions of (1.1) were discovered and discussed in detail [6,7,8], their original German names being "translatorische" and "oszillatorische Eigenbewegungen". From a historical view-point it is interesting to note that this preceded the discovery of the solitonic properties of the Korteweg-de Vries (KdV) equation by more than a decade and that in the original investigations of the Enneper and the KdV equation quite different techniques were involved (the above-mentioned Bäcklund transform in the case of (1.1), and numerical computations [9] and the so-called inverse scattering technique in the case of the KdV equation [10]).

Subsequently (1.1) was applied to many different physical problems, although one must realize that a simple equation such as (1.1) can describe physical phenomena only approximately and that sometimes drastic assumptions have to be made in order to arrive at (1.1). A field in which its use has been quite successful and where fairly close contact to experiments exist is that of kinks on dislocation lines [11,12]. Here x denotes the spatial coordinate along nearly straight dislocation lines, and $\Phi/2\pi$ the displacement of the dislocation lines from the bottoms of their "Peierls valleys", (for details on normalization see [12]).

In the formation of kink pairs (i.e., two kinks of opposite sign) it is of particular interest to consider the generalization

$$\frac{\partial^2\Phi}{\partial x^2} - \frac{\partial^2\Phi}{\partial t^2} + s = \sin\Phi \ , \tag{1.3}$$

where the time- and space-independent "external force" s represents the action of an applied shear stress on the dislocation lines. In the present normalization s is restricted to $[0,1]$.

The total energy of solutions $\Phi = \Phi(x,t)$ of (1.3) is given by

$$E = \int_{-\infty}^{\infty} \left[\frac{1}{2}\left(\frac{\partial\Phi}{\partial x}\right)^2 + \frac{1}{2}\left(\frac{\partial\Phi}{\partial t}\right)^2 + 1 - \cos\Phi - s\,\Phi \right] dx \ , \tag{1.4}$$

provided the integral converges. The units used in (1.4) are such that the energy of a single kink is $E_k = 8$.

Among the many remarkable properties of (1.1) is the existence of spatially localized time-periodic solutions [7]

$$\Phi_b(x,t) = 4\arctan\left\{ \tan\left(\frac{I}{16}\right) \frac{\sin\omega t}{\cosh\left[x\,\sin(I/16)\right]} \right\} \ , \tag{1.5}$$

now known as "breather at rest" (in the following for short "breather"). Eq.(1.5) represents a one-parameter family of solutions. The parameter I has been chosen in such a way that the amplitude of the breather is given by $I/4$ and its normalized circular frequency by

$$\omega = \cos(I/16) = \cos(I/2\,E_k) \ . \tag{1.6}$$

By evaluating (1.4) for $s = 0$, the breather energy was found to be

$$E_b = 2\,E_k\,\sin(I/2\,E_k) \ . \tag{1.7}$$

Elimination of I between (1.6) and (1.7) leads to [7][1]

$$E_b = 2\,E_k\left(1-\omega^2\right)^{\frac{1}{2}} \ . \tag{1.8}$$

[1] The original expression of Seeger, Donth, and Kochendörfer [7] was, in fact, more general and included the case of breathers moving with a constant velocity of translation.

As is well-known from the mechanics of systems of mass points, the frequency of a periodic motion equals the derivative of the total energy with respect to the action variable of that motion. Since in the present case

$$\partial E_b/\partial I = \omega \quad . \tag{1.9}$$

holds, we may identify $I/2\pi$ as the action variable associated with the breather mode. This is a result of great significance, since it allows us to map the breather motion (originally requiring a field description) on the one-dimensional motion of a mass point. Adopting the usage of Arnol'd [13] we shall call I "action variable". In the present notation the action variable I of a small-amplitude breather (which behaves as an harmonic oscillator) becomes equal to the breather energy.

In the case of nonvanishing s, (1.3) is no longer exactly integrable. This means that if we start with a breather solution at $t = 0$, its amplitude will gradually diminish. Since the total energy E of the system is preserved (as may be easily demonstrated from (1.4)), energy flows gradually from the breather mode into other modes. From physical arguments it is fairly clear that these modes will be the "phonons" of the system, i.e. the small-amplitude vibration following from the linearization of (1.3). The aim of the present paper is to calculate the rates of transfer of energy from the breather mode to phonon modes and to obtain analytical expressions for the "radiative damping" of the breather motion.

Our approach is based on the idea that for finite s, notwithstanding the coupling between different modes, breather-like solutions continue to exist on the time scales of physical interest. As a first step (Sect. 2) we disregard the existence of the phonon modes entirely and ask the question how the breather mode is affected by the external stress s in the absence of radiative damping. In this approximation the energy associated with the breather mode equals the energy of the entire system and may hence be taken as time-independent. Our treatment is analogous to the so-called adiabatic approximation in the mechanics of mass points and rigid bodies. As a second step the coupling of the phonon modes to the breather mode is calculated by a perturbation treatment based on the inverse scattering transformation (Sect. 3). From this analytical expressions correct to the order s^2 are derived for the radiative damping in the limiting cases of small and large breather amplitudes.

2. The adiabatic approximation

The breather-like solutions mentioned in Sect. 1 are obtained by putting

$$\Phi(x,t) = \Phi_b(x; I, \varphi) + \arcsin s \quad , \tag{2.1}$$

where $\Phi_b(x, ; I, \varphi)$ is the breather solution (1.5) with $\omega\, t$ replaced by φ, and calculating E from (1.4) under the conditions

$$\dot{I} = 0 \quad , \quad \dot{\varphi} = \omega \quad . \tag{2.2a, b}$$

(dots denote the differentiation with respect to the time t). The adiabatic approximation consists in taking E as time-invariant. We may then consider it as a Hamiltonian

$$H(I, \varphi) = 16 \sin(I/16) - 4\pi s \operatorname{cosec}(I/16) \cdot f(y) \tag{2.3}$$

with I and φ as canonically conjugate variables, hence obeying Hamilton's equations

$$\dot{I} = -\partial H(I,\varphi)/\partial\varphi \quad , \quad \dot{\varphi} = \partial H(I,\varphi)/\partial I \ . \tag{2.4a, b}$$

In (2.3) the following abbreviation have been introduced:

$$f(y) := \operatorname{arsinh} y - y\,(1+y^2)^{-\frac{3}{2}}$$
$$-(2/\pi s)\left(1-(1-s^2)^{\frac{1}{2}}\right)\left[y^2(1+y^2)^{-1} + y\,(1+y^2)^{-\frac{3}{2}}\operatorname{arsinh} y\right] \ , \tag{2.5a}$$
$$y := \sin\varphi\ \tan(I/16) \ . \tag{2.5b}$$

A simplified form of (2.3), correct to first-order terms in s, was first obtained by Karpman et. al. [14] using a perturbation procedure based on the inverse scattering technique.

The preceding treatment has a simple physical meaning. The "shape" of the "driven" breather remains the same as in the case $s = 0$ but the parameter I and φ have acquired different dependences on time, governed by

$$\dot{I} = 4\pi s\ \cos\varphi \sec(I/16)\, f'(y) \tag{2.6a}$$

$$\dot{\varphi} = \cos(I/16) + (\pi s/4)\Big\{\cos(I/16)\operatorname{cosec}^2(I/16)\, f(y)$$
$$-\sin\varphi \operatorname{cosec}(I/16)\sec^2(I/16)\, f'(y)\Big\} \tag{2.6b}$$

with

$$f'(y) := \mathrm{d}f/\mathrm{d}y = \left(y^2(4+y^2)(1+y^2)^{-\frac{3}{2}}\right)$$
$$+(2/\pi s)\left(1-(1-s^2)^{\frac{1}{2}}\right)\left\{3\,y(1+y^2)^{-2} + (1-2\,y^2)(1+y^2)^{-\frac{5}{2}}\operatorname{arsinh} y\right\} \ . \tag{2.6c}$$

A first integral of (2.6) is

$$E = H(I,\varphi) \ . \tag{2.7}$$

From (2.7) we may construct the I,φ-contours of constant energy. An example ($s = 0.05$) is shown in Fig.1. The energy diagrams are 2π-periodic and may hence be mapped on tori. Their saddle points (I^s,φ^s) are obtained from the conditions $\dot{I} = 0, \dot{\varphi} = 0$ and are given by

$$\varphi^s = (2m+1)\pi/2 \quad (m = 0, \pm 2, \pm 4, \ldots) \ , \tag{2.8a}$$

$$\sin^2(I^s/16)\ \cos^3(I^s/16) = (\pi s/4)\Big\{\sin(I^s/16)\, f'(\tan(I^s/16))$$
$$-\cos^3(I^s/16)\, f(\tan(I^s/16))\Big\} \ . \tag{2.8b}$$

The contour line $E^s := H(I^s,\varphi^s)$ connecting the saddle points is called the separatrix. Within the framework of the adiabatic approximation, time-periodic breather-type solutions are possible for $E < E^s$, whereas for $E > E^s$ we obtain unbound kink-antikink pairs.

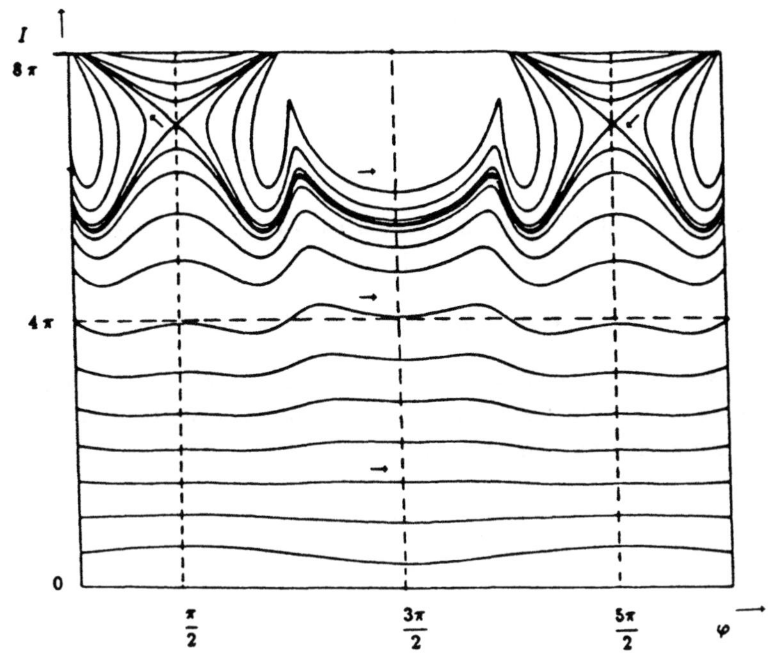

Figure 1: Contours of constant breather energy E in an (I, φ) plane (phase plane) according to the adiabatic approximation. The arrows indicate the flow pattern. The example pertains to $s = 0.05$. The E values associated with the contours may be obtained by inserting the I-value at $\varphi = 0$ into (1.7).

3. Perturbation treatment of breather-phonon coupling

The adiabatic approximation of Sect.2 disregards the coupling of the breather motion to the phonon modes. The coupling may be taken into account by a perturbation treatment based on the inverse scattering transformation, which is a generalization of the Hamilton-Jacobi theory of the mechanics of systems of mass-points.

The relationships between energy density, normalized frequency Ω, and normalized wave number k of running-wave solutions of (1.1) were derived very early [7]. In the present paper we confine ourselves to the harmonic approximation. This leads to the Yukawa-type dispersion relation

$$\Omega^2 = 1 + k^2 \ . \tag{3.1}$$

Eq.(3.1) justifies the name "heavy phonons" in order to distinguish the phonon solutions of (1.1) from the acoustic or "light" phonons of elasticity theory.

The Hamiltonian of the system "breather plus heavy phonons" may be written as

$$H = 16 \sin(I/16) + \int_0^\infty \Omega(\lambda)\, J(\lambda)\, \mathrm{d}\lambda \ , \tag{3.2}$$

where $J(\lambda)$ is the phonon "particle density" and, at the same time, the action variable associated with the heavy phonons. In terms of the parameter λ, which is introduced by the inverse scattering transformation, frequency and wavenumber read

$$\Omega(\lambda) = \left(\lambda + \frac{1}{4\lambda}\right) \quad , \quad k(\lambda) = \left(\lambda - \frac{1}{4\lambda}\right) \quad . \tag{3.3}$$

Both the action variable J and the spectral energy density $\epsilon(k,t)$ may be expressed in terms of the (in general complex-valued) Jost coefficient $b(\lambda,t)$:

$$J(\lambda) \approx (4/\pi\,\lambda)\,|b(\lambda,t)|^2 \quad , \quad \epsilon(k,t) \approx (4/\pi)\,|b(k,t)|^2 \quad . \tag{3.4a, b}$$

It is one of the fundamental properties of (1.1) that there is no energy exchange between the various solitonic modes (this holds even if the harmonic approximation for the phonons is not made). Hence for the unperturbed Enneper equation

$$|b(\lambda,t)| = |b(\lambda,0)| \tag{3.5}$$

is time independent.

The Jost coefficient obeys an evolution equation which in the case of (1.3) reads

$$\frac{\partial b(\lambda,t)}{\partial t} = i\Omega(\lambda)\,b(\lambda,t) + \frac{is}{4}\int_{-\infty}^{\infty}(1-\cos\Phi_b)\left\{\psi_2^{*2} - \psi_1^{*2}\right\}\,\mathrm{d}x \quad , \tag{3.6}$$

where ψ_i^* denotes the complex conjugate of the eigenfunction to eigenvalue λ of the so-called Zakharov–Shabat system [15]. If we expand the integrand in (3.6) in a Fourier series

$$(1-\cos\Phi_b)\left\{\psi_2^{*2} - \psi_1^{*2}\right\} = \sqrt{\frac{\omega}{2\pi}}\sum_{n=-\infty}^{\infty} C_n(x,\omega;\lambda;I)\,e^{in\omega t} \tag{3.7}$$

with Fourier coefficients

$$C_n(x,\omega;\lambda;I) = \sqrt{\frac{\omega}{2\pi}}\int_{-\pi/\omega}^{\pi/\omega}(1-\cos\Phi_b)\left\{\psi_2^{*2} - \psi_1^{*2}\right\}e^{-in\omega t}\,\mathrm{d}t \quad , \tag{3.8}$$

we obtain a formal solution of (3.6) in the form

$$b(\lambda,t;I)\,e^{-i\Omega(\lambda)t} = \frac{s}{4}\sqrt{\frac{\omega}{2\pi}}\sum_{n=-\infty}^{\infty}\frac{e^{i(n\omega-\Omega(\lambda))t} - 1}{n\omega - \Omega(\lambda)}\int_{-\infty}^{\infty} C_n(x,\omega;\lambda;I)\,\mathrm{d}x \quad . \tag{3.9}$$

However, since for $s \neq 0$ the eigenfunctions of the Zakharov–Shabat system are not known, we are unable to calculate the Fourier coefficients exactly. We have to resort to a perturbation treatment, in which we insert into (3.8) the eigenfunctions for $s = 0$. These are given by [14]

$$\psi_1^{*2}(x;\lambda;I;\varphi) = B\,\sin^2(I/8)\Bigg\{\frac{i}{4}\cosh(x\sin(I/16))\sin\varphi + \frac{\lambda}{2}\Big(\cos(I/16)\cosh(x\sin(I/16))\cos\varphi - \sin(I/16)\sinh(x\sin(I/16))\sin\varphi\Big)\Bigg\}^2 , \tag{3.10a}$$

$$\psi_2^{*2}(x;\lambda;I;\varphi) = B\Bigg\{\cos^2(I/16)\,(\lambda^2-(1/4))\cosh^2(x\sin(I/16)) + \sin^2(I/16)\,(\lambda^2+(1/4))\sin^2\varphi - \frac{i\lambda}{4}\sin(I/8)\Big[\cos(I/16)\sinh(2\,x\sin(I/16)) + \sin(I/16)\sin 2\varphi\Big]\Bigg\}^2 , \tag{3.10b}$$

$$B := \Bigg\{(\lambda+\lambda_1^*)^2(\lambda-\lambda_1^*)^2\cos^4(I/16)(\cosh^2(x\,\sin(I/16)) + \tan^2(I/16)\sin^2\varphi)\Bigg\}^{-1}\exp(ikx) , \tag{3.10c}$$

$$\lambda_1^* := \exp(-i\,I/16)/2 . \tag{3.10d}$$

As can easily be seen, the Fourier coefficients with odd n vanish whereas those with even n may be reduced to integrals of the type

$$\mathcal{I}_m(x;I) = \frac{(2a)^4}{\tan^8(I/16)}\int_0^{2\pi}\frac{\cos g(m)\tau}{(a^2+1-2a\cos\tau)^4}\,d\tau = \frac{32\pi}{\tan^8(I/16)}\frac{a^{10+g(m)}}{(1-a^2)^7}\sum_{j=0}^{3}\binom{3+g(m)}{j}\binom{6-j}{3}\left(\frac{1-a^2}{a^2}\right)^j , \quad |a|<1 . \tag{3.11}$$

with

$$\frac{1}{2}\left(a+\frac{1}{a}\right) = 1+2\cot^2(I/16)\cosh^2(x\sin(I/16)) , \tag{3.12}$$

$$m = \frac{n}{2} , \quad g(m) = m\,,\, m\pm 1\,,\, m\pm 2\,,\, m\pm 3. \tag{3.13}$$

Space limitations do not allow us to carry the general treatment further. Therefore we confine ourselves to two limiting cases.

α.) Radiation in the limit of small breather amplitudes:

In the limit of small breather amplitudes ($I \ll 16$, $E \approx I$, $\omega \approx 1$) we find

$$a \approx \frac{1}{4\cot^2(I/16)\cosh^2(x\,\sin(I/16))} , \tag{3.14}$$

$$\mathcal{I}_m(x;I) \approx \frac{32\pi}{\tan^8(I/16)} \sum_{j=0}^{3} \binom{3+g(m)}{j} \binom{6-j}{3} a^{10-2j+g(m)} , \tag{3.15}$$

$$b(k,t;I) = s\, e^{i\Omega(k)t} \sum_{n=2}^{\infty} A_n(k) \exp\left(-\frac{8\,k\pi}{I}\right) \left(\frac{e^{i(n\omega-\Omega(k))t}-1}{n\omega-\Omega(k)}\right) . \tag{3.16}$$

We see that the spectral energy density of the phonons posesses resonances at

$$\Omega_n = 2n\omega \ , \ k_n \pm \sqrt{4n^2\omega^2-1} \ , \ \omega \approx 1-(1/2)(I/16)^2 \tag{3.17}$$

and that the main phonon emission takes place at the wavenumber $k = \sqrt{3}$ [16,17]. The radiated power

$$L(I) = 2 \int_0^{\infty} \frac{\mathrm{d}\epsilon(k,t)}{\mathrm{d}t}\, \mathrm{d}k = 16\, s^2 \sum_{n=2}^{\infty} |A_n(k_n)|^2 \exp\left(-\frac{16k_n\pi}{I}\right) \tag{3.18}$$

is exponentially small with respect to I. For large n the intensity ratios of the emitted lines obey a Boltzmann distribution with E playing the rôle of a "temperature":

$$\frac{P_n}{P} \approx \exp - \left\{ \frac{16\pi \left(\Omega_n \sqrt{1-(1/\Omega_n^2)} - \sqrt{3}\right)}{E} \right\} . \tag{3.19}$$

The breather acts as the thermal reservoir of the emitted phonons; its "temperature" E decreases slowly because of radiative losses.

β.) Radiation in the limit of large breather amplitudes.

The case of large breather amplitudes is characterized in terms of the quantity

$$\zeta := 8\,\pi - I \tag{3.20}$$

by

$$\zeta \ll 1 \ , \quad \omega = \sin\zeta \approx \zeta \ . \tag{3.21}$$

We confine ourselves to the limit $\zeta \to 0$ and replace the Fourier series (3.7) by a Fourier integral. In the limit $t \to \infty$ the Fourier coefficients become

$$C(x,\Omega;\lambda) = \frac{1}{\sqrt{2\pi}} h(x,\lambda) \exp\left(ikx - \Omega\cosh x\right) \ , \tag{3.22}$$

where $h(x,\lambda)$ is a function located at the origin. In this limit the Jost coefficient evolves according to

$$\lim_{t\to\infty} b(\lambda,t)\, e^{-i\Omega t} = \frac{is}{4} \int_{-\infty}^{\infty} h(x,\lambda) \exp(ikx - \Omega\cosh x)\, \mathrm{d}x \ . \tag{3.23}$$

In the present approximation we are allowed to confine ourselves to x^2-terms in the exponent in (3.23). This gives us finally

$$\epsilon(k) = \frac{\pi s^2}{8} \frac{|h(0,k|^2}{k^2+1} \exp\left(-\frac{3k^2+2}{\sqrt{k^2+1}}\right) . \tag{3.24}$$

We see that in both limiting cases lowest-order contribution to the radiative damping is proportional to s^2.

References

[1] A.Enneper,'Über asymptotische Linien' in: Nachr. Königl. Gesellsch. d. Wiss., Göttingen, 1870, p.493

[2] A.V.Bäcklund, 'Om ytor med konstant negative krökning' in: Lunds Universitets Ars-skrift XIX, IV, p. 1, 1882-83

[3] L.Bianchi,: Rend. Acc. Naz. Lincei (5) **1** (2°em.), 1892, p. 3

[4] A.Seeger: Diploma thesis, Technische Hochschule Stuttgart 1948/49

[5] F.C.Frank, J.H.van der Merwe: Proc. Roy. Soc. London A, **201**, 261 (1950)

[6] A.Seeger, A.Kochendörfer: Z. Physik **130**, 321 (1951)

[7] A.Seeger, H.Donth, A.Kochendörfer: Z. Physik **134**, 173 (1953)

[8] A.Seeger: Z. Naturforschung **8a**, 246 (1953)

[9] N.J.Zabusky and M.D.Kruskal: Phys. Rev. Letters **15**, 240 (1965)

[10] C.S.Gardner, J.M.Greene, M.D.Kruskal , R.M.Miura: Phys. Rev. Letters **19**, 1095 (1967)

[11] A.Seeger, P.Schiller, 'Kinks in Dislocation Lines and Their Effects on the Internal Friction in Crystals' in: Physical Acoustics, Vol IIIA (W.P.Mason ed.), p.361, Academic Press, New-York and London 1966

[12] A.Seeger:'Structure and Diffusion of Kinks in: Monoatomic Crystals' in: Dislocation 1984 (Ed. Centre Nationale de la Recherche Scientific) p.141, Paris 1984

[13] V.I.Arnol'd: 'Mathematical Methods of Classical Mechanics, Springer Verlag, New York-Heidelberg-Berlin, 1978

[14] V.I.Karpman, E.M.Maslov, V.V.Solov'ev: Sov. Phys.-JETP **57**, 167 (1983)

[15] S.Novikov, S.V.Manakov, L.P.Pitaevskii, V.I.Zakharov: Theory of Solitons-The Inverse Scattering Method, Consultants Bureau, New-York and London 1984

[16] B.A.Malomed: Physica **27D**, 113 (1987)

[17] Y.S.Kivshar, B.A.Malomed: Europhys. Lett. **4**, 1215 (1987)

FLUXON TRAPPING BY INHOMOGENEITIES IN LONG JOSEPHSON JUNCTIONS

Tassos Bountis[a] , Stephanos Pnevmatikos[b]
Stavros Protogerakis[c] , George Sohos[a]

[a]Department of Mathematics, University of Patras, Patras 261 10, Greece.
[b]Research Center of Crete, P.O.B. 1527, 711 10 Heraklion, Crete, Greece.
[c]Physics Depart., University of Crete, P.O.B. 1470, 714 09 Heraklio, Crete, Greece.

ABSTRACT

Following the approach of collective coordinates for the location X(t), and speed U(t), of a fluxon travelling in a long Josephson junction (LJJ), we show that the conditions of fluxon trapping by an array of N point-like inhomogeneities are determined by the relative location of the separatrices of N unstable fixed points in the U,X phase plane. In the particular case of experimental interest in which the inhomogeneities are "microresistors", we find specific parameter values and initial conditions such that each "microresistor" will trap one fluxon thus turning the LJJ into a quantum flux "shuttle". We also solve for the same parameter values, the corresponding PDE numerically and verify the validity of the results obtained by the ODE's of the collective coordinate approach.

1. INTRODUCTION

The dynamics of Josephson junctions has been for many years, a very active and fruitful area of research [1,2]. From the point of view of applications, there has been renewed interest, since the recent discoveries of high T_c superconducting compounds [3]. On the other hand, advances in the theory and applications of solitons [4] and, more recently, exciting developments in the field of dynamical systems and chaos [5,6], have provided valuable new insight into a variety of Josephson junction phenomena [7,8].

In long (one dimensional) Josephson junctions (LJJ), it is well known that the phase difference $\varphi(x,t)$ of the superconducting current obeys the perturbed sine-Gordon equation written in the following dimensionless form [1,9]:

$$\varphi_{tt} - \varphi_{xx} + \sin\varphi = - \alpha\varphi_t + \beta\varphi_{xxt} - \gamma \qquad (1.1)$$

where $\alpha\varphi_t$ and $\beta\varphi_{xxt}$ are dissipation terms due to tunneling of normal electrons across the barrier, and the flow of normal electrons parallel to the barrier respectively, and γ is a distributed bias current.

For $\alpha=\beta=\gamma=0$, (1.1) becomes the sine-Gordon equation, which possesses an exact soliton solution of the kink (+) or antikink (-) type [4]:

$$\varphi_{\pm}(x,t) = 4\tan^{-1}\left[\exp\left(\pm\frac{x-vt-x_0}{(1-v^2)^{1/2}}\right)\right] \tag{1.2}$$

where the velocity v and the phase x_0 of the soliton are constant. It is remarkable that, if α,β,γ are nonzero, but small, a fluxon (+) or antifluxon (-) solution of the form (1.2) still appears to exist, provided its velocity and phase are taken to be time dependent. This fact has stimulated the development of a number of perturbation theories, which have been successful in calculating physically important quantities like fluxon radiation, the effect of fluxon-antifluxon collisions, etc. [10-13].

In this paper, we study the dynamics of a modulated fluxon

$$\varphi(x,t) = \varphi_+(x,t) = 4\tan^{-1}\left[\exp\left(\frac{x-X(t)}{(1-U^2(t))^{1/2}}\right)\right], \tag{1.3}$$

where X(t) and U(t) are the so called <u>collective coordinates</u>, in a LJJ with pointlike inhomogeneities of strength μ_i, at the locations $x=a_i$, $i=1,2,..,N$. The effect of these inhomogeneities can be included in (1.1) as follows [12,13].

$$\varphi_{tt} - \varphi_{xx} + \sin\varphi = -\alpha\varphi_t - \gamma - \sin\varphi \sum_{i=1}^{N} \mu_i\,\delta(x-a_i) \tag{1.4}$$

where we have decided, for simplicity, to ignore high order dissipation along the junction, setting $\beta=0$ [12].

If $\mu_i>0$ then the inhomogeneity is called a "<u>microshort</u>", since it can trap at a fixed stable point near it, a relatively energetic fluxon and thus impede the flow of current in the LJJ. On the other hand, if $\mu_i<0$, it is called a "<u>microresistor</u>", as relatively energetic fluxons can pass them by, not without suffering some radiative losses, however.

Now, under certain conditions, a microresistor can also trap fluxons, at a stable fixed point located almost exactly at the site of the microresistor. In fact, an array of such inhomogeneities can be used as a quantum flux "shuttle", measuring current pulses $A\delta(t)$, which would shift fluxons by one site to the right (or left), from the ith microresistor to the (i+1)st (or (i-1)st) [12].

It is very interesting that, recently, LJJ's with arrays of $\mu_i<0$ inhomogeneities have been constructed experimentally [14] and their I-V characteristics have been found to agree with the numerical results of simulation studies [15].

Our purpose is to approach this problem from the viewpoint of the <u>dynamics</u> of <u>collective coordinates</u>, X(t) and U(t). In a recent study [16], we showed, in the case of N=2 <u>microshorts</u>, that a simple analysis of the <u>unstable</u> fixed points and their invariant manifolds in the phase plane U,X, is very useful in determining parameter values and initial conditions for which fluxon trapping will or will not occur.

In section 2 of this paper we extend our analysis to the case of microresistors and obtain the corresponding conditions for the trapping of fluxons by an array of inhomogeneities with $\mu_i<0$ in (1.4). In section 3, we compare the predictions of our collective coordinate dynamics with the actual solution of the PDE (1.4) and find very satisfactory agreement.

Finally, we end, in section 4, with some remarks on the potential experimental usefulness of the method of collective coordinates in studying fluxon dynamics of long Josephson junctions.

2. COLLECTIVE COORDINATES DYNAMICS OF FLUXON TRAPPING

Following the analysis of the celebrated paper of McLaughlin and Scott [12], we substitute the modulated fluxon solution (1.3) in (1.4) and integrate out (from $-\infty$ to $+\infty$) the x-dependence, to derive finally two coupled ODE's

$$\frac{dU}{dt} = \frac{\pi\gamma}{4}(1-U^2)^{3/2} - \alpha\, U(1-U^2) + \frac{1}{2}(1-U^2)\sum_{i=1}^{N}\mu_i\, \mathrm{sech}^2\theta_i\, \tanh\theta_i \tag{2.1}$$

$$\frac{dX}{dt} = U - \frac{U}{2}\sum_{i=1}^{N}\mu_i\,(X-a_i)\,\mathrm{sech}^2\theta_i\,\tanh\theta_i \quad , \qquad \theta_i = \frac{X-a_i}{(1-U^2)^{1/2}}$$

describing the evolution of the "center" of the fluxon X and its velocity U in time.

Note that at $X=-\infty$, $U=U_\infty=\mathrm{const}$, $X=U_\infty t$, and there is an exact balance between energy loss and energy gain, where

$$U_\infty = [\,1 + (4\alpha/\pi\gamma)^2]^{-1/2} \tag{2.2}$$

is easily obtained from (2.1) as a fixed point solution. But we may now ask : What about other fixed point solutions of (2.1)? Where are they located in the U,X plane and what is their significance?

It turns out that all equilibrium solutions of (2.1) have U=0 and an X coordinate which satisfies

$$0 = \frac{\pi\gamma}{2} + \sum_{i=1}^{N}\mu_i\,\mathrm{sech}^2\,(X-a_i)\,\tanh\,(X-a_i) \tag{2.3}$$

This equation has in general 2N solutions, N of which represent <u>stable</u> fixed points X_i^s, while the remaining N, X_i^u, are <u>unstable.</u>

Thus, each inhomogeneity is accompanied by two fixed points : X_i^s and X_i^u. Let us suppose that all inhomogeneities are placed at equal distances from each other, i.e. $a_i=ia$, $i=1,2,..,N$. If all $\mu_i>0$, the unstable fixed points are very close to the microshorts, i.e. $X_i^u \approx ia$, while the stable ones are nearly at the middle of their separation distance, i.e. $X_i^s \approx ia-a/2$. On the other hand, in the case of microresistors, $\mu_i<0$, it is the stable point which is located near the inhomogeneity, i.e. $X_i^s \approx ia$, while $X_i^u \approx ia+a/2$.

Of central importance in the study of fluxon dynamics of collective coordinates is the location of the separatrices, or invariant manifolds $W_i^{\pm}$, associated with the unstable fixed points X_i^u in the U,X plane.

For example, in the case of 2 microshorts with $\mu_1=\mu_2=0.5$, $\alpha=0.033$, $\gamma=0.02$ and separation distance a=5, the invariant manifolds $W_i^{\pm}$, i=1,2, clearly forbid a fluxon coming from $X=-\infty$ to be trapped by the second impurity, see fig.1a. However, if we increase μ_2 to 0.6, fig.1b shows that W_1^+ and W_2^- "pass through" each other and thus make it possible for a fluxon with velocity within the appropriate area to be attracted by the stable fixed point at $X_2^s \approx 3$.

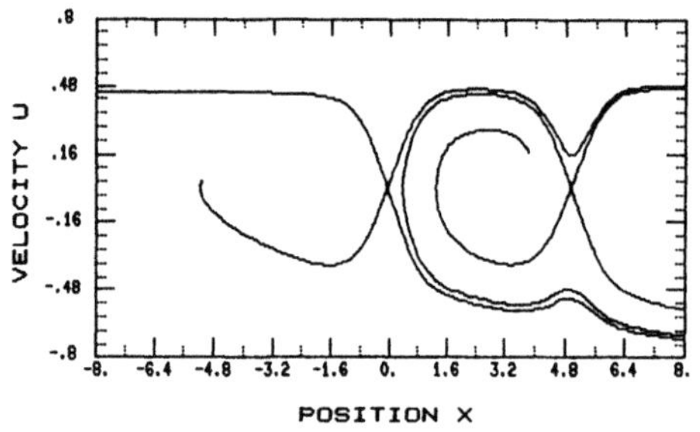

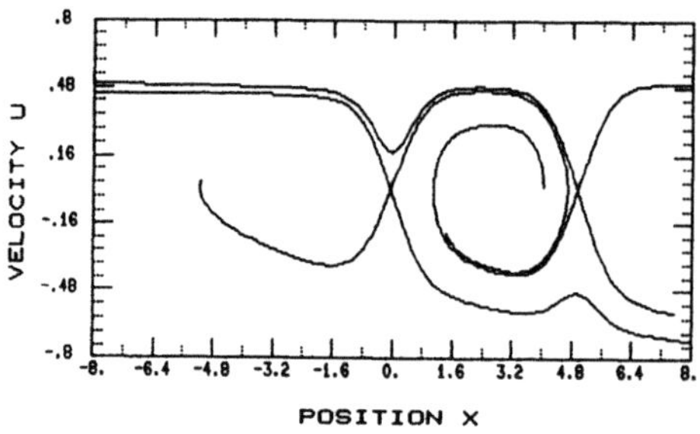

Figure 1 : (a) Stable and unstable invariant manifolds of the X_1^u and X_2^u unstable fixed points of (2.1) at a=5, $\gamma=0.02$, $\alpha=0.033$, $\mu_1=0.5$ and $\mu_2=0.5$. Note that, at this parameter values, orbits starting at $X=-\infty$ cannot be trapped by X_2^s . (b) Same as (a) but with $\mu_2=0.6$. Now orbits coming from $X(0)=-\infty$ can end up at X_2^s provided they lie within the appropriate region.

To understand, at least locally, how the invariant manifolds $W_i^{\pm}$ move around as we vary the μ_i's we can look at the eigenvalues of the linearized equations, near X_i^u, which are [16]

$$\lambda_i^{\pm} = [-\alpha \pm (\alpha^2+4A_iB_i)^{1/2}]/2 \quad , \tag{2.4}$$

$$A_i \equiv \frac{1}{2}\sum_{j=1}^{N} \mu_j \operatorname{sech}^2 (X_i^u-a_j)\,[3\operatorname{sech}^2 (X_i^u-a_j)-2]$$

$$B_i \equiv 1 - \frac{1}{2}\sum_{j=1}^{N} \mu_j\ (X_i^u-a_j)\operatorname{sech}^2 (X_i^u-a_j) \tanh (X_i^u-a_j) \tag{2.5}$$

i=1,2 with $a_1=0$, $a_2=a$. Note that, since $X_i^u \approx a_i$, and a=5 is large enough, $A_i \approx \mu_i/2$, $B_i \approx 1$ and to order α^2 the eigenvalues (2.4) may be approximated by

$$\lambda_i^{\pm} \approx [\ -\alpha \pm (2\mu_i)^{1/2}\ ]\ /2 \qquad i=1,2 \tag{2.6}$$

Calculating now the directions of the eigenvectors $u_i^{\pm}$ at the fixed points X_i^u we obtain

$$u_i^{\pm} = B_i/\lambda_i^{\pm} \approx 2 / [-\alpha \pm (2\mu_i)^{1/2}] \qquad (2.7)$$

which are accurate to within 2% of the exact values.

These formulas (2.7) clearly indicate that by increasing the value of μ_2, the invariant manifold W_2^- (at least close enough to X_2^u) turns in the clockwise direction! This explains why we may thus expect it to "pass through" W_1^+ at some value of $\mu_2>0.5$ and thus allow orbits to enter the basin of attraction of X_2^s, as is indeed observed to happen.

A similar analysis may be performed in the case of microresistors, where the μ_i's are negative. This is, in fact, a case of active experimental interest, as several properties of interactions between fluxons and their emitted plasma waves have been recently studied on Nb-NbO_x-Pb LJJ's with SiO inhomogeneities [14,15].

The main difference in the collective coordinate dynamics of this case (compared with $\mu_i>0$) is that the stable fixed points X_i^s are now located near the inhomogeneities, while the X_i^u have shifted in the space between (and to the right of) the microresistors, see fig.2. Moreover, as is clear from fig.2, in the microresistor case, relatively energetic fluxons, coming from large $X<0$, will not be trapped by the first microresistor at the origin, unless γ takes much smaller values than for the case of microshorts, e.g. $\gamma=0.002$.

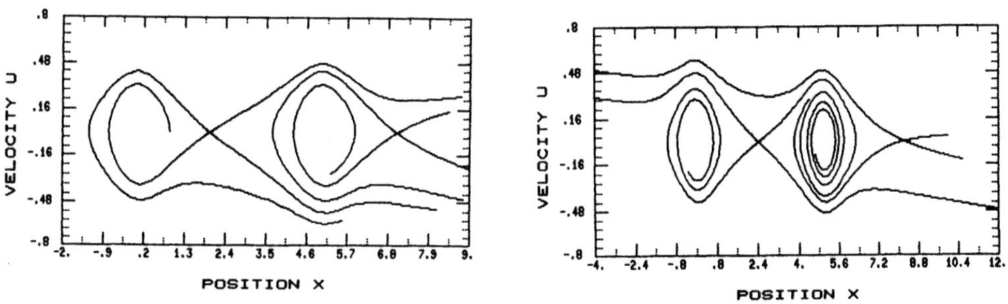

Figure 2 : Stable and unstable invariant manifolds of the X_1^u and X_2^u unstable fixed points of (2.1) at a=5, α=0.033, μ_1=-0.5, μ_2=-0.5 and (a) γ=0.02 , (b) γ=0.002.

In fact, in the case of a single microresistor at the origin, we can obtain an analytical expression for the separatrix, which bounds the trapping region around the origin, in the unforced,, conservative limit $\alpha=\gamma=0$. In that approximation, we can integrate eq. (2.1) to obtain

$$\mathrm{sech}^2[X/(1-U^2)^{1/2}] + 4/\mu(1-U^2)^{1/2} = C \qquad (2.8)$$

where C is an arbitrary constant.

Now, for oscillatory solutions, which intersect the U=0 axis at two points, $\pm X_o$, we have : $\mathrm{sech}^2 X_o + 4/\mu = C$. Since the separatrix is obtained in the limit $X_o \to \infty$, its equation follows directly from (2.8) with $C=4/\mu$:

$$\mu \, \mathrm{sech}^2 [X/(1-U^2)^{1/2}] + 4/(1-U^2)^{1/2} = 4 \qquad (2.9)$$

This formula is quite accurate in the vicinity of the microresistor (i.e for $-2 \le X \le 1$, with $\mu=-0.5$) but deviates significantly from the exact result, as we move further away from the origin.

To study, therefore, the trapping of fluxons by N>2 microresistors we proceed according to the following steps:

a) Place the N inhomogeneities with $\mu_i<0$, at the locations $a_i=ia$, $i=1,2,..,N$.

b) Solve (by a standard Newton scheme) eq.(2.3) for the 2N fixed points X_i^s and X_i^u.

c) Use the resulting X_i^u values to obtain the eigenvalues and eigendirections from (2.4) and (2.7).

d) Starting with initial conditions $U(0) = \pm 0.001$ and $X(0) = X_i^u + U(0)u_i^{\pm}$ integrate eq. (2.1) forward and backward in time to trace out in the U,X plane the invariant manifolds W_i^+ and W_i^- respectively.

Now, suppose we wish to inject fluxons from the $X=-\infty$ end of the junction and have them trapped by the ith microresistor. What we need to do first is to vary the parameters of the system so that the W_i^- traced above will form "corridors" in the U,X plane, through which, fluxons with the appropriate initial velocity U(0)>0 will reach the desired microresistor.

As we discussed earlier, in the case of microshorts, the values of the μ_i's significantly affect the location and "twisting around" of the separatrices. Thus, choosing for example

$$\mu_1=-0.9 \;, \quad \mu_2=-1.0 \;, \quad \mu_3=-1.5 \;, \qquad (2.10)$$

with $\alpha=0.033$ and $\gamma=0.002$, as in fig.3, we can create "corridors" in the U,X plane, through which fluxons (coming from $X(0)=-\infty$ and $U(0)>0$) can be trapped by any one of the 3 inhomogeneities.

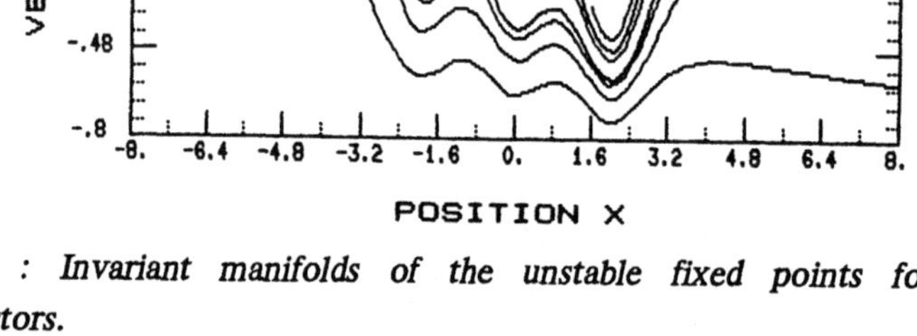

Figure 3 : Invariant manifolds of the unstable fixed points for the case of three microresistors.

Similarly, for N=5, analogous sequences of μ_i values can be found, for which fluxons can be trapped by the microresistors, after following a somewhat complicated path through a "labyrinth" of corridors, formed by the invariant manifolds of the unstable fixed points.

3. NUMERICAL RESULTS OF THE PDE

Naturally, the validity and usefulness of the collective coordinate approach can be estimated only when its predictions are compared against the solution of the PDE of the problem (1.4) (which also needs to be checked ultimately, of course, by performing the actual experiment!).

Thus, we have used a numerical scheme introduced by S. Pagano [17] to solve equation (1.4) for different parameter values α,γ and μ_i, taking as initial condition a fluxon of the form (1.2) approaching the inhomogeneities from the left end of the junction.

Starting, for example, with two microshorts $\mu_1=\mu_2=0.5$ located at a distance a=5 from each other, and setting $\alpha=0.033$ and $\gamma=0.02$, we have verified the predictions of fig.1 of this paper (and of [16]): Injecting a fluxon from the left, with U=0.4, trapping by the first microshort was indeed observed, while with U=0.45, the fluxon was seen to pass them both by, on its way to the right end of the junction.

On the other hand, setting $\mu_2=0.6$ and starting the fluxon with velocities $0.45\leq U\leq 0.52$ we found that it became trapped by the second microshort (while for $U\geq 0.53$ it escaped from both of them), exactly as predicted by our collective coordinate ODE solutions.

We then proceeded to test the microresistor case with $\mu_i<0$. One of our first observations was that the interaction between fluxon and inhomogeneities was a lot more dramatic than in the $\mu_i>0$ case : In particular, there was a considerable amount of radiation emitted from the fluxon in the form of "plasma waves", which kept going back and forth along the junction, dying away as $t\to\infty$. These waves became significantly more prominent as the values of the $|\mu_i|$'s were increased.

Furthermore, as is evident from fig.4a, the overall <u>shape</u> of the fluxon also gets distorted by developing smaller peaks centered at the sites of the inhomogeneities. Thus, even though, in fig.4a, its main part is located about the 3rd microresistor (as predicted by the collective coordinate dynamics of fig.3) significant parts of it have "spilled over" the separatrices and have been attracted by the 1st and 2nd microresistor.

We have performed a similar experiment with microresistors with smaller "strengths" $|\mu_i|$

$$\mu_1=-0.6 \quad , \quad \mu_2=-0.4 \quad , \quad \mu_3=-0.9 \tag{3.1}$$

Comparing with (2.10), the emitted plasma waves here were considerably weaker, while the peaks near the first and second inhomogeneity were significantly smaller, see fig. 4b. Again, there was reasonable agreement between the predictions of the collective coordinate approach and the time evolution of the fluxon as obtained from the solution of the PDE.

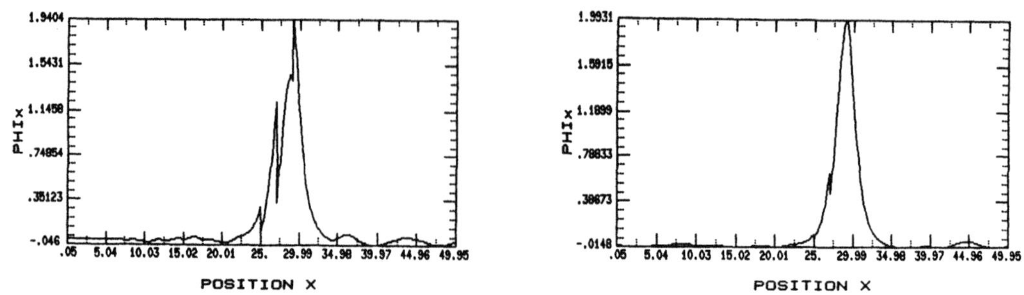

Figure 4 : The φ_x versus space for t=135 as results from the numerical integration of the PDE equation (1.1) for the three impurities case (see figure 3), located at X=25,27,29 in this figure, (a) μ_1=-0.9, μ_2=-1.0, μ_3=-1.5, (b) μ_1=-0.6, μ_2=-0.4 and μ_3=-0.9. In both (a) and (b) the fluxon has been trapped by the third impurity.

4. CONCLUDING REMARCS

In this paper we have studied the behavior of a single fluxon travelling in a long Josephson junction with inhomogeneities, from the point of view of the dynamics of its collective coordinates X(t) and U(t).

We have thus discovered that a simple analysis of unstable fixed points and their invariant manifolds in the U,X plane can provide us with very useful information concerning the different trapping possibilities of the fluxon. In particular, varying the inhomogeneity strengths μ_i in several examples, we were able to trace out "corridors" in the U,X plane though which fluxon trapping could occur near any one of an array of inhomogeneities.

Ultimately, of course, the important question is how much of this information represents an accurate description of what really happens to the fluxons as solutions of the PDE of the problem (1.4). To find out, we performed a number of computer experiments and obtained the following results:

First, in the case of two microshorts there was very satisfactory agreement between the ODE predictions and the corresponding numerical solution of the PDE. The situation was also helped by the fact that microshorts do not significantly distort the shape of the fluxon and do not cause serious radiation effects.

On the other hand, the case of microresistors was more complicated. Although the results of the ODE's, in the case of 3 inhomogeneities were generally verified by the PDE, there were instances where the fluxon was finally trapped by a different inhomogeneity than predicted by the collective coordinate approach.

The disagreement can be easily explained by the fact that the fluxon, due to its interaction with the microresistors, suffers not only radiation losses but also a significant distortion of its shape, causing parts of it to "spill over" the separatrix boundaries and form smaller peaks near several microresistors.

In conclusion, therefore, in order to make the predictive power of the ODE's as strong as possible, in the experimentally interesting case of trapping by microresistors, it seems appropriate to reduce first the "strengths" $|\mu_i|$ of the inhomogeneities just above the level of the existence of their stable fixed points. Then, spacing out the inhomogeneities and varying the μ_i in the manner indicated above, it should be possible to create large "corridors" in the U,X plane through which a fluxon can be trapped by any one of the available microresistors.

We are currently working on a more detailed quantitative comparison between the dynamics of collective coordinates and the solution of the PDE in the $\mu_i<0$ case. Further results in this direction, as well as a corresponding analysis of the effect of fluxon trapping on the I-V characteristics of the LJJ will appear in future publications [18].

5. ACKNOWLEDGEMENTS

Partial support for this work has been provided by a grant from the Ministry of Industry, Energy and Technology of Greece and the E.C. Stimulation grant ST2J-0267-J-C-(A). One of us (T.B.) wishes to acknowledge the hospitality and support of the Research Center of Crete, where the major part of this work was accomplished.

6. REFERENCES

[1] A, Barone and G. Paterno, "Physics and Application of the Josephson Effect" (Wiley, Interscience, New York, 1982).

[2] K.K. Likharev, "Dynamics of Josephson Junction and Circuits" (Gordon and Breach, New York 1986.

[3] J.G. Bednorz and K.A. Muller, Z.Phys. B64, 189 (1986); M.K. Wu et al, Phys. Rev. Lett. 58, 908 (1987).

[4] M.J. Ablowitz and H. Segur, "Solitons and the Inverse Scattering Transform" (SIAM Studies in Appl. Math. Philadelphia, 1981); "Solitons", S.E. Trullinger, V.E. Zakharov and V.L. Pokrovsky (eds), in Modern problems in Condensed Matter Sciences, Vol. 17, (North Holland, Amsterdam 1986).

[5] P. Cvitanovic, ed. "Universality in Chaos" (Hilger Bristol, 1984).

[6] H.G. Schuster, "Deterministic Chaos" (Physik Verlag, Weinheim, 1986).

[7] N.F. Pedersen, Physica Scripta, vol. T13, 129 (1986).

[8] P.L. Christiansen and N.E. Pedersen, in "Singular Behaviour and Nonlinear Dynamics" St. Pnevmatikos, T. Bountis and Sp. Pnevmatikos eds. (World Scientific Singapore, 1989).

[9] G. Reinisch and J.C. Fernandez, in same volume as ref.8.

[10] M.B. Fogel, S.E. Trullinger, A.R. Bishop and J.A. Krumhansl, Phys. Rev. Lett. 36 1411 (1976) ; see also Phys.Rev. B15, 1578 (1977).

[11] J.P. Keener and D.W. McLaughlin, Phys.Rev. A16, 777 (1977); see also J.Math.Phys. 18, 2008 (1977).

[12] D.W. McLaughlin and A.C. Scott, Phys. Rev. A18, 1652 (1978).

[13] Y.S. Kivshar and B.A. Malomed, Phys. Lett. 111A (8,9), 427 (1985); Phys. Lett. 118A(2), 85 (1986); Phys. Lett. 119A(5), 237 (1986); Phys. Lett. 129A(8,9), 443 (1988).

[14] I.L. Serpuchenko and A.V. Ustinov, JETP Lett. 46(11), 549 (1987).

[15] A.A. Golubov, I.L. Serpuchenko and A.V. Ustinov, JETP 94, 297, (1988).

[16] T. Bountis and St. Pnevmatikos, Phys. Lett. A, submitted, (1989).

[17] S. Pagano, private communication, (1989).

[18] St. Protogerakis, St. Pnevmatikos and T. Bountis, in preparation, (1989).

DYNAMICAL REGIMES PHASE-LOCKED TO AN EXTERNAL MICROWAVE FIELD IN A LONG, UNBIASED JOSEPHSON JUNCTION

J.C. Fernandez, R. Grauer* and G. Reinisch
Observatoire de la Cote d'Azur
Boîte Postale No. 139
F-06003 Nice Cédex, France

We demonstrate the existence of phase-locked limit cycles in inhomogenously driven sine-Gordon systems and point out their possible experimental verification by use of Josephson devices.

INTRODUCTION

Consider the following partial differential equation (PDE):

$$\phi_{tt} - \phi_{xx} + \sin\phi = \chi - \alpha\phi_t + \epsilon\cos\kappa x\cos\Omega t \tag{1}$$

with reflective boundary conditions $\phi_x(x) = 0$ at $x = \pm L/2$. A long Josephson junction (LJJ) allows the propagation of magnetic fluxons of soliton type, according to the above PDE with $\epsilon = 0$. Here ϕ means the phase difference between the two macroscopic (coherent) wave functions which describe the state of each superconducting layer – and therefore varies from zero to 2π (kink) or 2π to zero (antikink) modulo 2π, χ is a constant current flowing through the LJJ (the so called "bias current"), $\epsilon\cos\kappa x\cos\Omega t$ describes a spatially modulated ($\kappa = \pi/L$) a.c. external field of small amplitude ϵ and of (microwave) frequency Ω, and the last term in the r.h.s of equation (1) is an acceptable approximation of resistive damping effects present in the junction [1-3].

With the exception of the function ϕ itself, all variables and parameters of the PDE (1) are normalized to characteristic quantities related to the basic physical processes which occur in a LJJ. Hence the time t is given in units of the reciprocal LJJ plasma frequency, the (one-dimensional) space variable x and the LJJ length L are given in units of the Josephson penetration depth, the bias χ and the microwave field amplitudes ϵ are scaled with respect to the maximum Josephson current density, while the a.c. field frequency Ω and the wave vector κ are respectively given in units of the LJJ plasma frequency and of the reciprocal Josephson penetration depth. The parameter α^{-2} is the McCumber number (see ref. [1] for the equations defining the above characteristic physical quantities).

In absence of any external microwave field ($\epsilon = 0$), there exist well-known stationary shuttling dynamical regimes of the driven fluxons, which actually describe the power balance between energy feeding due to the bias and energy losses due to the damping [4]. Such regimes led to the first experimental evidence of the presence of fluxons effects inside a LJJ, as they result in the so-called "zero-field-steps" (ZFS) in the corresponding current-voltage-characteristics [5].

Quite naturally, one then has considered the additive effects of the a.c. field to the above ZFS regimes. The aim of such researches was the phase locking of two – and even more –

* Present address: Dept. of Mathematics, University of California of Santa Barbara, Santa Barbara, CA 93106, USA

LJJ's into an unique, coherent, "superradiant" regime characterized by a linewidth which is reduced by a factor n^{-1}, while the amplitude of the resulting radiation increases like n^2, if there are n LJJ's building the array [6-8].

Concerning such superradiant LJJ arrays, the present state of the experimental art is very promising [9]. To the contrary, the complete theoretical understanding of the precise phase-locking effects is far from being achieved. The reason might partially be due to the basic qualitative difference which exists between a d.c. driven – and an a.c. driven phase-locking regime. While a d.c. driven fluxon reverses its parity at each reflection occuring at either end of the LJJ – and therefore is driven by a newtonian force which always acts in the direction of the (anti)fluxon motion, such is no more the case if the external field is allowed to vary within one fluxon shuttling period.

One possible way to simplify this problem is to consider the a.c. driven field effect *alone*. Hence we assume $\chi = 0$. The aim of the present letter is therefore to give a preliminary description of such a phase-locked fluxon regime to an external, inhomogeneous, a.c. field in absence of any bias current, and to suggest its experimental verification. Actually, we first considered a homogeneous external a.c. field, corresponding $\kappa = 0$. It appeared that the limit "C"-cycles obtained in this situation were unstable, while such is no more the case if the external a.c. driving force is allowed to decrease from the LJJ center to its boundaries. Therefore we here assume a standing wave electric field pattern in the superconducting transmission line (1) such that the wave velocity Ω/κ equals the Swihart velocity (which is unity in the dimensionless units related to the PDE (1) [1].

THE NUMERICAL RESULTS

We have considered the following set of (reduced) parameters:

$$\chi = 0 \ , \quad \epsilon = 0.2 \ , \quad \Omega = 0.06 \ , \quad \alpha = 0.01 \ , \quad L = 52.4 \ , \quad \kappa = \pi L^{-1} = \Omega \ , \tag{2}$$

and performed numerical simulations of the PDE (1), using the method of characteristics. In a real experimental situation, it is impossible to control the initial data corresponding to the position and velocity of the fluxon when the a.c. field is switched on. Nor it is possible to control the parity of this initial fluxon. Hence we performed a series of numerical simulations which were supposed to cover a reasonably extensive range of such arbitrary initial data:

$$\phi(x,0) = 4\tan^{-1}\exp[-\gamma(x-x_0)] \tag{3}$$

$$\phi_t(x,0) = -v\phi_x(x,0) \ ; \quad \gamma = (1-v^2)^{-\frac{1}{2}} \tag{4}$$

$$v = -0.9 + n(0.2); 0 \le n \le 9 \ ; \quad x_0 = 14.5 \quad . \tag{5}$$

The PDE (1) was numerically integrated up to $t = 20000$ in all cases. Defining respectively the fluxon position $Y(t)$ and the velocity $\dot{Y}(t)$ according to the following formulas [10]:

$$\phi_x(Y,t) = \max_{-\frac{1}{2}L \le x \le +\frac{1}{2}L} [|\phi_x(x,t)|] \ , \ \dot{Y} = -\frac{\phi_t}{\phi_x}(x = Y, t) \quad , \tag{6}$$

and the fluxon parity σ ($\sigma = +1$: antikink; $\sigma = -1$: kink), then displaying the fluxon dynamics in its phase space $\{Y, \dot{Y}\}$ taking into consideration the parity, the numerical results clearly show the presence of two attractive limit "C-cycles" phase-locked to the external a.c. field. Each of these two C-cycles may be obtained from the other by the symmetry of the system $Y \to -Y$; $\dot{Y} \to -\dot{Y}$; $\sigma \to -\sigma$ (see next part). Moreover, each such C-cycle is actually described twice within a period of the external field, half a period π/Ω as a (say) antikink, and then as a kink. Therefore, if we only consider the (anti)soliton position and velocity, in the phase space $\{Y, \dot{Y}\}$

without taking into account the parity of the fluxon, the corresponding limit C-cycle displays a phase-locking of the system to the external a.c. driven frequency Ω with a ratio equal to 2. Then, the a.c. driven and damped (anti)fluxon is bouncing against a *single* LJJ end at each half-period of the external field. Therefore, the emission of the electromagnetic pulses by the (anti)fluxon at each reflection should occur at the frequency 2Ω. This property should lead to an unambigous experimental evidence of such an a.c. driven and damped (anti)fluxon phase-locked regime.

Let us now detail the numerical results obtained from the representative set of arbitrary initial conditons (3-5). For $n = 0$, we reach the "left" C-cycle ($Y \leq 0$), while for $n = 1$ to $n = 9$, we reach the "right" C-cycle ($Y \geq 0$). Note that, starting from other initial conditions, the fluxon may accidentally reach eather LJJ end with a kinetic energy below the annihilation threshold [11] during the transient phase (remember that the reflective boundary conditions mean a collision with a virtual antisoliton). As a result, the subsequent (half) breather would then decay to a final spatially flat state oscillating only with time.

The phase-locking of the system to the external driven frequency Ω with a ratio 2 in the $\{Y, \dot{Y}\}$ plane is displayed by the Poincaré sections of the numerical PDE results, starting at $t = 0$, at successive time intervals equal to π/Ω: see Figure 1.

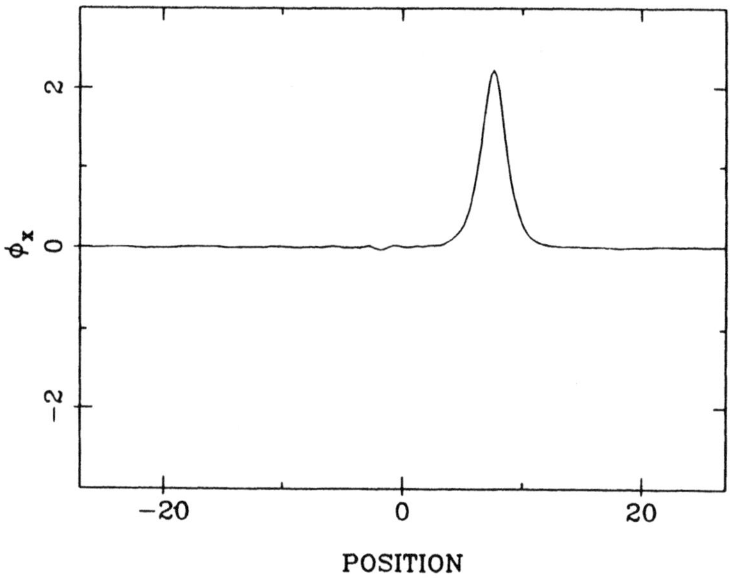

FIGURE 1

The asymptotic kink displayed from $t = 4000$ to $t = 20000$ in the Poincaré section of the PDE results corresponding to formulas (2–5) and $n = 1, 9$.

For $n = 0$, this mapping shows in the asymptotic regime ($t \geq 4500$) alternatively a kink and an antikink being both located at $Y_\infty = -0.7282\ldots$, with a velocity $\dot{Y}_\infty = -0.4569\ldots$. For $n = 1$ to $n = 9$, we obtain the corresponding sequence of kinks and antikinks with the *same* asymptotic fixed point given by its two coordinates: $Y_\infty = 0.7719\ldots$ and $\dot{Y}_\infty = 0.4552\ldots$. For each above series of results, we checked that the corresponding asymptotic Poincaré section points $\{Y_\infty, \dot{Y}_\infty\}$ remained invariant after a time of order 4000 up to the resolution of the computer (calculations were done in double precision).

These properties clearly prove that the final state of the system is periodic (phase-locked).

A FIRST THEORETICAL APPROACH

The ansatz of a one-degree of freedom (anti)kink

$$u(x,t) = 4\tan^{-1}\exp[-\sigma\frac{x - Y(t)}{\sqrt{1-\dot{Y}^2}}] \tag{7}$$

($\sigma = +1$: antikink ; $\sigma = -1$: kink) into the original PDE (1) and the projection of the resulting equation onto the soliton Goldstone mode as detailed in refs [10] and [12] leads to the following equation of motion:

$$\ddot{Y} = \frac{\pi\sigma}{4}(1-\dot{Y}^2)^{\frac{3}{2}}[\epsilon\cos\kappa Y\cos\Omega t + \chi] - (1-\dot{Y}^2)\alpha\dot{Y} \ . \tag{8}$$

This equation is also obtained by other techniques and actually describes the newtonian dynamics of a sine-Gordon kink [13].

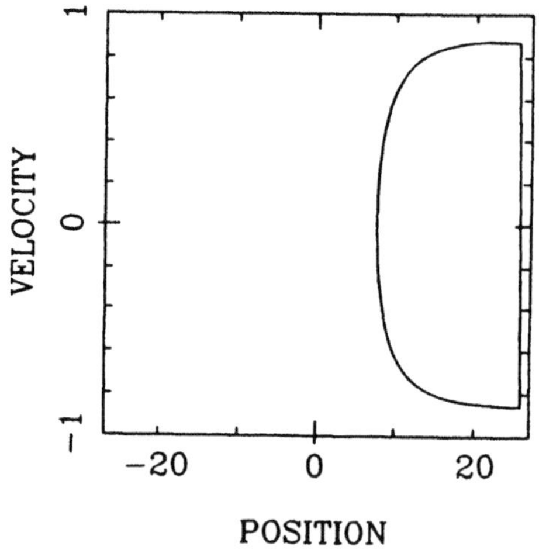

FIGURE 2

The right-sided limit C-cycle obtained from the initial conditions $n = 1$ to $n = 9$, after a transient time of order 4000, displayed in the $\{Y, \dot{Y}\}$ phase-space.

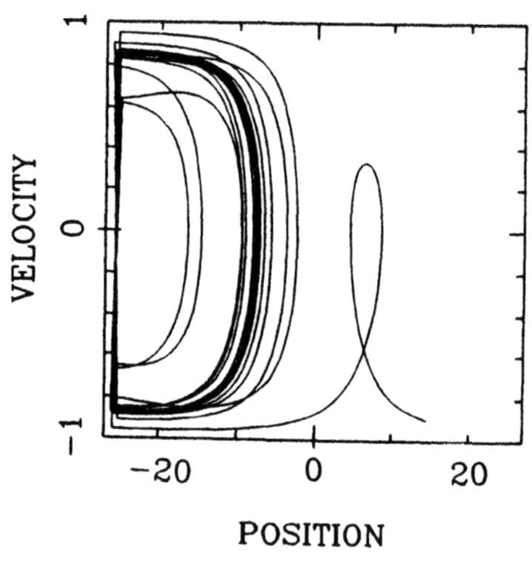

FIGURE 3

The whole fluxon dynamics corresponding to $n = 0$, displayed from $t = 0$ to $t = 5000$ in the $\{Y, \dot{Y}\}$ phase-space.

The ordinary differential equation (ODE) (8) allows to recover quite well the above 2Ω phase-locked C-trajectories (see Fig. 2) including also the transient behaviour (Fig. 3), once the following boundary conditions are used (− and + respectively mean the value of the function before and after the reflection at $Y = \pm L/2$):

$$\dot{Y}_+ = -\dot{Y}_- \left[1 - \frac{\pi^2 \alpha (1 - \dot{Y}^2)^{\frac{3}{2}}}{4\dot{Y}^2}\right] \tag{9}$$

$$Y_+ = sign(Y)[\frac{1}{2}L + 2(1 - \dot{Y}^2)^{\frac{1}{2}} \log |\dot{Y}|] \ . \tag{10}$$

The discrepancy between the PDE- and the corresponding ODE results is less than 3%. The conditions (9–10) respectively describe *in terms of the collective-coordinate* $Y(t)$ **both the energy loss experienced by an (anti)kink as it is reflected (i.e. while it inelastically collides with its virtual antikink) and the simultaneous phase shift [3],[13-15].**

FIGURE 4a

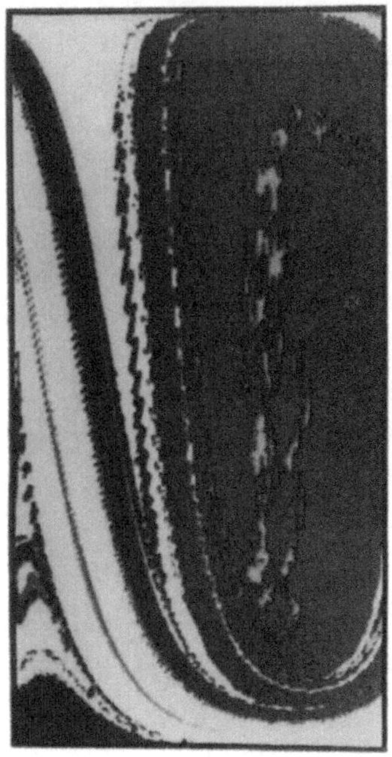

FIGURE 4b

The parameters are: $L = 52$; $\epsilon = 0.2$; $\kappa = 0.06$; $\Omega = 0.06$; $\alpha = 0.02$. **Starting with an initial condition given by, we reach either an asymptotic kink C-cycle if this initial condition is located in the dark area of the phase space, or an asymptotic antikink C-sycle if it is located in the pale area; the grey area represents the part of the phase space which lies outside any of the two above attraction basins.** $\chi = 0$ **for figure 4a and** $\chi = 0.02$ **for figure 4b.**

The interest of the above-described relevant reduction of the original PDE system into an ODE one allows to use the very powerfull "cell-mapping techniques", which basically consists in transforming a long-time range ODE numerical integration into a series of iterations, once the original phase-space $\{Y, \dot{Y}\}$ has first been approximated by a grid of elementary cells [16]. The huge gain of computing time then allows to consider all these cells – i.e., in our approximation, the whole phase-space – as possible initial conditions for the ODE system, and, hence, to obtain numerically a phase portrait of the system. Figures (4-5) displays the basins of attraction of the kink and antikink C-cycle, for two different values of the control parameter χ. Figure (6) displays the percentage (normalized to unity) of the whole phase-space covered by these attraction basins, as a function of χ for three different values of the damping α, and may therefore be considered as a probability of reaching a C-cycle attractor, when starting from random initial conditions.

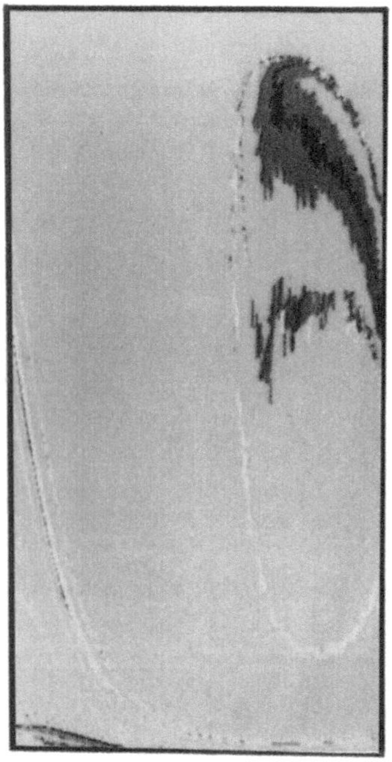

FIGURE 5

The parameters are the same as for figures 4, except that $\chi = 0.04$. Starting with a kink initial condition, the grey area of this phase portrait represents the attraction basin of the antikink C-cycle, the dark area is the attraction basin of the kink C-cycle, while a kink initial condition located in the (large) pale area does not lead to any C-cycle.

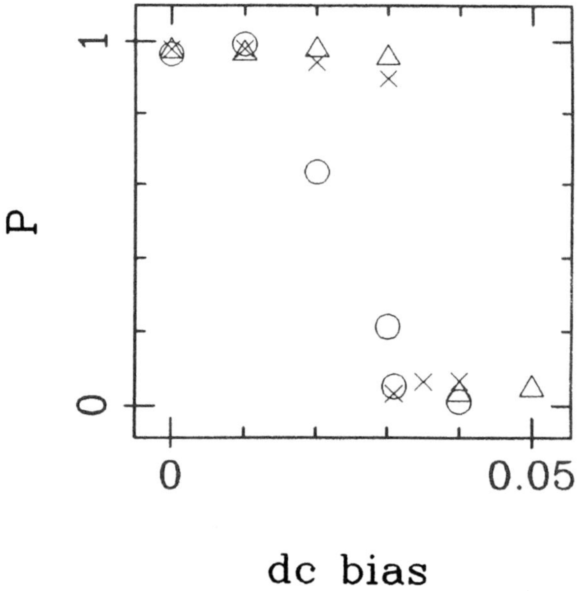

FIGURE 6

Percentage (normalized to unity) of the phase space area leading to a C-cycle versus dc bias value χ. The "o" markers correspond to $\alpha = 0.01$, the "x" markers to $\alpha = 0.02$ and the "$\triangle$" markers to $\alpha = 0.03$.

Finally, note the following symmetry which keeps the equation of motion (8) invariant:

$$\begin{aligned} Y \to \tilde{Y} = bY \ ; \ t \to \tilde{t} = bt \ ; \ L \to \tilde{L} = bL \ ; \ \kappa \to \tilde{\kappa} = \kappa b^{-1} \ ; \\ \Omega \to \tilde{\Omega} = \Omega b^{-1} \ ; \ \chi \to \tilde{\chi} = \chi b^{-1} \ ; \ \epsilon \to \tilde{\epsilon} = \epsilon b^{-1} \ ; \ \alpha \to \tilde{\alpha} = \alpha b^{-1} \ . \end{aligned} \tag{11}$$

Hence the kink velocity $\dot{Y}$ remains unchanged, and if an asymptotic limit cycle is obtained for, say, the range of parameters (2-5), the existence of other, similar limit cycles can be deduced from the above rescaling, provided $\tilde{\epsilon} \ll 1$. The particular case $b = -1$ is equivalent to the symmetry $Y \to -Y$; $\dot{Y} \to -\dot{Y}$; $\sigma \to -\sigma$ in equation (8), which was already mentioned in part 2 (cf. equation (8)).

This work was performed in the frame of the EEC "Stimulation" contract ST2-0267-J-C. Rainer Grauer was supported by this contract. It is a pleasure to acknowledge interesting discussions with N.F. Pedersen, M.J. Goupil and M.P. Soerensen.

REFERENCES

[1] A. Barone and G. Paterno, *Physics and applications of the Josephson effect*, (John Wiley, New York, 1982) and references therein.
[2] N.F. Pedersen, *Solitons in long Josephson junctions* in *Advances in Superconductivity*, edited by Deaver and Ruvolds (Plenum, New York, 1982).
[3] N.F. Pedersen, *Solitons in Josephson transmission lines* in *Solitons*, edited by Trullinger, Zakharov and Pokrovsky, (Elsevier Sc. Pub. 1986).
[4] T.A. Fulton and R.C. Dynes, Solid State Comm. *12*, 57 (1973).
[5] J.T. Chen, T.F. Finnegan and D.N. Langenberg, ICSS 413 (1969).
[6] T.F. Finnegan and S. Wahlsten, Appl. Phys. Lett. *21*, 541 (1972).
[7] R. Bonifacio, Phys. Lett. *A 101*, 427 (1984).
[8] M. Cirillo, S. Pace and S. Pagano, Phys. Lett. *A 125*, 20 (1987).
[9] R. Monaco, private communication (1988).
[10] J.C. Fernandez, M.J. Goupil, O. Legrand and G. Reinisch, Phys. Rev. *B 34*, 6207 (1986).
[11] N.F. Pedersen and D. Welner, Phys. Rev. *B 29*, 2551 (1984).
[12] O. Legrand, Phys. Rev. *A 36*, 5068 (1987).
[13] D.W. McLaughlin and A.C. Scott, Phys. Rev. *A 18*, 1652 (1978).
[14] N.F. Pedersen, M.R. Samuelsen and D. Welner, Phys. Rev. *B 30*, 4057 (1984).
[15] D. Welner, *Boundary effects in sine-Gordon systems*, Ph.D. Thesis, (Lyngby 1985).
[16] F. Varosi, C. Grebogi and J.A. Yorke, Phys. Lett. *A 124*, 59 (1987).

EXISTENCE AND STABILITY OF DRIVEN AND DAMPED PHASE-LOCKED NLS-SOLITONS AND SG-BREATHERS

K.H. SPATSCHEK, M. TAKI *, and TH. EICKERMANN

Institut für Theoretische Physik I
Heinrich-Heine-Universität Düsseldorf
D-4000 Düsseldorf
F.R. Germany

Abstract: The spatio-temporal behaviors of possible solutions of driven and damped nonlinear Schrödinger and sine-Gordon equations are analyzed. In the first part, the nonlinear Schrödinger equation is considered also as the limiting case corresponding to weak driving and damping in the sine-Gordon equation, with driving frequency close to one. It is shown that in this limit a phase-locked breather with $m/n = 1/1$ exists and undergoes period-doubling bifurcations. The close analogy to the dynamics of a phase-locked Schrödinger soliton is demonstrated in detail. In the second part, for smaller driving frequencies, the existence and stability of subharmonically phase-locked breathers is discussed. The quasi-periodic route to chaos is re-investigated and special attention is given to the length-dependence of the results. The findings are compared with predictions of simpler models in nonlinear dynamics. The whole paper consists of analytical and numerical results. For the latter, improved diagnostic tools have been developed.

1 INTRODUCTION

Within the fascinating area of nonlinear dynamics many unsolved problems exist. They reach from discrete mappings and nonlinear ordinary differential equations (ODE's) up to complicated partial differential equations (PDE's). This paper is concerned with a physically motivated problem of the latter type. In physics, very often complicated systems can be modelled by a relatively simple nonlinear PDE, e.g. the cubic nonlinear Schrödinger (NLS) equation for Langmuir waves, the Korteweg-deVries (KdV) equation for sound-waves, the sine-Gordon (SG) equation for the phase-difference in the Josephson transmission lines or the time-integrated electric field amplitude for light-pulse propagation in a 2-level atom system, respectively, etc. In the past, much work has been devoted to the integrable Hamiltonian systems [1]. A dynamical system is Hamiltonian, if it is possible to identify generalized coordinates (q) and momenta (p), and a Hamiltonian $H(q,p,t)$, such that the equations of motion of the system can be written in the form

$$\frac{\partial q}{\partial t} = \frac{\delta H}{\delta p} \quad , \quad \frac{\partial p}{\partial t} = -\frac{\delta H}{\delta q} \,. \tag{1.1}$$

Nontrivial integrable examples are the NLS equation $iq_t + q_{xx} \mp 2|q|^2 q = 0$ for $H = -i\int dx \, \{q_x p_x + (qp)^2\}$ and $p(x,0) = \pm q^*(x,0)$, as well as the SG equation $q_{tt} - q_{xx} + \sin q = 0$ for $H = \int dx \left\{\frac{1}{2}p^2 + \frac{1}{2}q_x^2 + 1 - \cos q\right\}$. PDE's of a certain general form (which we may call soliton equations in the AKNS form [1]) are integrable. They represent infinite dimensional Hamiltonian dynamical systems, in which q and p play the roles of conjugate variables, and which are solvable by the inverse scattering transform (IST). The latter makes use of a generalized scattering problem and the corresponding scattering data. There is a subset S of scattering data from which the rest of the scattering data can be reconstructed. It has been shown that the mapping $(q,p) \to S$ is a canonical transformation and that the conjugate variables in S (i.e. Q and P) are of action-angle type, i.e., $H = H(P)$ so that

$$\frac{\partial P}{\partial t} = 0 \quad , \quad \frac{\partial Q}{\partial t} = \frac{\delta H}{\delta P} = const. \tag{1.2}$$

However, when for physical reasons driving and damping terms are taken into account, the above examples turn over into non-integrable dissipative systems. Our understanding of such system in the near-conservative region (small driving and damping) is still low, however increasing during the past years [2]. In the near-conservative region the fascinating phenomenon of soliton appearence (at least well-understood in the integrable cases) helps us to construct a link between the complicated spatio-temporal developments and the conventional theory of nonlinear dynamical systems such as mappings and ODE's [3,4]. At this stage it is interesting to note that the existence of soliton solutions does not depend on the strength of a coupling parameter g (so long as the latter does not vanish; soliton solutions are singular when $g \to 0$). Thus, even in the case of weak coupling it is not possible to neglect the soliton solutions. This paper is devoted to that topic. To be more specific, we shall investigate the dynamical behavior of a driven and damped SG equation [5-7]

$$u_{tt} - u_{xx} + \sin u = -\alpha u_t + \Gamma \sin \Omega t \,, \tag{1.3}$$

where the driving frequency Ω is close to one, and the damping coefficient α as well as the driver amplitude Γ are small. We shall distinguish between two generic cases: First, the Schrödinger limit, when the first harmonic [$\sim \exp(i\Omega t)$] is dominating in the solution for u, and secondly, the still near-conservative state but with subharmonic phase-locking. Because of the concentration on vanishing boundary conditions – we are aware of the fact that there are many others – we anticipate the breather as the dominating nonlinear object. Of course, for the parameters of interest and the chosen initial conditions, we shall prove this conjecture.

2 The Nonlinear Schrödinger Limit

In this part we discuss the connection of the SG equation with the NLS equation in the presence of a driver and damping. Thus, we first summarize, and supplement by some new results, what is known for the NLS equation [8].

2.1 The driven and damped NLS equation

Let us start with the equation

$$iq_t + q_{xx} + 2 \mid q \mid^2 q = -i\gamma q - ia\ e^{i\omega t}, \tag{2.1}$$

which has many applications in plasma physics and nonlinear optics. First we note that the explicit time-dependence through the driving term $-ia\exp(i\omega t)$ can be transformed out; then without loss of generality we can set $\omega = 1$. For $\gamma \neq 0$ (dissipative case) or $a \neq 0$ (driven case), Eq.(2.1) is non-integrable. This can be checked by the so-called Painlevé test. Indeed, one can find a reduction so that the corresponding ODE has movable critical points. On the other hand, when we consider the space-independent ($\partial_x \equiv 0$) situation, we have effectively a two-dimensional autonomous system which cannot exhibit chaos. This is important since it means that any chaos in the system will not be induced by the boundaries when $q_x \to 0$ for $x \to \pm\infty$ is imposed.

For

$$\frac{2\gamma\omega^{1/2}}{\pi \mid a \mid} < 1 \tag{2.2}$$

Eq. (2.1) has a nontrivial phase-locked soliton-like solution. This has recently been shown numerically as well as analytically by perturbation theory [2]. (We shall come back to this procedure in part 3.) Here we present some plausibility argument for (2.2). When writing $q = (r + c)e^{i\phi_c}$ where $c\exp(i\phi_c)$ is the flat (x-independent) solution of Eq. (2.1), and treating γ, a, and c as perturbation terms, we find for the "mass" $I_0 = \int \mid r \mid^2 dx$ and the "momentum" $I_1 = \int (rr_x^* - r^* r_x)\, dx$ of the soliton-like part $r = \mid r \mid \exp(i\phi)$ the now violated "conservation laws"

$$I_{0t} = -2\gamma I_o + 4c \int \sin\phi \mid r \mid^3 dx + 4c^2 \int \sin 2\phi \mid r \mid^2 dx, \tag{2.3}$$

$$I_{1t} = -2\gamma I_1. \tag{2.4}$$

The second equation (2.4) tells us that the soliton-like state should have zero momentum, whereas from the time-averaged equation (2.3) the phase-looking condition $\omega = 4\eta^2$ follows. When performing the integrals on the r.h.s. of Eq. (2.3) by introducing

$$r \approx 2\eta \exp\left[-i\left(\psi_0 - \frac{\pi}{2}\right)\right] \operatorname{sech}[2\eta(x - x_0)] \tag{2.5}$$

we can conclude that only in the case (2.2) a soliton-like object with finite mass is possible.

Using a semi-implicit unitary Crank-Nicholson scheme, we have solved Eq. (2.1) numerically. The general picture, first reported by Nozaki and Bekki [8], has been confirmed. When (2.2) is satisfied,

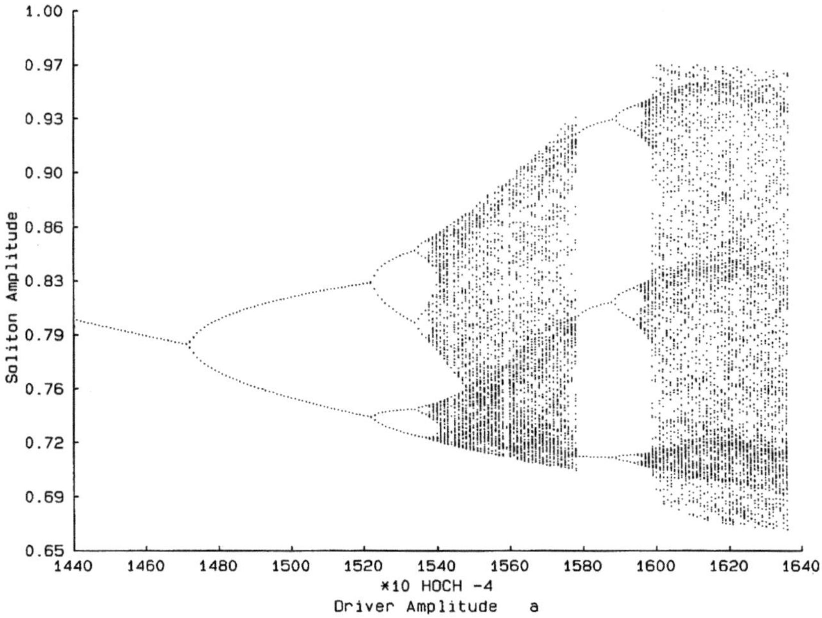

Fig.1 Bifurcation diagram for $\gamma = 0.11$ after the first instability (Hopf bifurcation). Note that the single line on the left corresponds to the limit cycle described in the text. Period-doubling, chaos and subharmonic windows are clearly shown.

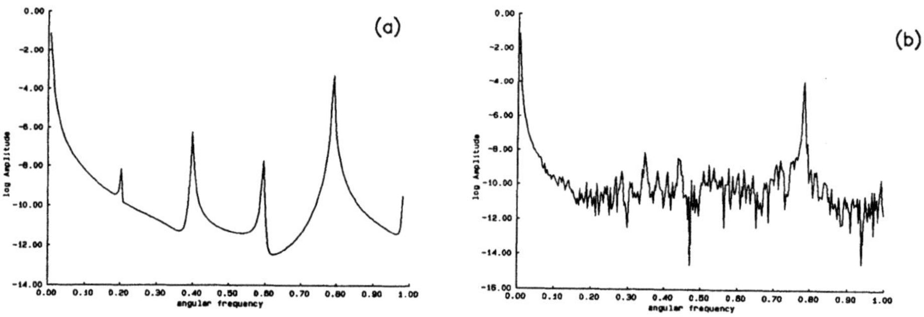

Fig.2 Power spectra for (a) the period-4 and (b) the chaotic state

a phase-locked soliton exists. E.g., when $\omega = 1$ and $\gamma = 0.11$ the phase-locked soliton exists in the region $0.07 \lesssim |a| \lesssim 0.1$. Then it becomes unstable via a Hopf bifurcation.

We call the new state a limit cycle. The reason is that it is actually an asymptotic attracting cycle in an appropriate phase-portrait. At a critical value of the amplitude (for fixed damping γ) the system undergoes a series of torus-doubling bifurcations. The scenario is similar to that known from the logistic map. More details are shown in Fig.1. When using the numerical values, $a_1 = 0.1472$, $a_2 = 0.15228$, $a_3 = 0.15350$, $a_4 = 0.15377, ...$ we can determine $\delta_n := (a_n - a_{n-1})/(a_{n+1} - a_n)$ which has the tendency to converge to the universal Feigenbaum constant $\delta_\infty = 4.6692...$ for the relative spacing between the bifurcation points. The numerically calculated values $\delta_2 = 4.16$ and $\delta_3 = 4.52$ are already quite good. Similarly, we can calculate the universal number $\alpha_\infty = 2.50291...$ for the relative widths, $\alpha_n = \Delta X_{n-1}/\Delta X_n$. We measured $\Delta X_1 = 0.0872$, $\Delta X_2 = 0.047$, $\Delta X_3 = 0.0205, ...$ leading to $\alpha_1 = 1.86$, $\alpha_2 = 2.29$; the tendency is quite correct. It is fascinating to recover the convergence tendency to the Feigenbaum constants for these torus-doubling cascades. In addition we have also determined the power spectrum for the various states. Typical examples are shown in Fig.2 which clearly prove the period-doubling.

We do not report here the analytical calculations which show the stability of the limit cycle (period 1) in a certain parameter regime. Also we have developed an analytical model (similar to that of Nozaki and Bekki [8]) which clearly explains the left part of the Feigenbaum tree in Fig.1 up to the window-3. In Fig.3 we show the period-doubling of the 3-cycle. Two diagnostic tools were used: First, as shown in Fig.3d, the Fourier-power-spectrum, and secondly the IST. By the latter we mean that we have used the numerically calculated distributions $q(x, t)$ as potentials in the scattering problem for periodic boundary conditions [9,10]. Two interesting aspects are worth mentioning: First, in finite systems instead of discrete eigenvalues bands appear. The widths of the bands depend on the length of the system. In our simulations the length of the system turns out to be large enough such that the band width becomes neglegible. This also supports our interpretation that the chaos is not induced (or dependent) on the boundaries if they are far enough. Secondly, the spectral diagnostics tell us which modes actually participate in the nonlinear dynamics. As we can see from Fig.3e, it is the one-soliton solution (with time-dependent parameters) which dominates. From time to time, very regularly a second but small soliton appears. What is not shown here is the $k = 0$ radiation mode which also appears. These findings form the basis for some analytical models which have been proposed elsewhere. In this respect it is also interesting to measure the dimension of the attractor. Following the concept of correlation dimension [11], we define

$$C(r) = \lim_{N\to\infty} \frac{1}{N^2} \sum_{i,j=1}^{N} \theta(r - |q^i - q^j|) \tag{2.6}$$

and determine the dimension ν from

$$C(r) \sim r^\nu \qquad \text{for} \quad r \to 0. \tag{2.7}$$

As is obvious, for the limit cycle the dimension should be 1 and close to the onset of chaos the dimension should approach 1.5.

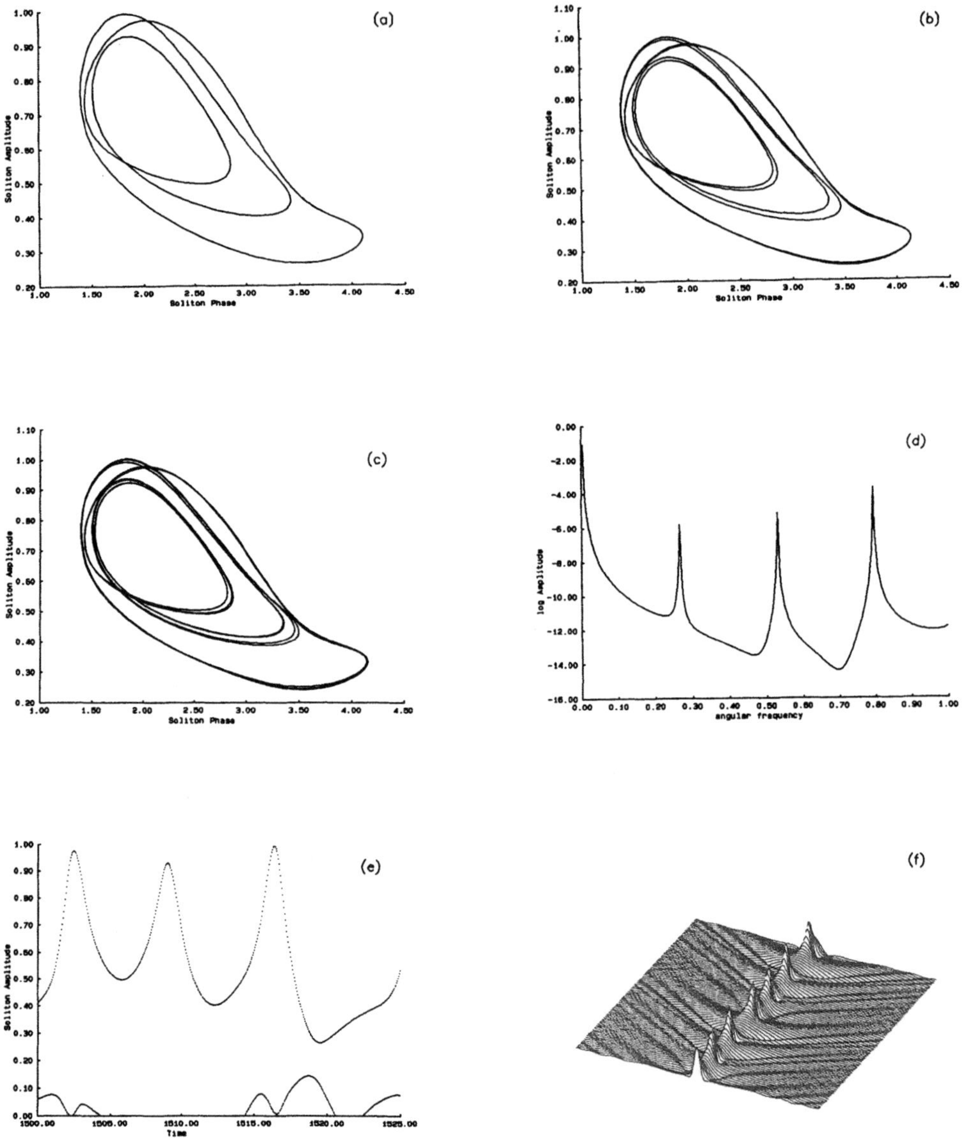

Fig.3 Period-doubling in the period-3 window: (a) 3-cycle for $\gamma = 0.11$ and $a = 0.158$; (b) the 6-cycle for $a = 0.159$; (c) the 12-cycle for $a = 0.1595$; (d) power spectrum for the 3-cycle with the peaks at the frequency $\omega_3 = 0.7977$ and the 1/3 and 2/3 values of it; (e) the development of the discrete eigenvalues; (f) $q(x, t)$ at different time steps.

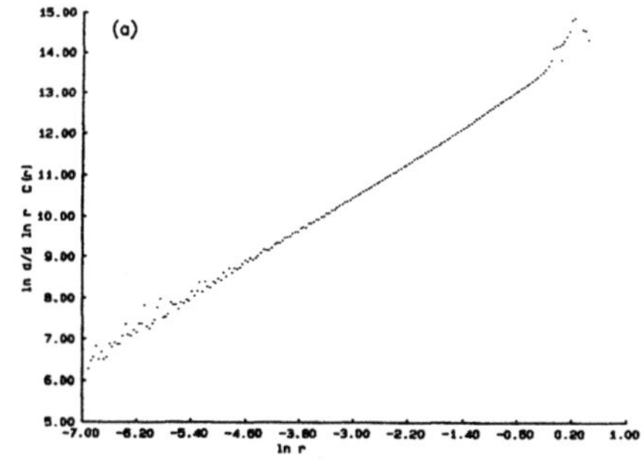

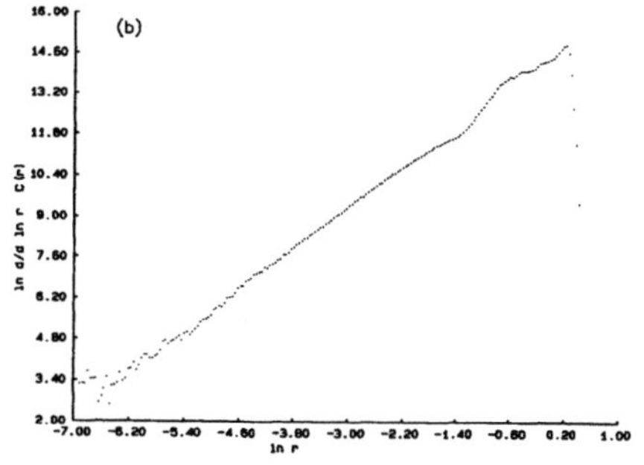

Fig.4 To determine the dimension of the attractor, we have plotted $\ln \frac{\partial}{\partial \ln r} C(r)$ vs. $\ln r$ for two cases (a) limit cycle ($\gamma = 0.111, a = 0.14$) leading to $\nu = 1.01$, and (b) chaos ($\gamma = 0.11, a = 0.1538$) leading to $\nu = 1.72$.

These predictions are confirmed by the numerical simulations of the PDE. In addition, for larger driving amplitudes we found that the dimension exceeds 2. This fact explains why simple models fail to work after the onset of chaos. We want to emphasize that even in the chaotic regime we have a spatially coherent phase-locked soliton as can be seen from Fig.5 where typical space-time plots are shown.

2.2 The link between the NLS and SG equation

As is well-known, in some parameter region, Eq. (2.1) can be considered as a simplified model for Eq. (1.3). This is important since Eq.(2.1) is sometimes easier to solve. The main idea is to expand the solution u of Eq. (1.3) in terms of harmonics,

$$u = \sum_{\nu=1}^{\infty} e^{i(2\nu-1)\tilde{\Omega}t} \epsilon^{2\nu-1} \, F^{(2\nu-1)}(X,T) + c.c., \tag{2.8}$$

where ϵ is a smallness parameter, and $X = \epsilon x$ and $T = \frac{1}{2}\epsilon^2 t$ are slow variables. When expanding $\cos u \approx 1 + a_2 u^2 + a_4 u^4 + \delta$, one finds by the Whitham method Eq. (2.1) for the first harmonic, when the following identifications are made: $\tilde{\Omega} = \sqrt{-2a_2} \approx 0.9967$, $\omega = 2\tilde{\Omega}(\tilde{\Omega} - \Omega)/\epsilon^2$, $t_{new} = T/\tilde{\Omega}$, $\gamma = \alpha\tilde{\Omega}/\epsilon^2$, $a = \sqrt{24a_4}\,\Gamma/4\epsilon^3$, and $F^{(1)} = 2q^*/\sqrt{24a_4}$. This shows us that for a certain scaling there is a one-to-one correspondence. Note that within our approach, the NLS approximation can be extended to breathers with amplitudes of order 1. However, any prediction is not trivial at all since higher harmonics are completely suppressed in the limit (2.1).

2.3 Period-Doubling for the SG equation

Guided by the above ideas and results for the NLS equation we looked for the existence of a phase-locked breather [as a solution of Eq. (1.3)], its stability, and nonlinear development when the control

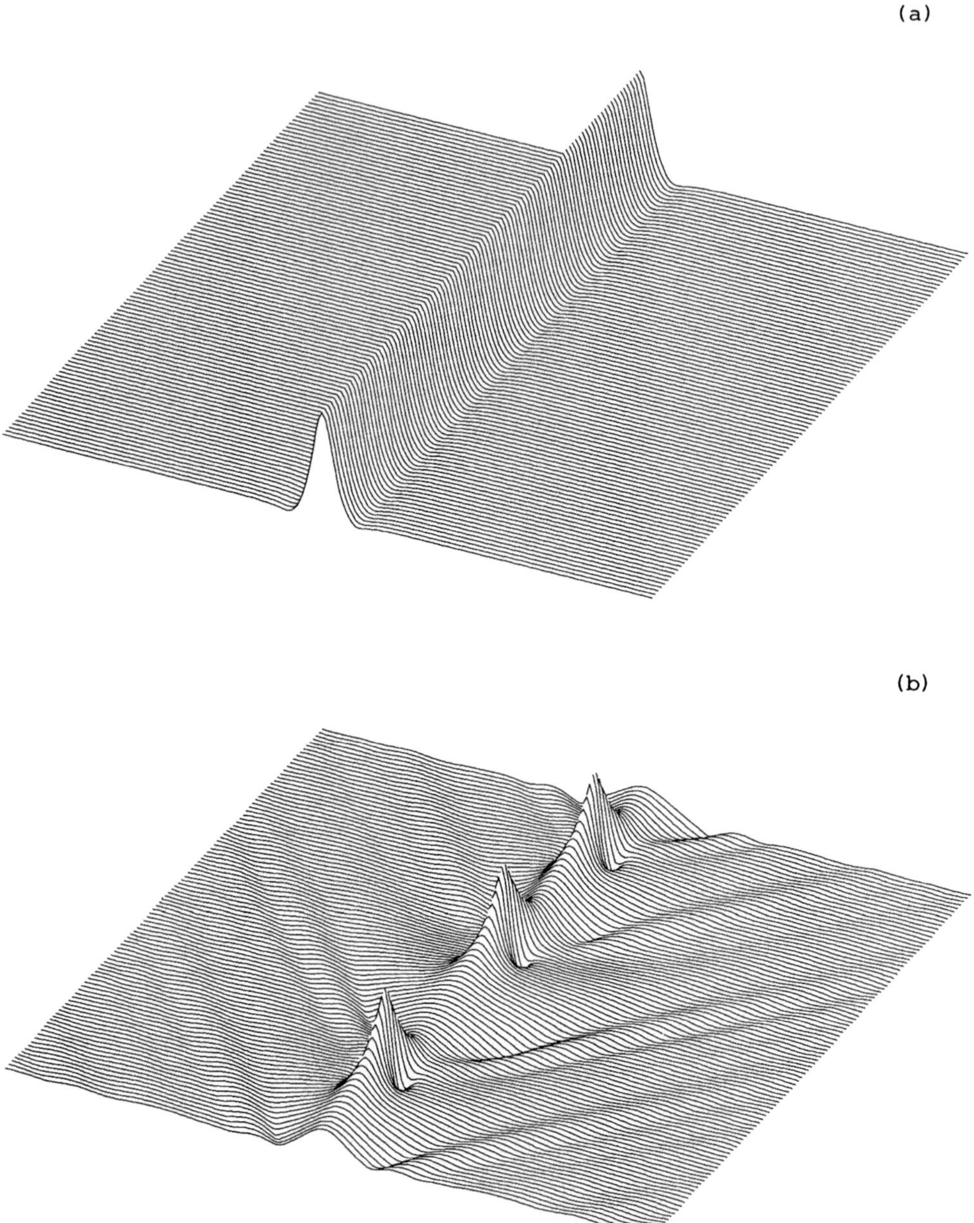

Fig.5 Space-time plots of spatially coherent structures: (a) a phase-locked-soliton for $\gamma = 0.11$ and $a = 0.1$ corresponding to a fixed point; (b) soliton-like structure in temporal chaotic evolution for $\gamma = 0.11$ and $a = 0.155$.

parameter Γ is increased. In this part we report results of the $m/n = 1/1$ phase-locked breather. All computations are performed in this part for a fixed length $L(\approx 80)$ of the system and constant damping α. A typical example is shown in Fig.6.

Let us discuss the existence of a $(m/n = 1/1)$ phase-locked breather-like solution of Eq. (1.3). We decompose the phase-locked state in two parts: the localized one (mainly a breather-like solution) and the x-independent $c(t)$ which solves the equation

$$c_{tt} + \sin c = -\alpha c_t + \Gamma \sin \Omega t. \tag{2.9}$$

Then the steady state solution of the linearized problem, in the lowest order, reads:

$$c(t) = \frac{\Gamma}{(1-\Omega^2)} \sin(\Omega t + \varphi_c). \tag{2.10}$$

For the localized part we take the breather expression at rest in the center of mass and centered about $x = 0$, i.e.

$$u_B = 4arc\ tan \left[\frac{k}{\Omega} \frac{\cos(\Omega t + \varphi_B)}{\cosh(kx)} \right]. \tag{2.11}$$

Here, $k = \sqrt{1-\Omega^2}$, the frequency of the breather is assumed to be locked to the external driver and φ_B stands for the asymptotic phase.

We write the total solution of Eq. (1.3) in the form $u = c + u_B + \delta u$, where δu obeys to lowest order the equation

$$H\delta u = c(1-\cos u_B) - \alpha u_{Bt} \equiv D; \tag{2.12}$$

$H := \partial_t^2 - \partial_x^2 + \cos u_B$ is an operator which possesses the kernel functions u_{Bx} and u_{Bt}, i.e. $Hu_{Bt} = Hu_{Bx} = 0$.

Although the two eigenfunctions u_{Bx} and u_{Bt} are linearly independent the solvability condition for the Eq. (2.12) reduces to

$$\langle u_{Bt}, D\rangle = \int_{-\infty}^{-\infty} dx \int_0^{2\pi/\Omega} dt \left[-\alpha\left(u_{Bt}\right)^2 + (1-\cos u_B)\, u_{Bt} c\right] = 0. \tag{2.13}$$

In fact D is orthogonal to the eigenfunction u_{Bx}, thus $\langle u_{Bx}, D\rangle \equiv 0$ is trivially satisfied.

This result can be easily understood from the SG equation symmetries and has a physical interpretation. Indeed, the SG equation is invariant under the transformations $t \leftrightarrow \pm t \pm \Delta t$ and $x \leftrightarrow \pm x \pm \Delta x$. They imply for the linearization operator H the kernel functions u_{Bt} and u_{Bx}, respectively. However, the perturbative term [r.h.s. in Eq. (1.3)] breaks only the first symmetry. Then one expects that the projection on u_{Bx} is trivially satisfied. This means that the phase-locked breather remains at rest and can have no translation at all under such perturbations. This result can also be of interest

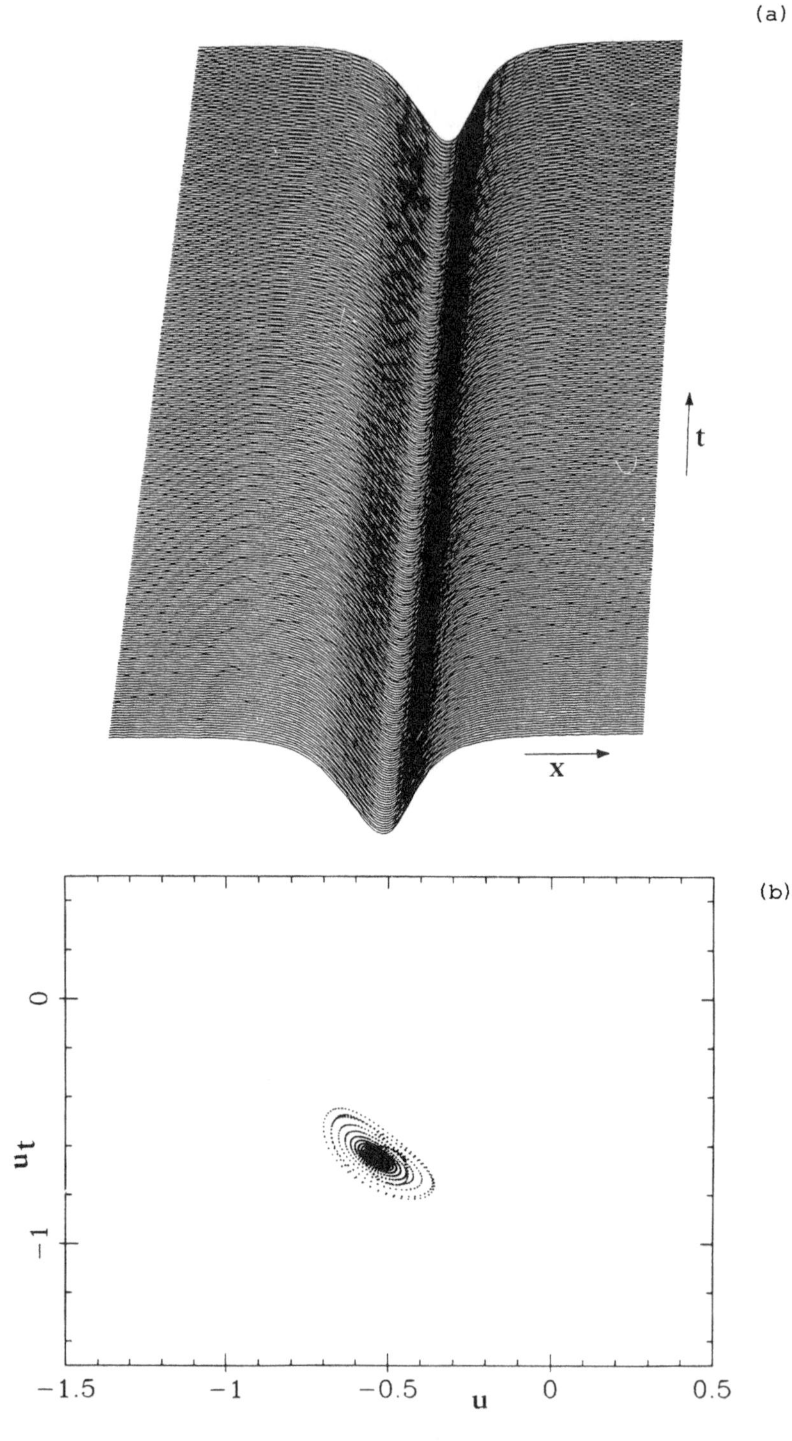

Fig.6 The phase-locked breather (a) space-time-plot of the amplitude u; (b) approach to the fixed point (phase-locked breather) in the reduced phase-space.

for the kink solution of SG equation. Since the above symmetries are characteristic for the pure SG equation, the procedure and the results may be extended to any perturbations provided that it is periodic in time. When evaluating the integrals which appear on the l.h.s. of Eq. (2.12) one obtains

$$2\alpha \, arctan\left(\frac{k}{\Omega}\right) + \Gamma\frac{\cos\Delta\varphi}{1-\Omega^2}\left[2K(k)-\pi\right] = 0, \tag{2.14}$$

where $K(k)$ is the complete elliptic integral of the first kind and $\Delta\varphi = \varphi_B - \varphi_c$. We first note that in the small-amplitude-limit ($\Omega \approx 1, k \ll 1$) we have $K(k) \approx \frac{\pi}{2} + \frac{\pi}{8}k^2$ and a simple calculation shows that then Eq. (2.13) leads exactly the result for the existence of a phase-locked soliton of the NLS equation. Next, Eq. (2.13) can be viewed in two respects. First, a necessary condition for the existence is obtained by setting $\cos\varphi_B = -1$. Then we can plot the critical amplitude $\Gamma_c(\Omega)$ of the driver versus the external frequency (for fixed damping α, as always in this paper). The result is shown in Fig.7a.; above the critical amplitude phase-locking is possible.

On the other hand, Eq. (2.13) shows the important dependence of the driver amplitude Γ upon the phase-difference $\Delta\varphi$ between the localized phase-locked state and the x-independent locked mode ($k = 0$ mode or background). This means that for different values of Γ one can find different asymptotic phase-locked states. The most interesting fact is that they are described by the above phase-difference variable $\Delta\varphi$. Fig.7b shows the Γ_c curve versus $\Delta\varphi$ for $0 \leq \Delta\varphi < 2\pi$ and α, Ω fixed. From Fig.7b, four observations must be noted: (i) A band between the $\Gamma = \pm\Gamma_c(\Omega)$ curves exists where no phase-locked state can occur. (ii) For $|\Gamma| \gtrsim |\Gamma_c(\Omega)|$ there exist two symmetric states about π. (iii) No phase-locked state is possible for $0 \leq \Delta\varphi \leq \frac{\pi}{2}$ and $\frac{3\pi}{2} \leq \Delta\varphi \leq 2\pi$. The reason is that the phase-locking phenomenon does not depend on the sign of the driver amplitude Γ. This is clear from the transformation $\Gamma \leftrightarrow -\Gamma$ and $u \leftrightarrow -u$ for which Eq. (1.3) is invariant. (iv) The values of Γ become quite large when $\Delta\varphi$ tends to $(2k+1)\frac{\pi}{2}$.

Finally we mention that in [12], the authors showed the crucial role of the phase-difference $\Delta\varphi$ (θ in their notation) in the instability mechanism of the phase-locked state (they termed this phenomenon "phase-pulling"). They also determined numerically the critical instability curve $\alpha_c(\theta)$ (where $\alpha = \Gamma(2\epsilon)^{-3/2}$ and $\epsilon = 1 - \Omega = 0.1$, see Fig.7 in [12]) which presents a close similarity to the above critical existence curve.

An analytical calculation of the stability of a phase-locked state is in progress. Here we report numerical results. First, we note that the theoretically predicted existence curve [Fig.7] for a phase-locked breather has been confirmed numerically. Secondly, as is shown in Fig.8b, the phase-locked breather (fix-point) undergoes a Hopf bifurcation when the amplitude of the driver is further increased. In Fig.8 we report results which correspond exactly to the values obtained from the proposed link between the NLS and the SG equation; see Sec.2.2. In this region, the period-doubling route to chaos is clearly demonstrated. We have also computed the power spectra corresponding to the various cases. The halfening of the characteristic frequency of the limit cycle is then obtained for all cases of interest.

We have also analyzed the time behavior by the inverse spectral diagnostics. Following Forest and McLaughlin [9], a program has been written [7] to compute the discrete and continuous eigenvalues of the SG scattering problem with periodic boundary conditions.

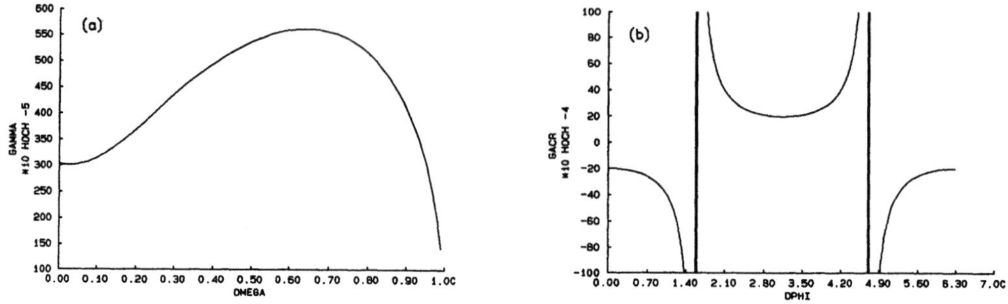

Fig.7 Evaluation of Eq.(2.13): (a) Critical value Γ_c of the driver amplitude vs. Ω for fixed $\alpha = 0.004$. Below Γ_c no phase locking is possible. (b) The phenomenon of phase pulling. The critical curve Γ_c (for $\Omega = 0.98$ and $\alpha = 0.004$) vs $\Delta\varphi$; the phase-difference between the driver and the breather.

We remind the reader that to the SG equation belongs the scattering problem

$$\left[\begin{pmatrix} 0 & -1 \\ 1 & 0 \end{pmatrix}\frac{d}{dx} + \frac{i}{4}w\begin{pmatrix} 0 & 1 \\ 1 & 0 \end{pmatrix} + \frac{1}{16\sqrt{E}}\begin{pmatrix} e^{iu} & 0 \\ 0 & e^{-iu} \end{pmatrix} - \sqrt{E}\right]\psi = 0, \qquad (2.15)$$

together with the time evolution equation

$$\left[\begin{pmatrix} 0 & -1 \\ 1 & 0 \end{pmatrix}\frac{d}{dt} + \frac{i}{4}w\begin{pmatrix} 0 & 1 \\ 1 & 0 \end{pmatrix} - \frac{1}{16\sqrt{E}}\begin{pmatrix} e^{iu} & 0 \\ 0 & e^{-iu} \end{pmatrix} - \sqrt{E}\right]\psi = 0, \qquad (2.16)$$

where $w = u_x + u_t$, $\psi = (\psi_1, \psi_2)^T$; the SG equation follows from the compatibility condition $\psi_{xt} = \psi_{tx}$.

In the case of periodic boundary conditions, the E-spectrum is continuous and bands appear. We observe growing of "spines" from the positive real E axis, corresponding to "linear" radiation modes, and bands in the complex E-plane, e.g. close to the breather cycle. By the Floquet theory, it follows that the bands are determined by the requirement that the eigenvalues $\rho(E)$ of the transfer matrix T obey $\mid \rho(E) \mid = 1$. As for the NLS case, it turns out (because of the relatively long widths used in the simulations) that the bands actually degenerate (approximately) to points. Three typical results for the motion of the discrete eigenvalues (bands) in a 1-, 2-, and 4-cycle, respectively, are shown in Fig. 9. For the 2-cycle the eigenvalue splits every second time when it comes close to the real axis. At the same time, a spine grows from the positive real axis and comes close to the discrete eigenvalue. For a 4-cycle, a similar behavior occurs, but only every second splitting is of the same form.

These results clearly support our interpretation that a phase-locked breather is the dominating nonlinear object. We succeeded to create a reduced model of four ODE's which reveals the basic phenomena in a satisfactory manner. The details are being published elsewhere [7].

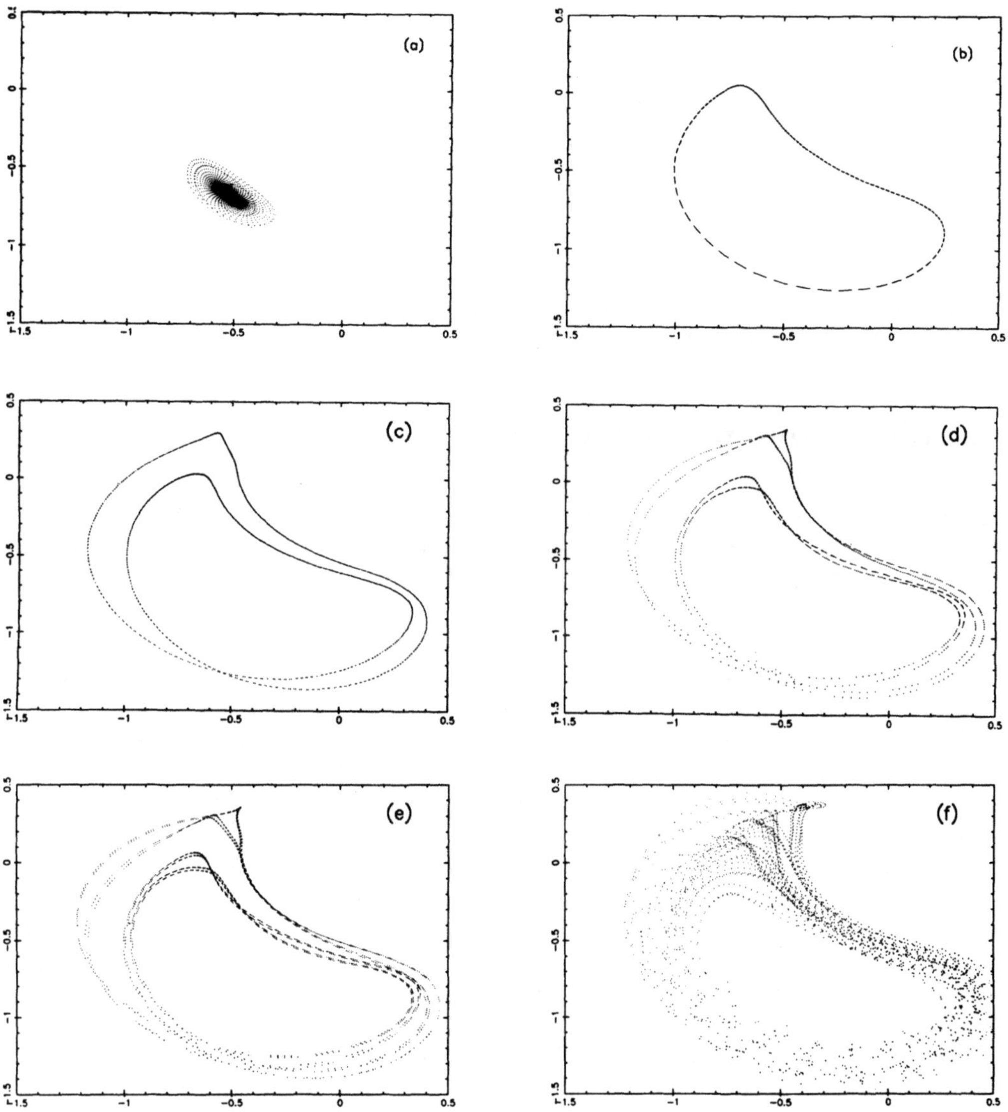

Fig.8 Phase portraits displaying $u_t(x = 0, T = nP)$ vs. $u(x = 0, T = nP)$, where P is the period of the external driver. In all cases the damping coefficient is $\alpha = 0.004$ and the driver frequency is $\Omega = 0.98$. The driver strength varies: (a) $\Gamma = 0.0032$, (b) $\Gamma = 0.0038$, (c) $\Gamma = 0.00432$, (d) $\Gamma = 0.00435$, (e) $\Gamma = .0045$, (f) $\Gamma = 0.00456$.

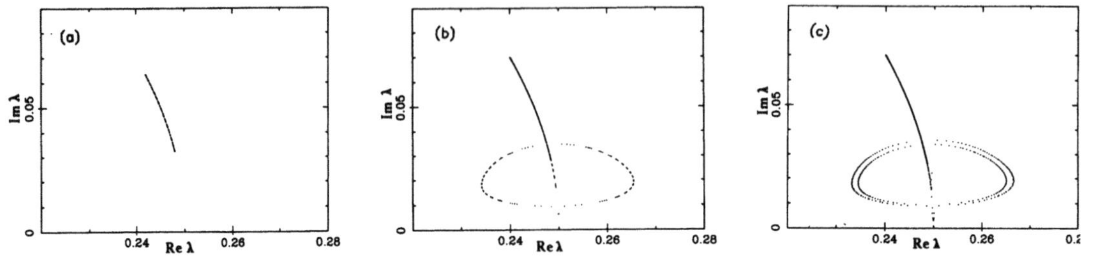

Fig.9 Motion of the discrete eigenvalue $\lambda = \sqrt{E}$ in the complex plane, as found from the nonlinear spectral transform for the (a) limit cycle, (b) 2-cycle, and (c) 4-cycle.

3 Subharmonic phase-locking

So far we have reported new results in a parameter regime which is different from those investigated in literature previously. Bishop et al. [5] found for larger amplitudes (the parameter $\alpha = 0.04$ and $\Gamma \approx 0.1$ chosen by them are one order of magnitude larger than those used in Sec. 2) and smaller driving frequency $\Omega(= 0.87$ in contrast to $\Omega = 0.98$ chosen by us in Sec. 2) the quasi-periodic route to chaos. When taking their values for a length $L = 24$ we have confirmed exactly their result. However, we were interested in modifications when small changes in the parameters are introduced. The reason for this procedure is to understand better the similarity to simpler models in nonlinear dynamics, especially, since Jensen, Bak and Bohr [13] were able to demonstrate that the driven damped pendulum has a strong connection with the circle map. The question we want to pose is whether such a similarity also holds with respect to the driven and damped SG equation. For fixed damping three parameters (Γ, Ω and L) enter our problem. Because we have some applicatons to Josephson junctions in mind, we put special attention on the L-dependence of the results. First we looked for the existence of m/n-phase-locking which should occur in addition to the 1/1-phase-locking. We do not report analytical calculations here which follow exactly the procedure outlined in Sec. 2.3. Instead, in Fig.10 we show a typical 1/8-phase-locking in the region of $L \approx 38$. These results demonstrate that we do not have only the scenario reported by Bishop et al. [5] where the two frequencies appearing in a limit-cycle (corresponding to Fig.10a) are irrational in ratio. The system is even richer: we can have an, e.g., 1/8-phase-locking with subsequent islands formation. The latter proves the appearence of a third frequency (via a second Hopf bifurcation) which can be either irrational (Fig.10b) or rational (Fig.10f) with respect to the previous ones. After the third frequency appeared, the system can go to chaos. In that transition, with decreasing L, the islands first grow and then nearly disappear when chaos sets in.

It is interesting to note that in accordance with the phenomena known from the circle map and other two-parameter families of maps of the plane we have observed in addition:

(a) 1/8-phase-locking with subsequent period-doubling, and

(b) 1/8-phase-locking with a tendency to intermittend behavior.

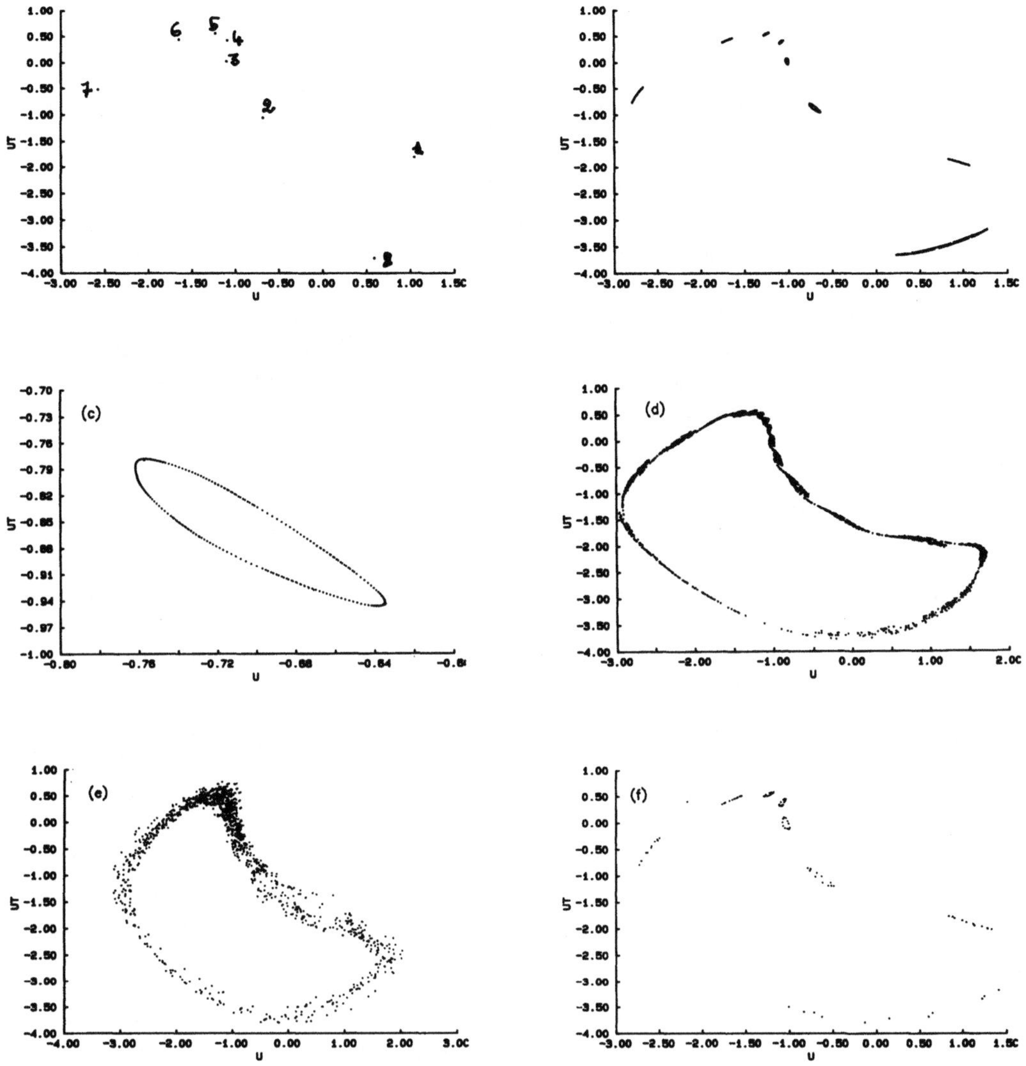

Fig.10 Subharmonic 1/8-phase-locking and subsequent transition to chaos. All simulations are for $\Gamma = 0.1051, \alpha = 0.04$, and $\Omega = 0.87$ whereas L is varied: (a) $L = 38$, (b) $L = 36.2$, (c) enlargement of one island shown in (b), (d) $L = 36$, (e) $L = 34.8$, (f) three frequencies with rational ratio appear in between the irrartional ones. Each one of the eight groups contains exactly 11 points.

Both phenomena are briefly shown in Fig.11 for the same parameters α, Ω and Γ as in the previous figures.

Of course, other phase-locked states have also been observed: 1/7 for $L = 77$, 2/13 for $L = 45$, 2/15 for $L = 39$, and so on.

Even though the solutions lie in an infinite dimensional phase space, it is well-known that attractors can be low-dimensional. However, our observations in the numerical study permit us to believe that the different phase-locking behaviors and the subsequent transitions to chaos may be well-understood in lights of bifurcation theory of maps [3]. Indeed, typical results are shown, and interpreted, in Fig.12 according to the study of the two-parameter families of maps of the plane [14]. Note that, in particular, chaos in the latter reference appears from the 1/8-phase-locking and not, e.g., from 1/7-phase-locking (no overlapping at all in this Arnold's tongue for these maps, see Figs. 3.4 and 5.1 in [14]). That is exactly what we observe in the numerical experiments.

4 Conclusion

Two aspects we want to emphasize in this report: Firstly, a driven and damped PDE possesses a rich variety of phenomena and secondly the dynamical behavior is generic even if we compare with the relatively simple and standard results from maps and ODE's. Let us discuss these conclusions now after we have obtained all the detailed results.

The period-doubling via torus-doubling route to chaos is now well-established for the NLS and SG equation. By comparing all these results we can conclude that the torus-doubling is generic in the true NLS limit. Here, we have shown for the first time the Feigenbaum scaling for the NLS equation.

We would like to remind the reader that all the points in Fig.1 were obtained by solving the PDE which requires a huge amount of computer time. The similarity to the Feigenbaum tree of the logistic map is fascinating and the convergence to the universal Feigenbaum values (known from mappings) after a few iterations is convincing. We were even able to observe the 3-torus cascade for the period-3 window. When performing a Poincaré mapping, the quadratic nature of the reduced model of a mapping becomes obvious; in the same typical manner the mapping changes its behavior when chaos sets in. The correlation dimension of the chaotic attractor increases and is larger than 2 in the region of well-developed chaos. This throws some light on the coherent structure undergoing temporal chaos. Within the period-doubling route, the low-dimensional nature of the attractor suggests a simple model with a few degrees of freedom for the interpretation of the results. In the NLS case, a one-soliton solution, together with long-wavelength radiation (k=0 mode) reveals in a satisfactory manner the dynamical behavior in time and space. This is not true anymore when the chaos is fully developed. The dimension of the PDE attractor is bigger than 2 whereas the just mentioned simplified model has a significantly lower-dimensional attractor. These conclusions have been confirmed by a nonlinear spectral diagnostics. In the latter the more complicated case of periodic boundary conditions was applied. The results show the appearence of a second (but small) discrete eigenvalue which belongs to a second soliton.

One of the main results of our previous work is that there is a nice link between the SG equation and the NLS equation. In a certain parameter regime, the SG breather also shows clear torus-doubling. All the tools, just discussed for the NLS euqation, can also be developed for the SG equation.

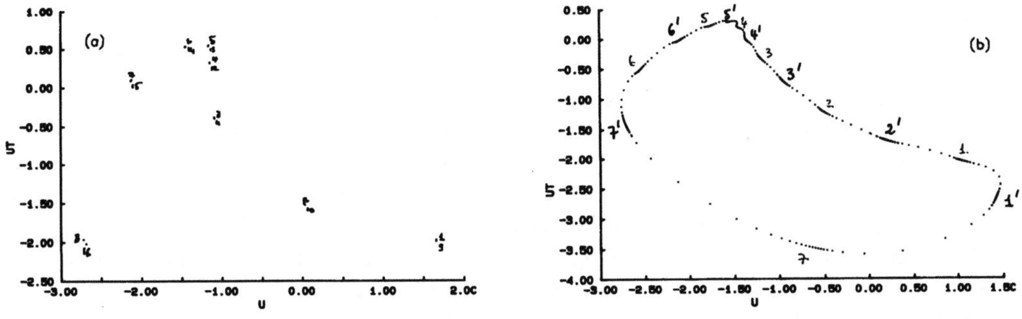

Fig.11 (a) When increasing L from 38 to 38.02 a period-doubling occurs and stops. The 2/15-phase-locked state appears immediately. (b) In the region $37 \leq L \leq 37.5$ an intermittend behavior occurs.

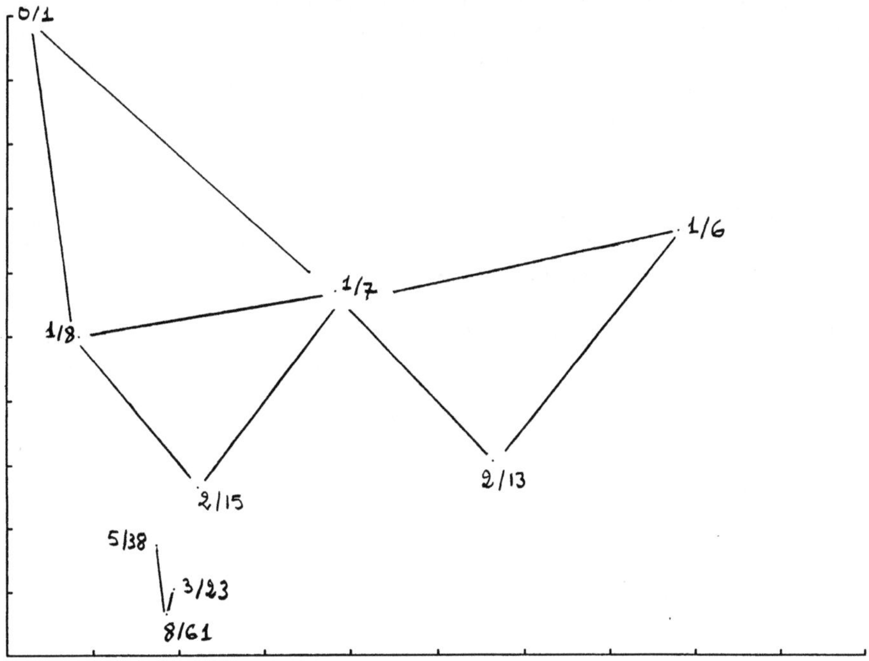

Fig.12 Plot of the Farey tree as observed for the SG-equation.

Especially, the nonlinear spectral transform is very useful. The movement of the breather eigenvalue as depicted in Fig.9 is very enlightening when comparing with results of other dynamical systems.

Here, we have shown analytically that a phase-locked breather can exist. The phenomenon of phase pulling is easy to understand now. However, the SG breather is even more rich in its time-behavior. We can have subharmonic phase-locking in the "non-true" NLS limit which appears at larger amplitudes (of the breather, the driver, etc.). Starting from that, different scenarios appear in different parameter regions: Appearence of a second Hopf bifurcation, period-doubling for a finite number of times, intermittend behavior, and so on. Two of the main conclusions of this paper are that, first, this behavior is also generic since we have found a strong link to the two-parameter families of maps of the plane. There, very similar results can be observed. Second, the interesting link between the appearence of the second Hopf bifurcation (islands) and the three-frequency quasiperiodicity. It is well-known from the theorem by Newhouse, Ruelle and Takens [15] that a three-frequency quasiperiodic attractor may be converted to a strange attractor under a certain small perturbation. Then one may conjecture that, for flows on 3-torus, three-frequency quasiperiodicity is unlikely and would not exist in real physical systems (there is always small perturbation). Here we are interested in an infinite-dimensional system and altough the islands disappear, they are persistent and clearly observable from our numerical simulations. In the case of 1/8-phase-locking they disappear in a complicated manner giving rise to a temporal chaotic state. The same qualitative behavior of mode-locking and chaos including islands has been observed recently in a Rayleigh-Benard convection system by Haucke and Ecke [16].

The most important aspect we want to emphasize again is that all these results for the time development were obtained with a quite simple spatially coherent structure. This shows that the soliton concept has really a brought region of application.

Acknowledgements: This work was supported by the DFG through SFB 237. Discussions with J.C. Fernandez and G. Reinisch within the PROCOPE collaboration are gratefully acknowledged.

(*) New address: Laboratoire de Spectroscopie Hertzienne de l'Université de Lille, Unité Fondamentale de Recherche de Physique, 59655 Villeneuve d'Ascq Cédex, France.

References

1. M.J. Ablowitz and H. Segur, Solitons and the Inverse scattering Transform (Siam, Philadelphia 1981).

2. K.H. Spatschek, H. Pietsch, E.W. Laedke, and Th. Eickermann, in Singular Behavior and Nonlinear Dynamics, eds. St.Pnevmatikos, T. Bountis, and Sp. Pnevmatikos (World Scientific, Singapore 1989), Vol.2, p.555.

3. G. Iooss, Bifurcation of Maps and Applications (Mathematical Studies, Vol. 36, North-Holland, Amsterdam 1979)

4. J. Guckenheimer and P. Holmes, Nonlinear Oscillations, Dynamical Systems, and Bifurcations of Vector Fields (Springer, Heidelberg 1986).

5. A.R. Bishop, M.G. Forest, D.W. McLaughlin, and E.A. Overman II, Physica D 23, 293(1986).

6. M. Taki and K.H. Spatschek, J. de Physique C3, Tome 50, 77(1989).

7. M. Taki, K.H. Spatschek, J.C. Fernandez, R. Grauer, and G. Reinisch, Physica D (1989), in press.

8. K. Nozaki and N. Bekki, Physica D 21, 381(1986).

9. M.G. Forest and D.W. McLaughlin, J. Math. Phys. 23, 1248 (1982).

10. E.R. Tracey and H.H. Chen, Phys. Rev. A 37, 815 (1988).

11. P. Grassberger and I. Procaccia, Phys. Rev. Lett. 50, 346 (1983).

12. A. Mazor and A.R. Bishop, Physica D 27, 269 (1987).

13. M.H. Jensen, P. Bak, and T.Bohr, Phys.Rev. A 30, 1960(1984) ibid. 1970(1984).

14. D.G. Aronson, M.A. Chory, G.R. Hall, and R.P. McGehee, Commun. Math. Phys. 83,303 (1982).

15. S. Newhouse, D. Ruelle, and F. Takens, Commun. Math. Phys. 64, 35 (1978)

16. H. Haucke and R. Ecke, Physica D 25, 307 (1987).

SPECTRAL CHARACTERIZATION OF 2-D NONLINEAR COHERENT STRUCTURES

J . LEON+, M. BOITI++, F. PEMPINELLI++

+Laboratoire de Physique Mathématique, U.S.T.L.
34060 Montpellier Cedex 01 (FRANCE)

++Dipartimento di Fisica, Università di Lecce,
73100 Lecce (ITALIA)

ABSTRACT

It is shown that a convenient choice of the definition of the spectral transform of the solutions of the Davey-Stewartson equation (or two-dimensional nonlinear Schrödinger equation) allows to characterize the 2-D nonlinear coherent structures by a point spectrum like in the one-dimensional case.

INTRODUCTION

The purpose of this note is to point out the essential features of the spectrum (also referred as nonlinear Fourier spectrum) characterizing the two-dimensional solitons recently discovered [1]. We shall present the results in parallel with the well known one-dimensional case and will avoid as much as possible the technical details which will be found in [2] [3]. Our goal here is only to show that like in one dimension :

i) the spectrum of the solitons consists in isolated points in a complex plane,

ii) these points are the poles of a solution of a linear non homogeneous integral equation.

This result is essential in order to recognize in a (measured) field whose small amplitude limit obeys some prescribed linear wave equation, the existence and number of true two-dimensional solitons. Physicists now use more and more the nonlinear spectral analysis (numerically) to identify, count and follow in time the nonlinear coherent structures in one-dimensional problems. This is possible because the soliton spectrum (points) is easily recognizable from the radiation spectrum (lines) and everything is unambiguously defined in 1-D problems.

Conversely, in 2-D problems the spectrum is not in general uniquely defined and the nature of the soliton spectrum crucially depends on the choice of the "good" spectral transform, more precisely the choice of the "good" non homogeneous term of the fundamental integral equation to be solved.

To obtain this "good" spectrum (by "good" we mean : of practical use for the spectral analysis of 2-D nonlinear coherent structures) we have been led to reconsider in [2] the direct and inverse spectral problems in the light of the discovery that the 2-D solitons of [1] actually result from a mixed initial / boundary value problem [4]. It is worth remarking that if one keeps the previously known spectral transform (see f.i. [5] [9]) like it is done in [4], then the solitons are not distinguishable from the radiation and have not a simple readable characterization [6].

To fix the ideas, we choose here as a representative example the nonlinear Schrödinger equation (NLS) in one and two spatial dimensions (in 2-D, it is also known as the Davey-Stewartson equation for hydrodynamics [7]) :

$$\begin{cases} iq_t = -\frac{1}{2} q_{xx} - wq & \text{(1 a)} \\ w_x = -|q|^2_x & \text{(1 b)} \end{cases}$$

in one dimension and

$$iq_t = -\frac{1}{2}(q_{xx} + q_{yy}) - (w_1 + w_2)q \tag{2 a}$$

$$w_{1x} + w_{1y} = -\frac{1}{2}|q|^2_x + \frac{1}{2}|q|^2_y \tag{2 b}$$

$$w_{2x} - w_{2y} = -\frac{1}{2}|q|^2_x - \frac{1}{2}|q|^2_y \tag{2 c}$$

in two dimensions (if the fields do not depend on y, we recover (1) from (2)).

It is essential to remark here that the constant of integration of (1b) can always be scaled off by a time-dependent phase shift of q(x,t). This is not anymore true for (2b) and (2c) where the "constant of integration" are arbitrary functions of x-y and x+y respectively (and of time t of course). We shall see that these boundaries play an essential role in the creation of 2-D solitons [4].

THE SPECTRAL TRANSFORM IN ONE DIMENSION

We briefly recall the method for the determination of the spectral transform of some 1-D field q(x,t) (at any fixed time) whose small amplitude limit $\tilde{q}$ would obey

$$i\tilde{q}_t + \frac{1}{2}\tilde{q}_{xx} = 0. \tag{3}$$

The spectral transform of q(x,t) is obtained for any fixed t by solving the two following Volterra integral equations for the 2x2 matrices ϕ and ψ

$$\phi(k,x,t) = e^{-ik\sigma_3 x} + \int_{-\infty}^{x} d\xi \exp[-ik\sigma_3(x-\xi)]Q(\xi,t)\phi(k,\xi,t) \tag{4}$$

$$\psi(k,x,t) = e^{-ik\sigma_3 x} - \int_{x}^{\infty} d\xi \exp[-ik\sigma_3(x-\xi)]Q(\xi,t)\psi(k,\xi,t) \tag{5}$$

where we have defined

$$Q(x,t) = \begin{pmatrix} 0 & q(x,t) \\ q^*(x,t) & 0 \end{pmatrix} \tag{6}$$

and where k is a complex spectral parameter.

Then we define the transfer matrix by

$$S(k,t) = \psi^{-1}(k,x,t)\phi(k,x,t) \tag{7}$$

which in turn can be written [8]

$$S(k,t) = e^{ik^2\sigma_3 t}\begin{pmatrix} \dfrac{1}{\beta(k)} & \dfrac{\alpha(k)}{\beta(k)} \\ -\dfrac{\alpha^*(k^*)}{\beta^*(k^*)} & \dfrac{1}{\beta^*(k^*)} \end{pmatrix} e^{-ik^2\sigma_3 t} \tag{8}$$

This is enough to obtain the nonlinear spectrum (9) of Q(x,t) which consists in the set

$$\{\alpha(k),\ k\in\mathbb{R}\ ;\ k_n,\ \ C_n,\ \ n = 1,\ldots, N\} \tag{9}$$

where the k_n's are the N simple poles of $\beta(k)$ in $\text{Im}k > 0$, and the C_n some constants given by

$$C_n = \lim_{k\to k_n} (k-k_n)\alpha(k) \tag{10}$$

To the continuoum part ($k\in\mathbb{R}$) of the spectrum corresponds the radiative part of q(x,t) which asymptotically vanishes in time. To the discrete part of the spectrum (the N points k_n) corresponds the soliton part of q(x,t) which asymptotically, separates into N individual solitons.

Therefore the nonlinear spectral analysis of a one-dimensional field q(x,t) (remember t is a parameter) requires only solving two Volterra linear non homogeneous matrix integral equations with complex parameter k. The singular points, if any, characterize the presence of solitons.

The spectral transform can be written in a more convenient form by using the $\bar{\partial}$-formalism which actually will be essential to solve 2-D problems as it was originally stated in [9].

The $\bar{\partial}$ operator

$$\frac{\partial}{\partial\bar{k}} = \frac{\partial}{\partial k_R} + i\,\frac{\partial}{\partial k_I}\quad ,\qquad k = k_R + i\,k_I\quad , \tag{11}$$

is the distributional derivative in the complex plane and we have for instance

$$\frac{\partial}{\partial\bar{k}}\left(\frac{1}{k-\lambda}\right) = -\,2i\pi\,\delta(k-\lambda) = -\,\pi\,\delta(k_R-\lambda_R)\delta(k_I-\lambda_I). \tag{12}$$

The problem of the spectral transform of q(x,t) can be formulated as the following $\bar{\partial}$ problem for the 2x2 matrix μ :

$$\frac{\partial}{\partial \bar{k}} \mu(k,x,t) = \mu(k,x,t)\, R(k,x,t) \quad , \quad k \in \mathbb{C} \tag{13 a}$$

$$\mu(k,x,t) = \mathbb{1} + O(\tfrac{1}{k}), \quad |k| \to \infty \tag{13 b}$$

where the x and t-dependence of the matrix distribution R(k) are explicitly given by [10]

$$i R_x = [R,\Lambda], \quad \Lambda = k\,\sigma_3 \quad , \tag{14 a}$$

$$i R_t = [R,\Omega], \quad \Omega = k^2 \sigma_3 \quad . \tag{14 b}$$

The eq. (14 a) implies that the function $\mu e^{-ik\sigma_3 x}$ is actually a linear combination of the solutions ϕ and ψ of (4) and (5) [10]. The analytical properties of ϕ and ψ lead us to choose

$$\mu = (\beta(k)\phi_1,\ \psi_2)e^{ik\sigma_3 x} \quad \text{for} \quad \mathrm{Im}k > 0 \tag{15 a}$$

$$= (\psi_1, \beta^*(k^*)\phi_2)e^{ik\sigma_3 x} \quad \text{for} \quad \mathrm{Im}k < 0 \tag{15 b}$$

Then it is clear that μ solves for $\mathrm{Im}k > 0$:

$$\mu_{11}(k,x,t) = \beta(k) + \int_{-\infty}^{x} d\xi\ q(\xi,t)\ \mu_{21}(k,\xi,t)$$

$$\mu_{21}(k,x,t) = \int_{-\infty}^{x} d\xi\ q^*(\xi,t)\ \mu_{11}(k,\xi,t)e^{2ik(x-\xi)}$$

$$\mu_{12}(k,x,t) = -\int_{x}^{\infty} d\xi\ q(\xi,t)\ \mu_{22}(k,\xi,t)e^{-2ik(x-\xi)}$$

$$\mu_{22}(k,x,t) = 1 - \int_{x}^{\infty} d\xi\ q^*(\xi,t)\ \mu_{12}(k,\xi,t) \tag{16 a}$$

and for $\mathrm{Im}k < 0$:

$$\mu_{11}(k,x,t) = 1 - \int_{x}^{\infty} d\xi\ q(\xi,t)\ \mu_{21}(k,\xi,t)$$

$$\mu_{21}(k,x,t) = -\int_{x}^{\infty} d\xi\ q^*(\xi,t)\ \mu_{11}(k,\xi,t)e^{2ik(x-\xi)}$$

$$\mu_{12}(k,x,t) = \int_{-\infty}^{x} d\xi\ q(\xi,t)\ \mu_{22}(k,\xi,t)e^{-2ik(x-\xi)} \quad ,$$

$$\mu_{22}(k,x,t) = \beta^*(k^*) + \int_{-\infty}^{x} d\xi\ q^*(\xi,t)\ \mu_{12}(k,\xi,t). \tag{16 b}$$

Consequently the function $\mu(k)$ is discontinuous on the real axis and has the 2N singular points k_n and k_n^* in the complex plane coming from the non homogeneous termes $\beta(k)$ and $\beta^*(k^*)$. This can be summarized with (13a) for the following characterization of the SPECTRAL TRANSFORM R(k) :

$$R(k,x,t) = \exp[i\sigma_3(kx + k^2t)] \left\{ \frac{i}{2} \begin{pmatrix} 0 & \alpha^*(k^*)\,\delta(k_I + 0) \\ \alpha(k)\,\delta(k_I - 0) & 0 \end{pmatrix} + \right.$$

$$\left. + 2i\pi \sum_1^N \begin{pmatrix} 0 & C_n\,\delta(k-k_n) \\ C_n^*\,\delta(k-k_n^*) & 0 \end{pmatrix} \right\} \exp[-i\sigma_3(kx + k^2t)] \tag{17}$$

which contains in a very clear form the spectrum of $q(x,t)$ defined in (9).

THE SPECTRAL TRANSFORM IN TWO DIMENSIONS

As noted in the introduction, the asymptotic behaviours of the potentials w_1 and w_2 (namely the functions a_1 and a_2 defined below) play a fundamental role in the equation (2) and its solution.

Making the change of variables

$$u = x + y \qquad v = x - y \tag{18}$$

the 2-D NLS equation becomes

$$iq_t = -(q_{uu} + q_{vv}) - (w_1 + w_2)q \tag{19 a}$$

$$w_{1u} = -\frac{1}{2}\,|q|_v^2 \quad , \quad w_{2v} = -\frac{1}{2}\,|q|_u^2 \tag{19 b}$$

or, by integrating (19b)

$$w_1 = a_1(v,t) - \frac{1}{2}\int_{-\infty}^{u} |q|_v^2 \quad , \quad w_2 = a_2(u,t) - \frac{1}{2}\int_{-\infty}^{v} |q|_u^2 \quad , \tag{19 c}$$

where a_j are ARBITRARY FUNCTIONS. Therefore the question of finding the spectral transform of $q(u,v,t)$ is meaningless unless a_1 and a_2 are given, and hence we will be concerned with the spectral transform of the set

$$\{q(u,v,t)\ ,\ a_1(v,t)\ ,\ a_2(u,t)\} \quad , \tag{20}$$

where the small amplitude limit $\tilde{q}$ of q obeys (from (19))

$$\tilde{q}(u,v,t) = r(u,t)\,s(v,t)\ ,$$

$$ir_t + r_{uu} + a_2 r = 0 \quad , \qquad is_t + s_{vv} + a_1 s = 0. \tag{21}$$

The above Schrödinger equations with the potentials a_1 and a_2 replace in 2-D the free Schrödinger equation (3) in 1-D.

The spectral transform of (20) is now performed along the following successive steps :

Solve the following scalar integral equation for the function $\phi_{o1}(k,v,t)$ from the datum of $a_1(v,t)$

$$\phi_{o1}(k,v,t) = e^{ikv} + \frac{1}{2i\pi}\int_{-\infty}^{+\infty} d\eta \, \{\int_{-\infty}^{t} d\tau \int_{-\infty}^{k} d\ell - \int_{t}^{\infty} d\tau \int_{k}^{\infty} d\ell\} \times$$

$$\times \exp[i\ell(v-\eta) - i(\ell^2-k^2)(t-\tau)]a_1(\eta,\tau)\phi_{o1}(k,\eta,\tau), \text{ for } \mathrm{Im}k > 0,$$

$$\phi_{o1}(k,v,t) = e^{ikv} + \frac{1}{2i\pi}\int_{-\infty}^{+\infty} d\eta \, \{\int_{-\infty}^{t} d\tau \int_{k}^{\infty} d\ell - \int_{t}^{\infty} d\tau \int_{-\infty}^{k} d\ell\} \times$$

$$\times \exp[i\ell(v-\eta) - i(\ell^2-k^2)(t-\tau)]a_1(\eta,\tau)\phi_{o1}(k,\eta,\tau), \text{ for } \mathrm{Im}k < 0, \tag{22a}$$

and the analogous one for ϕ_{o2} :

$$\phi_{o2}(k,u,t) = e^{-iku} + \frac{1}{2i\pi}\int_{-\infty}^{+\infty} d\xi \, \{\int_{t}^{\infty} d\tau \int_{-\infty}^{k} d\ell - \int_{-\infty}^{t} d\tau \int_{k}^{\infty} d\ell\} \times$$

$$\times \exp[-i\ell(u-\xi) + i(\ell^2-k^2)(t-\tau)]a_2(\xi,\tau)\phi_{o2}(k,\xi,\tau), \text{ for } \mathrm{Im}k > 0,$$

$$\phi_{o2}(k,u,t) = e^{-iku} + \frac{1}{2i\pi}\int_{-\infty}^{+\infty} d\xi \, \{\int_{t}^{\infty} d\tau \int_{k}^{\infty} d\ell - \int_{-\infty}^{t} d\tau \int_{-\infty}^{k} d\ell\} \times$$

$$\times \exp[-i\ell(u-\xi) + i(\ell^2-k^2)(t-\tau)]a_2(\xi,\tau)\phi_{o2}(k,\xi,\tau), \text{ for } \mathrm{Im}k < 0. \tag{22b}$$

We shall see later that this step is actually the only one really necessary if we are merely interested in the soliton part of the solution.

Measure the analytical properties of $\phi_{oj}(k)$ by computing the $\bar\partial$ derivative which gives [3] :

$$\frac{\partial}{\partial\bar k}\phi_{o1}(k,v,t) = \iint_{\mathbb{C}} d\ell \wedge d\bar\ell \; r_1(k,\ell,t)\phi_{o1}(\ell,v,t) \tag{23a}$$

$$\frac{\partial}{\partial\bar k}\phi_{o2}(k,u,t) = \iint_{\mathbb{C}} d\ell \wedge d\bar\ell \; r_2(k,\ell,t)\phi_{o2}(\ell,u,t) \tag{23b}$$

where r_1 and r_2 have like in (17) for the one dimensional case, two parts : the Radiation part

$$[r_1(k,\ell,t)]_R = -\frac{1}{4}\, e^{i(k^2-\ell^2)t} f_{o1}(k,\ell)\, \delta(k_I+0)\, \delta(\ell_I+0) \tag{24a}$$

$$[r_2(k,\ell,t)]_R = -\frac{1}{4}\, e^{-i(k^2-\ell^2)t} f_{o2}(k,\ell)\, \delta(k_I-0)\, \delta(\ell_I-0) \tag{24b}$$

and the Soliton part :

$$[r_1(k,\ell,t)]_S = 4\pi\, e^{i(k^2-\ell^2)t} \sum_{1}^{N} \lambda_{nI} \exp[2\lambda_{nI}v_{on}]\, \delta(k-\lambda_n)\, \delta(\ell-\bar\lambda_n) \tag{25a}$$

$$[r_2(k,\ell,t)]_S = 4\pi\, e^{-i(k^2-\ell^2)t} \sum_{1}^{M} \mu_{mI} \exp[-2\mu_{mI}u_{om}]\, \delta(k-\mu_m)\, \delta(\ell-\bar\mu_m) \tag{25b}$$

The above mentioned "measure" of the analytical properties consists precisely in computing $f_{oj}(k,\ell)$ $(k,\ell \in \mathbb{R})$, the positions of the N singular points λ_n and the M singular points μ_m, and finally the constants v_{on} and u_{om}.

The above result means that ϕ_{o1} and ϕ_{o2} are menomorphic functions of k with i) a discontinuity on the real axis, ii) N and M simple poles in the complex plane.

With ϕ_{o1} and ϕ_{o2} in hands, compute now the matrix solution $\phi(k,u,v,t)$ of the following integral equation (forgetting the parametric t-dependence) :

$$\begin{cases}\phi_{11}(k;u,v) = \phi_{o1}(k;v) - \frac{1}{2}\int_{-\infty}^{u} du' q(u',v)\, \phi_{21}(k;u',v) \\ \phi_{21}(k;u,v) = \frac{1}{2}\int_{v}^{\infty} dv' q^{*}(u,v')\, \phi_{11}(k;u,v')\end{cases}$$

$$\begin{cases}\phi_{12}(k;u,v) = -\frac{1}{2}\int_{-\infty}^{u} du' q(u',v)\, \phi_{22}(k;u',v) \\ \phi_{22}(k;u,v) = \phi_{o2}(k;u) - \frac{1}{2}\int_{-\infty}^{v} dv' q^{*}(u,v')\, \phi_{12}(k;u,v')\end{cases} \tag{26a}$$

for Imk >0, and of

$$\begin{cases}\phi_{11}(k;u,v) = \phi_{o1}(k;v) - \frac{1}{2}\int_{-\infty}^{u} du' q(u',v)\, \phi_{21}(k;u',v) \\ \phi_{21}(k;u,v) = -\frac{1}{2}\int_{-\infty}^{v} dv' q^{*}(u,v')\, \phi_{11}(k;u,v')\end{cases}$$

$$\begin{cases}\phi_{12}(k;u,v) = \frac{1}{2}\int_{u}^{\infty} du' q(u',v)\, \phi_{22}(k;u',v) \\ \phi_{22}(k;u,v) = \phi_{o2}(k;u) - \frac{1}{2}\int_{-\infty}^{v} dv' q^{*}(u,v')\, \phi_{12}(k;u,v')\end{cases} \tag{26b}$$

for Imk <0.

Like in the preceding step we measure the analytical properties of ϕ and we get [2].

$$\frac{\partial}{\partial \bar{k}}\, \phi(k,u,v,t) = \iint_{\mathbb{C}} d\ell \wedge d\bar{\ell}\, \phi(\ell,u,v,t)\, R(k,\ell,t) \tag{27}$$

Using that $\phi_o \exp[-ik(\sigma_3 x-y)]$ is everywhere bounded in the plane, the above Volterra equations can easily be rescaled to prove that their Neumann series are absolutely convergent provided

$$q(u,v) \in L^1(\mathbb{R}^2), \quad \forall t. \tag{28}$$

Therefore the related homogeneous equations have only the vanishing solution and the only singularities of ϕ in the complex k-plane are those of ϕ_o and of the sectionally holomorphic Green function of the integral equations. In other words the discrete spectrum in $R(k,\ell)$ is entirely due to the boundaries a_1 and a_2 but not to the initial condition q(x,y,0). This result has been obtained in [4] for a_1 and a_2 time-independent, and finds here a very natural interpretation.

We have therefore in R, a radiative part which reads [2]

$$[R(k,\ell,t)]_R = -\frac{1}{4}\begin{pmatrix} \delta(\ell_I+0)\delta(k_I+0) & 0 \\ 0 & \delta(\ell_I-0)\delta(k_I-0) \end{pmatrix} \exp[-i\ell^2\sigma_3 t]\ \times$$

$$\times \begin{pmatrix} f_{o1}(k,\ell) & s_2(k,\ell) \\ s_1(k,\ell) & f_{o2}(k,\ell) \end{pmatrix} \exp\,[ik^2\sigma_3 t] \tag{29a}$$

and a soliton part [2] :

$$[R(k,\ell,t)]_S = \exp[-i\ell^2\sigma_3 t] \begin{pmatrix} 0 & \sum_1^N \beta_m\ \delta(k-\mu_m)\ [\delta(\ell-\lambda_m) + \delta(\ell-\lambda_m^*)](\ell-\lambda_m) \\ \sum_1^N \alpha_n\ \delta(k-\lambda_n)[\delta(\ell-\mu_n) + \delta(\ell-\mu_n^*)](\ell-\mu_n) & 0 \end{pmatrix} \exp[ik^2\sigma_3 t] \tag{29b}$$

where α_n and β_m are some constants that have an explicit expression in terms of q and ψ [2].

We remark that the explicit time dependence corresponds only to exact solutions of the 2-D NLS equation (19) and therefore a spectral analysis of a function q(u,v,t) with unprescribed evolution should be made at any fixed t (just forget then the "exp" in (19)).

The formula (29) is the main resultand we see that, like in 1-D, the soliton part of the function q(u,v,t) has a point spectrum. But now this spectrum is entirely due to the boundaries $a_1(v,t)$ and $a_2(u,t)$ of the potentials w_1 and w_2.

Consequently, if one is interested only in finding (at some fixed t) the number and velocities of the solitons present in some two dimensional function q(u,v) which has the propertie (21), it is enough to solve (22) and localize the poles of $\psi_{o1}(k)$ and $\psi_{o2}(k)$.

To close this section we want to stress that the 2-D spectral transform of a solution of (19) is not unique. Indeed, in the fundamental integral equation (26) the non homogeneous terms ψ_{o1} and ψ_{o2} can be chosen arbitrarily. Actually in the work [4] they have been taken respectively as exp[ikv] and exp[-iku]. But only for the choice made here would we have both a linear (integrable explicitely) time evolution of R(k,ℓ,t) and a point spectrum for characterizing the solitons.

THE TWO-DIMENSIONAL NLS-SOLITON

As an example we give hereafter the one-soliton solution of the 2-D NLS equation (2) and its spectral transform. For $\{\lambda_R, \lambda_I, \mu_R, \mu_I, \eta, \rho\}$ arbitrary real constants, we have [1] [6] :

$$q(x,y,t) = \frac{\lambda_R \eta \exp[-i\varphi]}{\gamma \cosh \xi_1 + [\gamma(1+\gamma)]^{\frac{1}{2}} \cosh \xi_2} \tag{30}$$

with $\gamma = \frac{1}{4}\eta\rho$, $\lambda_R \eta = \mu_R \rho$ and

$$\varphi = (\mu_R + \lambda_R)x + (\mu_R - \lambda_R)y + [\lambda_R^2 + \mu_R^2 - \lambda_I^2 - \mu_I^2]t$$

$$\xi_1 = (\mu_I + \lambda_I)x + (\mu_I - \lambda_I)y + 2(\lambda_R \lambda_I + \mu_R \mu_I)t$$

$$\xi_2 = (\mu_I - \lambda_I)x + (\mu_I + \lambda_I)y + 2(\mu_R \mu_I - \lambda_R \lambda_I)t - \frac{1}{2} \ell n \frac{1+\gamma}{\gamma} .$$

The related boundaries of the potential are

$$a_1(x-y,t) = \frac{4\lambda_I \gamma}{\gamma + [\gamma + \theta(\mu_I)] \exp[\xi_1 + \xi_2]}$$

$$a_2(x+y,t) = \frac{-4\mu_I \gamma}{\gamma + [\gamma + \theta(-\lambda_I)] \exp[\xi_2 - \xi_1]} \tag{31}$$

(θ is the step function).

The soliton (30) moves in the plane (x,y) with velocity

$$\vec{v} = \begin{pmatrix} \lambda_R + \mu_R \\ \mu_R - \lambda_R \end{pmatrix} . \tag{32}$$

Its spectral transform reads

$$R(k,\ell,t) = e^{-i\ell^2 \sigma_3 t} \begin{pmatrix} 0 & \eta\, \delta(k-\mu)[\delta(\ell-\lambda) + \delta(\ell-\lambda^*)](\ell-\lambda) \\ \rho\, \delta(k-\lambda)[\delta(\ell-\mu) + \delta(\ell-\mu^*)](\ell-\mu) & 0 \end{pmatrix} e^{ik^2\sigma_3 t} \tag{33}$$

where $\lambda = \lambda_R + i\lambda_I$ and $\mu = \mu_R + i\mu_I$.

CONCLUSION

We have focus our attention on the SPECTRAL TRANSFORM of some measured field and seen that the problem is that of solving an integral equation, either in one or two dimensions. The solutions of these integral equations have some poles which give all the necessary informations about the number, energy... of the solitons present in the measured field.

We have not examined the question of solving the NLS equation for given initial (plus boundaries in 2 D) conditions. This is another problem (the INVERSE SPECTRAL TRANSFORM) and the interested reader may refer to [2] [3] and [6] for the discussion of this question in the present formalism, and to [4] [5] for a different approach.

Another question is also of interest : the interaction of these travelling 2-D solitons and we refer to [1] for this problem.

REFERENCES

[1] M. BOITI, J.JP. LEON, L. MARTINA, F. PEMPINELLI, "Scattering of localized solitons in the plane" Phys. Lett. A 132, 432 (1988)

[2] M. BOITI, J.JP. LEON, F. PEMPINELLI "A new spectral transform for the Davey-Stewartson eq. I" Preprint PM 89/10, (Montpellier Feb. 1989) to appear in Phys. Lett. A

[3] M. BOITI, J.JP. LEON, F. PEMPINELLI "A new spectral transform for the Kadomtsev-Petviashvili eq. I" Preprint PM 89/9 (Montpellier Feb 1989) to appear in Phys. Lett. A.

[4] A.S. FOKAS, P.M. SANTINI "Solitons in multidimensions" Preprint INS # 106, Clarkson (Nov. 1988).

[5] A.S. FOKAS, M.J. ABLOWITZ J. Math. Phys. 25, 2494 (1984)

[6] M. BOITI, J.JP. LEON, F. PEMPINELLI "Multidimensional solitons and their spectral transform", Preprint PM 88/44 (Montpellier, October 1988). See also : M. BOITI, J.JP. LEON, L. MARTINA, F. PEMPINELLI "Localized solitons in the plane" in "Nonlinear Evolution Equations : Integrability and Spectral Methods" Eds : A. DEGASPERIS, A.P. FORDY, M. LAKSHMANAN, Manchester Univ. Press (1989) ; and "Solitons in two dimensions" in "Integrable Systems and Applications" Eds : M. BALABANE, P. LOCHAK, D.W. Mc LAUGHLIN, C. SULEM in Lecture Notes in Physics (1989).

[7] A. DAVEY, K. STEWARTSON Proc. Roy. Soc. London A 338, 101 (1974)

[8] F. CALOGERO, A. DEGASPERIS, Nuovo Limento B 32, 201 (1976)

[9] S.V. MANAKOV, Physica D 3, 420 (1981)

[10] J.JP. LEON : "On the nonlinear evolutions having nonanalytic dispersion relations" in "Some topics on inverse problems" Ed : P.C. SABATIER, World Scientific (Singapore 1988) and
J.JP. LEON, F. PEMPINELLI : "Singular general evolutions in 1+1 and 2+1 dimensions" in "Nonlinear evolution equations : integrability and spectral methods" Eds. : A. DEGASPERIS, A.P. FORDY, M. LAKSHMANAN, Manchester University Press (1989).

The stochastic ϕ^4 model

Angel Sánchez and Luis Vázquez

Departamento de Física Teórica I
Facultad de Ciencias Físicas
Universidad Complutense
28040 Madrid (Spain)

Abstract

Some techniques for the study of the stochastic ϕ^4 model are described. These techniques are applied to a certain perturbation of the ϕ^4 potential and compared with numerical simulations. Results are reported concerning the accuracy of the stochastic numerical scheme, the evolution of the kink mean parameters with lose of brownian motion under certain conditions, and the modifications suffered by kinks, as well as their instability threshold.

1 Introduction

A great deal of attention has been devoted in the last years to study solitary-wave motion in one dimensional systems to modelate dislocations, domain walls, structural phase transitions, magnetic flux in Josephson junctions, charge density waves and many other phenomena. More recently, some simple models are being studied to clarify how these nonlinear excitations and the presence of disorder affect each other. Among them, the ϕ^4 chain is very suitable to incorporate different forms of disorder to its intrinsic nonlinearity, and so it can exhibit most of the features of these kind of problems.

In principle, the unperturbed (non disordered) model is a bistable monoatomic chain of N atoms described by the Hamiltonian

$$H = \sum_{n=1}^{N} \left\{ \frac{1}{2} m \left(\frac{dy_n}{dt} \right)^2 + \frac{1}{2} K (y_{n+1} - y_n)^2 + \varepsilon_0 \left(1 - \frac{y_n^2}{y_0^2} \right)^2 \right\}, \tag{1}$$

where $y_n(t)$ is the longitudinal displacement of the n-th particle; each of them interacts only with its two nearest neighbours and it is under the influence of an on-site, bistable potential. From the Hamiltonian (1) the equation of motion for each atom can be obtained and, adimensionalized for convenience, written as follows:

$$\frac{d^2 y_n}{dt^2} = y_{n+1} - 2y_n + y_{n-1} + p y_n - q y_n^3, \tag{2}$$

where p and q are some constants depending on the parameter y_0 and the height ε_0 of the potential barrier; as we can always change variables in (2) in such a way that $p = q = 1$, from now on we will only consider that choice and perturbations of it. The continuum description of the chain follows from the replacement of the discrete coordinate $y_n(t)$ by the displacement field $\phi(x,t)$, that leads to the well known ϕ^4 equation,

$$\phi_{tt} - \phi_{xx} - \phi + \phi^3 = 0. \tag{3}$$

To date, several ways of introducing disorder in these systems have been considered, involving the change of some of the parameters (mass, coupling constants, barrier parameters) of one or more atoms. The one we deal with is to assume that ε_0 is a random time-dependent function that introduces random coefficients into our equation; more precisely, we choose our system to be

$$\phi_{tt} - \phi_{xx} + \left[1 + \chi_{[a,b]}(x)\, V(t)\right] \left[-\sqrt{2D_1}\phi + \sqrt{2D_3}\phi^3\right] + \alpha\, \phi_t = 0, \tag{4}$$

where we limit the zone of influence of noise introducing $\chi_{[a,b]}(x) = \frac{1}{2}\left(\theta(x-a) - \theta(x-b)\right)$, θ being the Heaviside step function, and we allow dissipation through the term $\alpha\phi_t$. With respect to $V(t)$, its statistical properties are

$$< V(t) >= 0,\ \forall\, t \in [0,\infty],\ < V(t)V(t') >= \delta(t-t')\ \forall\, t,t' \in [0,\infty]. \tag{5}$$

The purpose of this kind of perturbation is twofold. Firstly, it may be used as a model for the potential arising from the interaction of noisy external fields with the regular background due to the other atoms, or for thermal movement caused by a heat bath. Secondly, as it admits further analytical study than the somehow more realistic spatial case, it makes possible to settle the accuracy of the numerical scheme that will be used for, and to get an idea of what is to be expected in, the case of disorder in space.

2 Analytical Results

2.1 Perturbative method

This technique has been widely employed to study nonlinear Klein-Gordon equations under weak perturbations [1,2,3]. The effect of such perturbations is supposed small enough to write the solution to (4) as a kink with velocity v plus a small deviation, i.e.

$$\phi(x,t) = \phi_k(x - vt) + \psi(x,t),\ \mid \psi(x,t) \mid \ll 1. \tag{6}$$

We insert *ansatz* (6) in equation (3)in order to learn what the excitations of the chain are, and after Lorentz transforming to the reference frame in which the kink is at rest at the origin, linearizing in the small quantity $\psi(x,t)$, and assuming for it a harmonic time dependence given by $\psi(x,t) = f(x)\, e^{-i\omega t}$, we obtain the following Schrödinger equation for the "modes" of the kink distortion:

$$- f''(x) + [2 - 3\cosh^{-2}\frac{\gamma}{\sqrt{2}}x]\, f(x) = \omega^2 f(x), \tag{7}$$

where $\gamma = (1-v^2)^{-1/2}$.

The spectrum of the operator in the l.h.s. of (7) contains a discrete part $\{\omega = 0, \sqrt{3/2}\}$ and a continuum of collision states $\{\omega = \sqrt{2+k^2}, k \in [0,\infty]\}$. From this spectrum, it can be seen that the effect of weak perturbations is a translation of the kink due to the $\omega = 0$ mode, and a deformation of its shape by phonon-like overimposed oscillations.

We can use this basis of eigenfunctions in two ways: on the one hand, it is possible to obtain information about the radiation spectrum originated by the perturbation; on the other hand, if the noise strength is not so large, the solution to (4) can be expanded using dispersions as small parameters and the coefficients in this expansion projected along the eigenfunctions of (7) to remove unbounded contributions. By this doing we have found a compatibility condition for the first order coefficient to exist; it imposes that, *in absence of dissipation, the kink speed is kept constant.* Succesive terms and conditions can be worked out without any problem but the rather cumbersome look of the equations. We only mention the previous result as a confirmation of the conclusions reported below.

2.2 Adiabatic approximation

In the previous section we have showed that the effect of small perturbations is a translation and a distortion of the kink. Now, we neglect deformations and suppose that the only influence of noise on the kink is to change its center and its velocity, that become functions of time. So we look for a solution to (4) of the form

$$\phi(x,t) = \tanh \psi(x,t), \;\; \psi(x,t) = \frac{\gamma}{\sqrt{2}}(x - z(t)), \tag{8}$$

that verifies $\psi_x = \gamma/\sqrt{2}$, and $\psi_t = -\gamma v/\sqrt{2}$; moreover, we decompose the effect of the perturbation by writing $z'(t) = v(t) + z_0'(t)$, to separate the contributions of the center and velocity. This is the foundation of the so called adiabatic approximation.

Next, we must recall that there are two quantities conserved in the evolution prescribed by the unperturbed equation (3). These quantities are the energy and the momentum; their expressions are

$$E \equiv \int_{-\infty}^{\infty} dx \, \frac{1}{2}(\phi_t^2 + \phi_x^2) + \frac{1}{4}(\phi^2 - 1)^2 = \frac{4\gamma}{3\sqrt{2}}, \tag{9}$$

$$P \equiv -\int_{-\infty}^{\infty} dx \, \phi_t \, \phi_x = \frac{4\gamma v}{3\sqrt{2}}, \tag{10}$$

the last expressions being valid only if ϕ is a kink. When we perturbe (3), both E and P are not conserved anymore but change in a precise way depending on ϕ, ϕ_t, ϕ_x and, of course, on the precise form of the perturbation considered. If we use this fact, take derivatives with respect to time in the particularization for kinks of (9) and (10) and compare the formulae so obtained, we get two nonlinear, coupled, stochastic differential equations for v and z_0 or equivalently z, that are quite unpleasant. To deal with them we assume that the interval $[a,b]$ where the noise acts is very much greater than the one in which the kink energy is concentrated. In this limit we arrive at

$$\begin{aligned} v'(t) &= -\alpha\, v\, (1 - v^2) \\ z'(t) &= v - \frac{3}{2} v\, (1 - v^2)\, V(t) \left[\sqrt{2D_1} - \frac{1}{3}\sqrt{2D_3}\right]. \end{aligned} \tag{11}$$

These are the Langevin equations that we must solve to obtain the mean properties of the kink parameters. However, the nonlinearity does not allow us to do it. An alternative way to find the main moments of z and v is to write the associated Fokker-Planck equation for their joint probability density W, namely

$$\frac{\partial W}{\partial t} = \left\{ -\frac{\partial}{\partial z} v + \frac{\partial}{\partial v}\left[\alpha v (1 - v^2)\right] + \frac{\partial^2}{\partial z^2}\left[\frac{9}{4}\left(D_1 - \frac{D_3}{9}\right) v^2 (1 - v^2)^2\right]\right\} W, \tag{12}$$

from which we can obtain an infinite hierarchy of equations for $< v^n(t) >$ and $< z^n(t) >$ $n = 1, 2, \ldots$. To get something out of it we could close and cut it by making further hypothesis, e.g. $< v^n(t) > = < z^n(t) > = 0 \; \forall n \geq 3$. This is the same as take the "non-relativistic" limit in the Langevin equations (11) supposing $v \ll 1$. We then linearize and solve them, and find

$$< v(t) > = v_0, \tag{13}$$

$$< z(t) > = v_0 t, \tag{14}$$

$$< v^2(t) > = v_0^2, \tag{15}$$

$$< z^2(t) > = \frac{9}{2} v_0^2 t \left(D_1 - \frac{D_3}{9}\right) + v_0^2 t^2, \tag{16}$$

$$< z(t) v(t) > = v_0^2 t; \tag{17}$$

where initial conditions $v(0) = v_0$, $z(0) = 0$ have been taken into account. If there is dissipation, $\alpha \neq 0$, these equations become slightly more complicated, because some coefficients that contain essentially $e^{-\alpha t}$ and powers of it appear, but there are no more noise contributions.

The main conclusion we can draw from these expressions is that *weak perturbations spatially extended do not affect seriously the evolution in time of non-relativistic kinks even if dissipation is present.* They only introduce a brownian contribution, i.e., a term containing t in $< z^2(t) >$. *However, and this is a remarkable fact, there are special cases, when $9D_1 = D_3$, in which even this contribution is absent, and there is no brownian motion at all.*

2.3 Exact results

Aside from the two approaches so far described, it is possible to calculate the evolution of the energy and momentum defined in (9) and (10). Using geometrical techniques to construct several Fokker-Planck-like equations, specially suitable to be applied to Hamiltoniam systems because of their associated symplectic structure, ordinary integro-differential equations can be found for E and P [4]. We do not go into the detail of their derivation but only mention the final expressions:

$$< \frac{dP}{dt} > \; = \; 0 \tag{18}$$

$$< \frac{dE}{dt} > \; = \; \int_{-\infty}^{\infty} dx \; \chi_{[a,b]}(x) \; < (-D_1\phi + D_3\phi^3)^2 > . \tag{19}$$

We must emphasize that these formulae are exact and do not hide any approximation at all, and hence they provide an excellent test for our numerical simulations.

3 Numerical simulation details

In order to achieve some knowledge of the system (4) in general situations, where the approximations may fail to account for the kink dynamics, we have performed some numerical simulations. We integrate (4) using a straightforward generalization of the Strauss-Vázquez scheme [5]:

$$\frac{\phi_j^{n+1} - 2\phi_j^n + \phi_j^{n-1}}{\Delta t^2} - \frac{\phi_{j+1}^n - 2\phi_j^n + \phi_{j-1}^n}{\Delta x^2} + \alpha \frac{\phi_j^{n+1} - \phi_j^{n-1}}{2\Delta t} +$$

$$+ \frac{1}{4}(1+V^n) \frac{[\sqrt[4]{2D_3}(\phi_j^{n+1})^2 - \sqrt{2D_1}/\sqrt[4]{2D_3}]^2 - [\sqrt[4]{2D_3}(\phi_j^{n-1})^2 - \sqrt{2D_1}/\sqrt[4]{2D_3}]^2}{\phi_j^{n+1} - \phi_j^{n-1}} \; = \; 0, \tag{20}$$

where $\phi_j^n \equiv \phi(j\Delta x, n\Delta t)$, Δx and Δt are respectively the spatial and temporal steps and V^n is the discretization of $V(n\Delta t)$.

When there is neither dissipation nor perturbations, the stability and convergence of the scheme can be proved [6] using its conservative character: it has a discrete energy given by

$$E^n \; \equiv \; \Delta x \sum_{j=-\infty}^{+\infty} \frac{1}{2}\left(\frac{\phi_j^n - \phi_j^{n-1}}{\Delta t}\right)^2 + \frac{1}{2}\left(\frac{\phi_j^n - \phi_{j-1}^n}{\Delta x}\right)\left(\frac{\phi_j^{n-1} - \phi_{j-1}^{n-1}}{\Delta x}\right) +$$

$$+ \frac{1}{8}\{[(\phi_j^n)^2 - 1]^2 + [(\phi_j^{n-1})^2 - 1]^2\}, \tag{21}$$

which remains invariant in time evolution.

To carry out calculations in the perturbed case, the noise $V(t)$, interpreted in the Stratonovič sense, can be simulated using a generator of gaussian distributed random numbers such that their dispersion is $\sigma^2 = 2/\Delta t$ [7,8]. What we did was to compute 30 particular time evolutions or "realizations" for ϕ and average over them to obtain the relevant quantities. We have always used $\Delta x = 2\,\Delta t = 0.05$. The numerical lattice was the spatial interval [-20,20] and the time interval [0,20], with null time derivative boundary conditions, $\phi_J^{n+1} = \phi_J^n$, $\phi_{-J}^{n+1} = \phi_{-J}^n$, $J = 1 + (20/\Delta x)$. Finally, we have chosen $D_1 = D_3$ varying from 10^{-3} to 1, and restricted to the interval $[-10, 10]$ and exceptionally $[0, 10]$, with initial conditions $v_0 = 0, 0.2$.

4 Main results and conclusions

We can summarize our main preliminary findings on the stochastic ϕ^4 chain as follows:

1. The stochastic Strauss-Vázquez numerical scheme is very accurate: the exact formulae (18) and (19) are perfectly consistent with our simulations. As far as we know, this is the first result concerning, and in some sense verifying, the stability and convergence of a numerical scheme for the integration of stochastic partial differential equations.

2. Kinks are stable under noises of strength 0.1 or less in adimensional units. Around that order of magnitude radiation begins to appear and kinks seem to be sensitively distorted by energy spreading to its surroundings. Over dispersions greater than 1 they are quickly destroyed, so we can take this value as the instability threshold. This must be compared to the situation in which $D_3 = 0$ [8], in which kinks are only stable under much weaker noises, of dispersion, say, 10^{-4}.

3. Far from the non-relativistic limit, kinks suffer a slowing down via energy conversion into phonon-like radiation whose rate is increased by nonlinear contributions. We can consider this fact important when v is around 0.1 or over.

4. We have also studied a case where our first approximation for widely spatially extended noises fails: incidence of a kink from an unperturbed to a stochastic zone. We have observed that kinks are more sensitive to noise, the instability threshold value is decreased by approximately an order of magnitude, but energy is much less converted into radiation and it diminishes practically only through kink slowing down.

Acknowledgements

We gratefully acknowledge the Centro de Investigaciones Energéticas, Medio Ambientales y Tecnológicas (C.I.E.M.A.T., Madrid, Spain) for allowing us to use their IBM 3090 computer in which all our computations have been done. We thank the financial support of Comisión Interministerial de Ciencia y Tecnología (C.I.C.Y.T., Spain) under grant PB86-0005. One of us (A.S.) wants to thank also the support from a fellowship of the Universidad Complutense de Madrid.

References

[1] M.B. Fogel, S.E. Trullinger, A.R. Bishop and J.A. Krumhansl, *Phys. Rev.* **B15**, 1578 (1977).

[2] J.F .Currie, J.A. Krumhansl, A.R. Bishop and S.E. Trullinger, *Phys. Rev.* **B22**, 477 (1980), and references therein.

[3] D.W. McLaughlin and A.C. Scott, *Phys. Rev.* **A18**, 1652 (1978), and references therein.

[4] J.M.R. Parrondo, M. Mañas and F.J. de la Rubia, preprint (1989).

[5] W. Strauss and L. Vázquez, *J. Comp. Phys.* **28**, 271 (1978).

[6] Guo Ben Yu and L. Vázquez, *J. App. Sci. (China)* **1**, 25 (1983).

[7] P.J. Pascual and L. Vázquez, *Phys. Rev.* **B32**, 8305 (1985).

[8] M.J. Rodríguez, Ph.D.Thesis, Universidad Complutense de Madrid (1988).

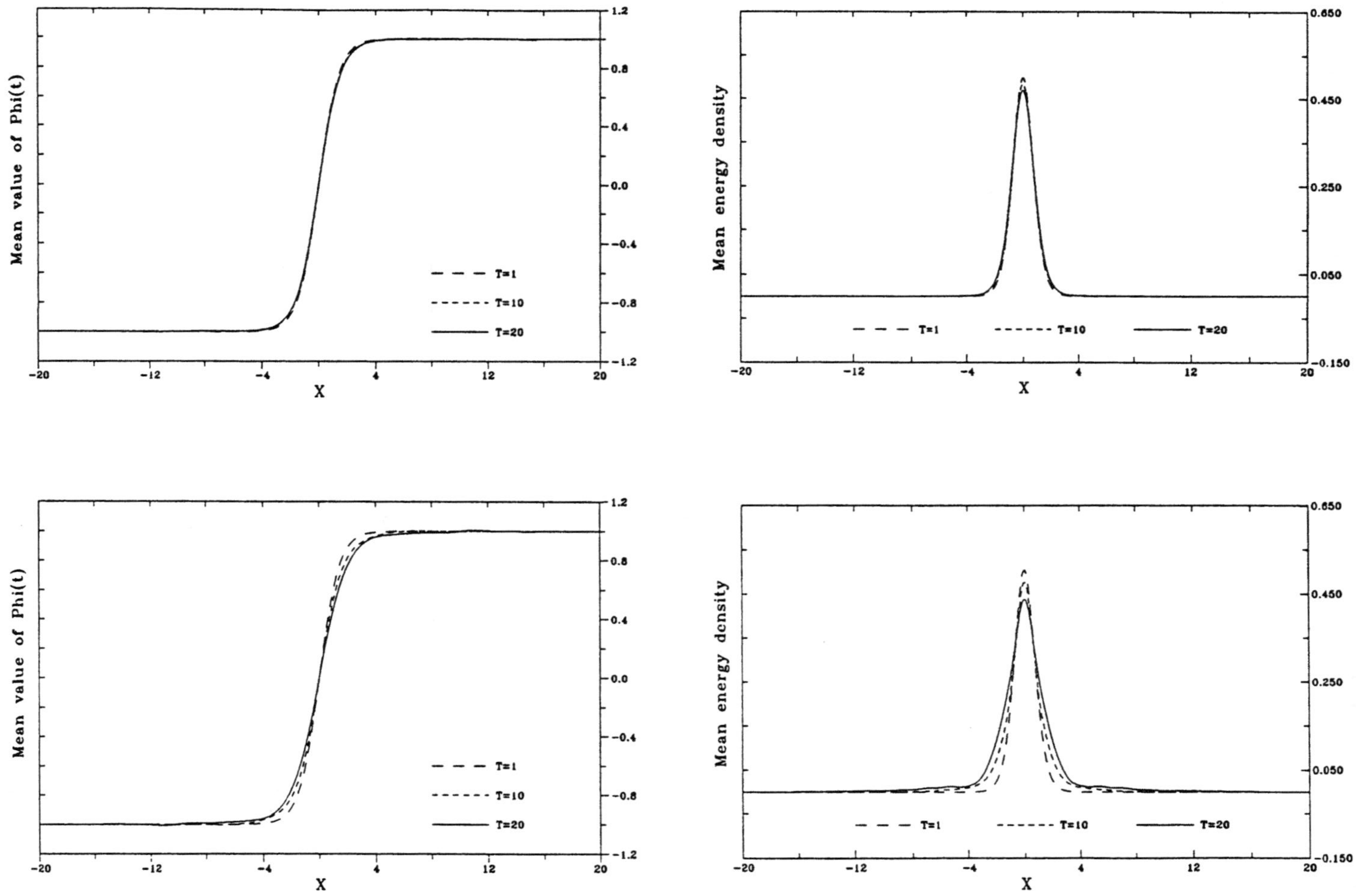

Figure 1 Mean kink shape and mean energy density at three different times for $v_0 = 0$. The effect of two noise intensities $2D = 0.01$ (up) and 0.1 (down) is shown. Notice that for the stronger one kink is clearly affected and phonon-like oscillations appear distinctly propagating outwards.

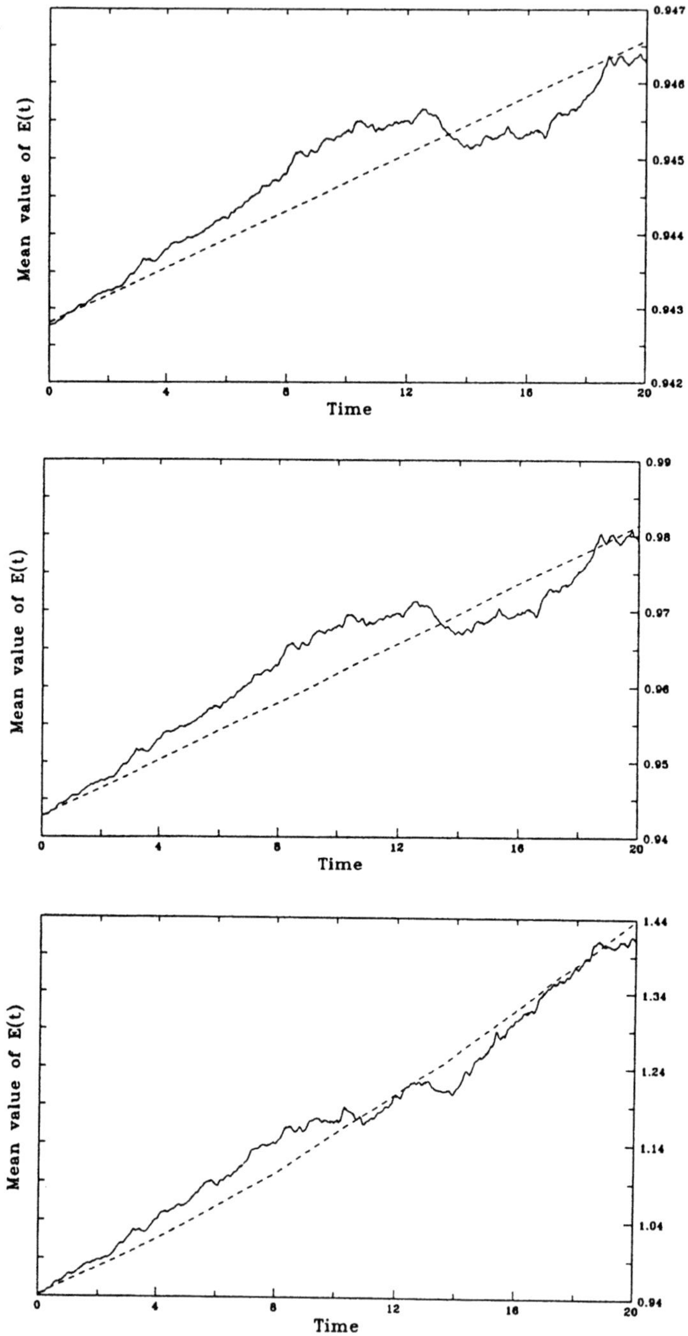

Figure 2 Mean value of the total kink energy as a function of time. Solid line is calculated directly from the numerical simulation, dashed line is obtained integrating equation (19). Dispersions are $2D = 0.001$, (up), 0.01 (middle) and 0.1 (down). Initial velocity is always $v_0 = 0$. Different scales are used for the vertical axis in each plot to emphasize the very good fitting of both results.

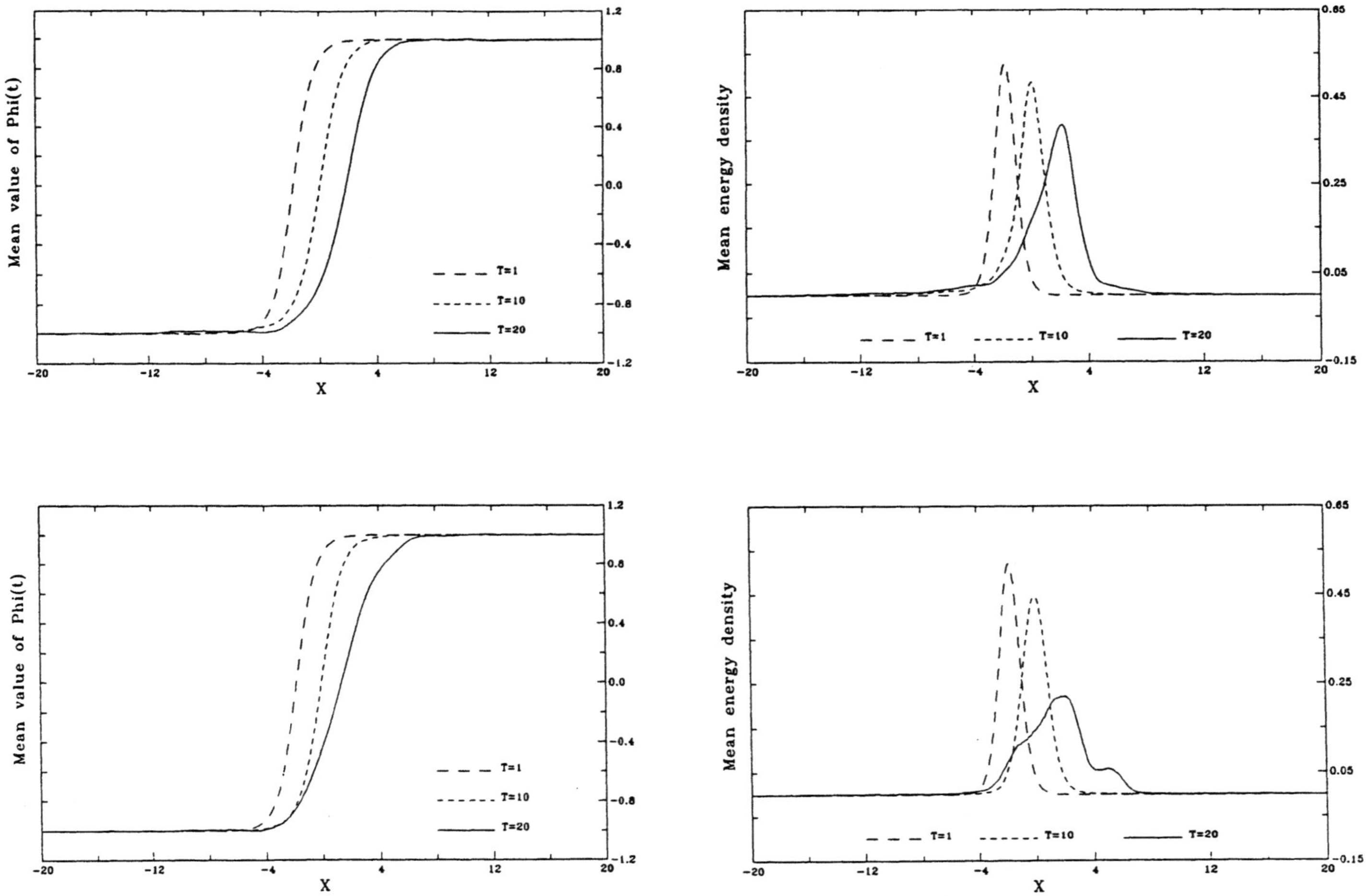

Figure 3 Simulation of a kink with $v_0 = 0.2$ when initially embedded in the noise zone (up) or incident on it from the unperturbed left zone (down; in this last case noise acts on $[0, 10]$). Separation from the adiabatic formulae (14) to (17) can be observed. Besides, the difference in radiation emitting in these two cases are also appreciated.

INVARIANTS FOR THE GENERALIZED LOTKA-VOLTERRA EQUATIONS

Laurent CAIRÓ and Marc R. FEIX
PMMS/CNRS - Université d'Orléans

Joao GOEDERT
Universidade Federal do Rio Grande do Sul
Porto Alegre (Brazil)

ABSTRACT

A generalisation of Lotka-Volterra System is given when self limiting terms are introduced in the model. We use a modification of the Carleman embedding method to find invariants for this system of equations. The position and stability of the equilibrium point and the regression of system under invariant conditions are studied.

INTRODUCTION

The main object of this work is to find the invariants (first integrals) of the nonlinear system of equations :

$$\dot{x}_i = x_i \left(a_i + \sum_{j=1}^{N} b_{ij} x_j \right) \quad \forall\, i \in \{1, \ldots, N\} \tag{1}$$

where the parameters a_i and b_{ij} are representative respectively of the linear growth rate (malthusian terms) and the interaction between the variables x_i. The dot means time derivative. This system has been introduced by Lotka [1] in the frame of chemical reactions, then by Volterra [2] to study the interaction between biological species (prey-predator system). Among the terms of the matrix $\| b_{ij} \|$, those of the main diagonal play a particular role as they control the growth of species having the malthusian terms a_i positif. They have been introduced by Verhulst [3]. Recently, the system has been employed in the dynamical study of gaz mixtures [4] [5]. Actually, it has other applications including turbulence [6].

I. THE VOLTERRA INVARIANT

Volterra [2] obtained a general invariant for this system in the particular case, where the Verhulst terms are zero. This invariant implies the antisymetry of matrix $\| a_{ij} \|$ such that :

$$a_{ij} = \beta_i\; b_{ij} \tag{2}$$

where the β_i called "equivalence" numbers by Volterra are representatif of an equivalent "mass" of species i. The antisymetry conditions lead to the mass conservation law when the malthusian terms are zero. In a previous paper [7], we show that in order to be able to write the general matrix in the Volterra form, we need the N relations $b_{ii} = 0$ plus (N-1) (N-2)/2 additional relations which make a total of N(N-1)/2 + 1 conditions. The aspect of Volterra invariant differs on the parity of N. So, for N even it writes :

$$\mathfrak{J} = \left(\frac{e^{x_1}}{x_1^{q_1}}\right)^{\beta_1} \left(\frac{e^{x_2}}{x_2^{q_2}}\right)^{\beta_2} \cdots \left(\frac{e^{x_N}}{x_N^{q_N}}\right)^{\beta_N} \tag{3}$$

Where the q_i are the coordinates of the central equilibrium point i.e. the solution of the system :

$$a_i + \sum_{j=1}^{N} b_{ij} x_j = 0 \qquad \forall\, i \in \{1, \ldots, N\} \tag{4}$$

In the N odd case, the invariant writes :

$$\mathfrak{J} = x_1^{R_1\beta_1} x_2^{R_2\beta_2} \cdots x_N^{R_N\beta_N} e^{st} \tag{5}$$

where the R_i are the minors of the matrix $\|a_{ij}\|$ and $s = \sum_{i=1}^{N} a_i \beta_i R_i$. (6)

The temporal dependance of this invariant implies that if the solutions to (1) must be considered bounded then there is evolution to a system of N - 1 variables, as the disappearance of at least one species is necessary to balance the effect of the exponential time dependent term. Volterra also shows that in the N = 2 case, the solutions satisfying (3) have the particularity of being periodic so that in phase space, they describe cyclic orbits.

II. THE CARLEMAN EMBEDDING TO FIND INVARIANTS

In the general case, written in (1) where the Verhulst terms are not zero we use a method which is a modification to the Carleman embedding [8] [9] and which has been used for the Lorenz model of turbulence [10] to [12]. It consists in a formal linearisation of the problem of finding invariants of the general type :

$$\mathfrak{J} = \prod_{j=1}^{N} x_j^{\alpha_j} \left[\sum_{k_1=0}^{L} \cdots \sum_{k_N=0}^{L} A_{k_1 \ldots k_N} \prod_{j=1}^{N} x_j^{k_j}\right] e^{st} \tag{7}$$

where the α_j and s are real numbers, the k_j are integers and L is taken generaly small. Then (7) is embedded in system (1) after applying the fundamental relation of invariance i.e.

$$\frac{d\mathcal{I}}{dt} = \frac{\partial \mathcal{I}}{\partial t} + \sum_{i=1}^{N} \frac{\partial \mathcal{I}}{\partial x_i} \dot{x}_i = 0 \tag{8}$$

A similar but less general ansatz was used in [12] to the Lorenz system. It differs from the present on the absence of terms in α_j. In fact, these terms take particular relevance in the Lotka-Volterra problem because of the presence of the factor x_k in the expression $\dot{x}_k$ preventing the cancellation of the α_j. Straightforward algebra using (7) and (1) transforms (8) as follows :

$$\frac{d\mathcal{I}}{dt} = \prod_{j=1}^{N} x_j^{\alpha_j} \left\{ \sum_{K_1=0}^{L} \sum_{k_N=0}^{L} [K_1, K_2 \ldots K_N] \prod_{j=1}^{N} x_j^{K_j} \right\} e^{st} \tag{9}$$

where

$$[K_1, K_2 \ldots K_N] \equiv \left[\underline{0}A_{k_1 \ldots k_N} + \underline{1}A_{k_1-1 \ldots k_N} + \underline{2}A_{k_1 k_2-1 \ldots k_N} + \underline{N}A_{k_1 \ldots k_N-1} \right] \tag{10}$$

and $\underline{0}$, $\underline{1}$, $\underline{2}$, ... $\underline{N}$ are operators acting on the polynomial coefficients $A_{k_1 \ldots k_N}$ of (7). If we introduce the notation $(k_1 \ldots k_N)$ for $A_{k_1 \ldots k_N}$, they write :

$$\underline{0}\ (k_1 \ldots k_N) \equiv \left[s + \sum_{i=1}^{N} (k_i + \alpha_i)\, a_i \right] (k_1 \ldots k_N)$$

$$\underline{1}\ (k_1 \ldots k_N) \equiv \left[\sum_{i=1}^{N} (k_i + \alpha_i)\, b_{i1} \right] (k_1 \ldots k_N)$$

$$\underline{2}\ (k_1 \ldots k_N) \equiv \left[\sum_{i=1}^{N} (k_i + \alpha_i)\, b_{i2} \right] (k_1 \ldots k_N) \tag{11}$$

$$\cdots\cdots\cdots\cdots\cdots\cdots$$

$$\underline{N}\ (k_1 \ldots k_N) \equiv \left[\sum_{i=1}^{N} (k_i + \alpha_i)\, b_{iN} \right] (k_1 \ldots k_N)$$

Now, the invariance condition (8) implies :

$$[K_1\ K_2 \ldots K_N] = 0 \tag{12}$$

Then the condition (12) can be written :

$$[K_1\ K_2 \ldots K_N] = \sum_{k_1} \ldots \sum_{k_N} M(K_1\ K_2 \ldots K_N\ ;\ k_1\ k_2 \ldots k_N)\ (k_1 \ldots k_N) = 0 \tag{13}$$

and the matrix M writes :

	(0...000)	(0...001)	(0...010)	(0...100)	....
[0...000]	0				
[0...001]	N				
[0...010]	N-1				
[0...100]	N-2				
......	⋮				
[01...00]	2				
[1...000]	1				
[0...002]		N			
[0...011]		N-1	N		
[0...020]			N-1		
[0...101]		N-2		N	
[0...110]			N-2	N-1	
[0...200]				N-2	
....					

Figure 1

Here each line of the matrix gives an equation to be fulfilled. The organisation of Figure 1 follows a certain pattern. First of all, we consider the coefficients $(k_1 \ldots k_N)$ for which $k_1 + \ldots + k_N \leqslant p$ (actually we take $p = 1$, as in the $N = 2$ case no other invariant is obtained for $p = 2$ and conjecture that the interesting results are limited to small p, otherwise we have too many relations to satisfy). If $p = 1$, we see that we have to cancel terms $[K_1 \ldots K_N]$ with $K_1 + \ldots + K_N \leqslant 2$. This is due to the nonlinear part of equation (1) which raises by one unit the degree of the term associated to $(k_1 \ldots k_N)$. As a consequence, we have more equations (lines) than unknowns (columns). This will lead to satisfy relations among the parameters of system (1) which are called the invariance conditions.

III. THE THREE INVARIANTS

a) Invariant I

The simplest choice of the coefficients $(k_1 \ldots k_N)$ is to consider only the first one i.e. (0...0). Now, from Figure 1, we see that this implies :

$$[0\ldots01] = 0 \ \ldots\ [1\ldots0] = 0 \tag{14}$$

i.e.

$$\mathrm{Det}\ \| b_{ij} \| = 0 \tag{15}$$

in order to obtain a non trivial solution for α_i. Condition (15) states that the equilibrium point given by (5) does not exist, and as a consequence, it has less interest, except in the particular case, where all the malthusian terms are zero.

b) Invariant II

If we take the N columns (0...01) to (1...00), we determine the α_i from [0...2] = 0 to [2...0] = 0 provided Det $\|b_{ij}\| \neq 0$. Then s is computed from relation [0...0] = 0. As for the rest, we obtain the invariance conditions namely : N - 1 conditions from [0...1] = 0 to [1...0] = 0 giving $a_i = a_j = a$ and (N-1)(N-2)/2 conditions resulting from relations [0...11] = 0 to [11...0] = 0 in order to obtain the N coefficients (0...1) to (1...0) of the problem. These last conditions write :

$$\bar{R}_{ijk} = (b_{ki} - b_{ii})(b_{ij} - b_{jj})(b_{jk} - b_{kk}) + (b_{ik} - b_{kk})(b_{ji} - b_{ii})(b_{kj} - b_{jj}) = 0 \quad (16)$$

Notice that all together, we have N(N-1)/2 conditions. The invariant writes :

$$\mathfrak{I} = \prod_j x_j^{\alpha_j} \, [(1...0)\, x_1 + (01...0)\, x_2 + \ldots + (0...1)\, x_N]\, e^{st} \quad (17)$$

where s has the particularity of being zero for N odd.

c) Invariant III

If we consider the N + 1 first columns, the N + 1 coefficients (0...0) to (1...0) will be perfectly determined without any condition but lines [0...11] = 0 to [11...0] = 0 will give the N(N-1)/2 relations $R_{ij} = 0$ defined by :

$$R_{ij} = a_i\, b_{ji}\, b_{jj} + a_j\, b_{ij}\, b_{ii} - (a_i + a_j)\, b_{ii}\, b_{jj} \quad (18)$$

and the invariant writes :

$$\mathfrak{I} = \prod_j x_j^{\alpha_j} \left[1 + \frac{b_{11}}{a_1} x_1 + \ldots + \frac{b_{NN}}{a_N} x_N\right] e^{st} \quad (19)$$

Where now s is zero for N even. For N = 2, this invariant produces cyclic motions [13].

IV. THE CENTRAL EQUILIBRIUM POINT

Under invariant II, the central equilibrium point Q is over the hyperplane $\bar{P}$, defined as :

$$\bar{P} = (1...0)\, x_1 + (01...0)\, x_2 + ... + (0...1)\, x_N = 0 \qquad (20)$$

if and only if $s \neq 0$ which is true for N even. This follows immediately from the definition of equilibrium point and the form (17) of the invariant which has then an exponential time dependance. Similarly and if invariant III conditions are satisfied, the central equilibrium point Q is over the hyperplane P :

$$P = 1 + \frac{b_{11}}{a_1} x_1 + ... + \frac{b_{NN}}{a_N} x_N \qquad (21)$$

if and only if $s \neq 0$, that happens now for N odd, as is deduced from (19). Notice that for invariant II and N odd, Q is also over P. Actually analytic calculations have been carried on up to N = 5.

An important point is the stability of the solutions near the point Q. Linearizing the system (1) around Q one can see that when the invariant III conditions are satisfied and the equilibrium point is in R^+, the equilibrium is marginally stable (real parts of imaginary eigenvalues are zero).

V. THE REGRESSION

a) The assymptotic regression

As for the equilibrium point, we must distinguish between invariant II and invariant III and N odd or even. Moreover, we assume to deal with bounded solutions. So, under invariant II and for even values of N, the expression (17) of the invariant tells us that there is an asymptotic regression either to the disappearance of one species like in the Volterra case, or to a drift to the limit hyperplane $\bar{P}$ defined above.

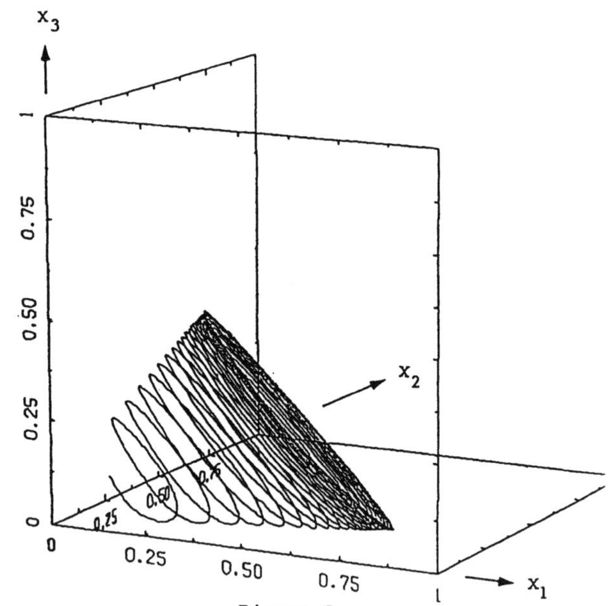

Figure 2

Under invariant III conditions, and for N odd, we have the same phenomena, except that now, there is evolution to the hyperplane P (see figure 2). Moreover, one can see that over those hyperplanes, the invariant character is preserved, that is to say we have regression to a N-1 system with conservation of the invariant type [7].

b) <u>The classical regression</u>

When all the a_i are of the same sign, we are faced with the competition problem. If morever all the a_i are equal, then we can regress from N to N - 1(see [14] and [15]). Now, if in addition, we satisfy the $\bar{R}_{ijk} = 0$ relations (invariant II) then, the reduced system satisfies the $R_{ij} = 0$ relations of invariant III. This comes from the particular form of the reduced N - 1 system of equations. From [15], in the N = 3 case, we have, distinguishing with bars the reduced system :

$$\bar{a}_1 = (b_{13} - b_{33})\, c \qquad \bar{a}_2 = (b_{23} - b_{33})\, c$$
$$\bar{b}_{11} = b_{11} - b_{31} \qquad \bar{b}_{12} = b_{12} - b_{32} \qquad \bar{b}_{21} = b_{21} - b_{31} \qquad \bar{b}_{22} = b_{22} - b_{32} \tag{22}$$

Where c is a constant which represents the value of the third coordinate. Taking (22) and replacing them in $\bar{R}_{123} = 0$, we easily obtain :

$$\bar{a}_1 \left(- \bar{b}_{22}\right)\left(\bar{b}_{21} - \bar{b}_{11}\right) + \bar{a}_2\left(- \bar{b}_{11}\right)\left(\bar{b}_{12} - \bar{b}_{22}\right) = 0$$

that is to say $R_{12} = 0$.

c) <u>The regression N to N - 2</u>

It takes place if N is even under invariant II conditions and follows from the properties described preceedingly. Consequently (see Figure 3) one can go from A to D via two routes namely ABD and ACD. The first one is characterised by a horizontal path AB, done under invariant II conditions by asymptotic regression (let's call this : horizontal regression) and an oblique path BD done by classical regression (oblique regression). As for the ACD route, we first apply the "oblique" regression on path AC and then the "horizontal" one on path CD. The two routes ABD and ACD commute provided that the species taken as constant in the "oblique" regression be the same in both routes.

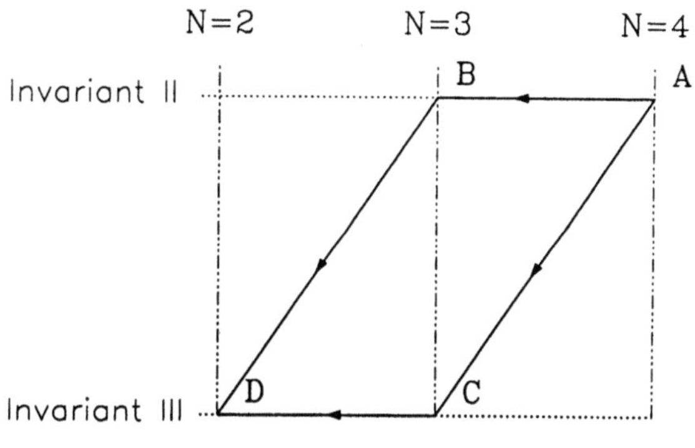

<u>Figure 3</u>

d) <u>Application : cyclic orbits for N = 4</u>

If N = 4 in Figure 3, we may proceed like it follows. On route ABD, the path AB is done, for instance, eliminating species 3 and the path BD by taking species 4 as a scale constant. On route ACD, the path AC is covered taking species 4 as constant, and finaly we eliminate species 3 on path CD. Now at D we are under invariant III conditions and consequently the orbits are cyclic [13]. As this cyclic character must be conserved through the multiple regressions considered, it turns out immediately that it must also be present for N = 4 under invariant II conditions (see Figure 4).

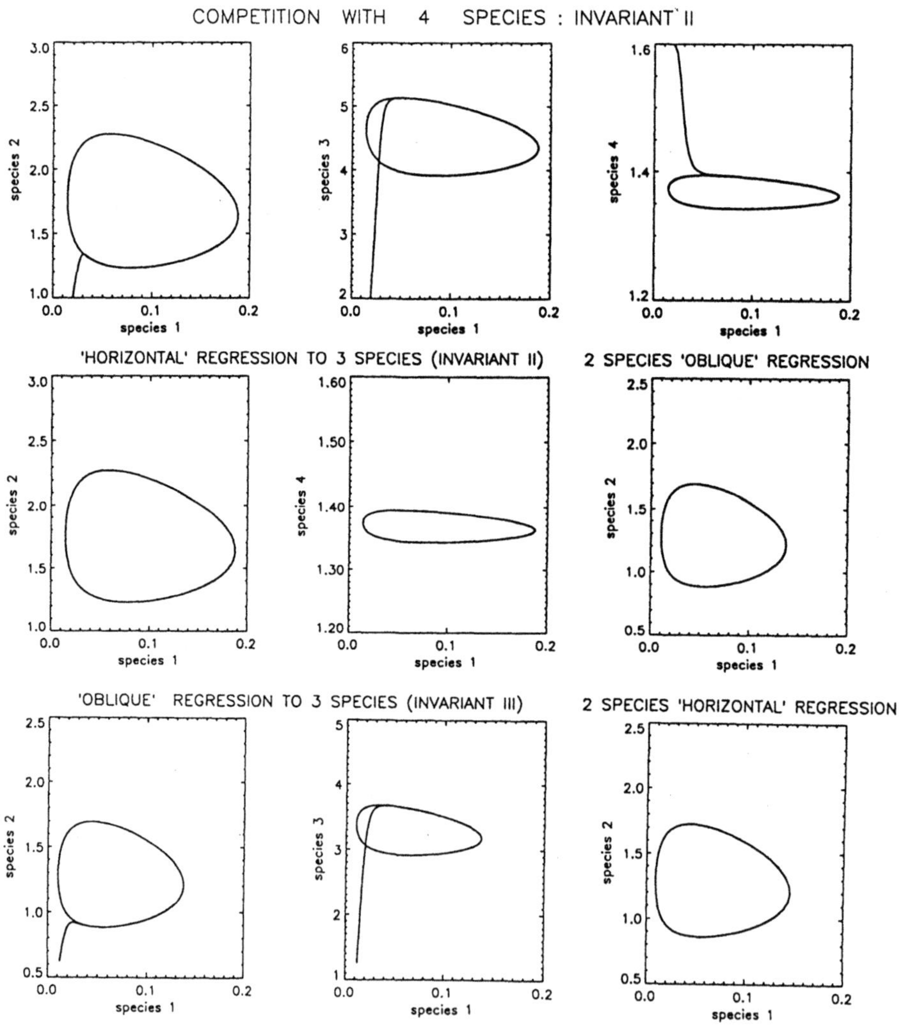

Figure 4

CONCLUSION

This work is a generalisation to an arbitrary matrix $\|b_{ij}\|$ of the theory done by Volterra for a special one. Among the most interesting results are the two "horizontal" asymptotic regressions for type II invariant (N even) and type III invariant (N odd) and the two "oblique" ones from a type II to a type III invariant. As a consequence, we can explain the asymptotic cyclic orbits for the N = 4 case under invariant II conditions.

BIBLIOGRAPHY

[1] **LOTKA A.J.** "Contribution to the theory of Periodic Reactions", Journal of Physics Chemistry 14, (1910), 271-274.

[2] **VOLTERRA V.** "Leçons sur la théorie mathématique de la lutte pour la vie" Gauthier Villars, Paris, 1931.

[3] **VERHULST P.F.** Nuov. Mem. Acad. Roy. Bruxelles 18 (1845), 1. See also 20 (1847), 1.

[4] **BOFFI V.C., FRANCESCHINI V. and SPIGA G.** "Dynamics of a Gaz Mixture in an Extended Kinetic Theory", Phys. of Fluids 28 (1985), 3232-3236.

[5] **LUPINI R. and SPIGA G.** "Chaotic Dynamics of Spatially Homogeneous Gaz", Phys. of Fluids 31 (1988), 2048-2051.

[6] **GOEL N., MITRA S. and MONTROLL E.W.** "On the Volterra and other Nonlinear Models of Interacting Populations", Rev. of Mod. Phys. 43 (1971) 231-276.

[7] **CAIRO L., FEIX M.R. and GOEDERT J.** "Invariants for Models of Interacting Populations", accepted for publication in Physics Letters A, 1989.

[8] **CARLEMAN T.** "Application de la théorie des équations intégrales linéaires aux systèmes d'équations différentielles non linéaires", Acta Mathematica 59, (1932), 63-87.

[9] **MONTROLL E.W. and HELLEMAN R.H.G.** "Topics of Statistical Mechanics and Biophysics", R.A. Piccirelli Ed. AIP Conf. Proc. 27 (1976), 75.

[10] **ANDRADE R.F.S. and RAUCH A.** "The Lorenz Model and the Method of Carleman Embedding", Phys. Letters 82 A, (1981), 276-278.

[11] **STEEB W.H. and WILHELM F.**, "Non-linear Autonomous Systems of Differential Equations and Carleman Linearization Procedure", J. Math. Anal. Appl. 77 (1980) 601-611.

[12] **KUS M.** "Integrals of Motion for the Lorenz System", J. Phys. A. Math. Gen. 16 (1983), L 689-691.

[13] **GUCKENHEIMER J. and HOLMES P.** "Nonlinear Oscillations, Dynamical Systems and Bifurcations of Vector Fields", Chapter 7.5, Springer Verlag, 1985.

[14] **MAY R.M., LEONARD W.J.** "Nonlinear Aspects of Competition between three Species", SIAM J. Appl. Math. 29 (1975) 243-253.

[15] **COSTE J., PEYRAUD J. and COULLET P.** "Asymptotic Behaviours in the Dynamics of Competing Species", SIAM J. Appl. Math 36 (1979) 516-543.

SLOW RELAXATION EFFECTS IN THE EVOLUTION OF NONLINEAR QUANTUM SYSTEMS

V. M. Kenkre
Department of Physics and Astronomy, University of New Mexico
Albuquerque, NM 87131, USA

1. Introduction

Recent approaches to the study of one of the fundamental problems of condensed matter physics, viz., the description of the influence of strong interactions of the lattice on quasiparticles such as electrons or excitons moving in solids, have been based on discrete nonlinear quantum evolution equations[1-11]. Here we present new results regarding the coupled evolution of such quasiparticles and of the lattice oscillators with which they interact, obtained without making the usual "adiabatic" approximation. There are three parts to the results we have recently obtained: basic nonadiabatic nonlinear effects[12], interplay of nonlinearity and quantum phases in nonadiabatic systems[13,14], and the effects of finite temperature[15]. We describe only the first two in the following.

The essential feature of the systems of interest to the present investigation is a strong interaction between a quasiparticle which moves on a lattice in keeping with a quantum evolution equation, and oscillators whose displacements modulate the quasiparticle parameters such as the site energy. Several different sets of coupled equations for the moving quasiparticle and the lattice vibrations have been arrived at from microscopic analysis or postulated[1-5]. Consider the set used by Scott[2,3] and his collaborators[4,5]. Of their two coupled equations, one for the moving quasiparticle and the other for the vibrations which interact with the quasiparticle, we simplify the second and write,

$$i\,dc_m/dt = V(c_{m+1} + c_{m-1}) + E\,x_m\,c_m \tag{1.1}$$

$$d^2x_m/dt^2 + \alpha(dx_m/dt) + \omega^2 x_m = -(\text{constant})|c_m|^2 \tag{1.2}$$

Equation (1.1) describes the evolution of the quasiparticle: its amplitude for occupation of the site m is $c_m(t)$ and its (nearest-neighbour) intersite transfer matrix element is V. Equation (1.2) describes the evolution of the lattice vibrations with which the quasiparticle interacts: the oscillator at site m has displacement $x_m(t)$ and frequency ω, and is damped at a rate α. The strong interaction between the quasiparticle and the lattice oscillators is represented by the last terms in each of (1.1) and (1.2): the site energy of the quasiparticle is proportional to the displacement of the oscillator at site m; and the equilibrium position of the oscillator is shifted by the presence of the quasiparticle at site m by an amount proportional to the probability of site occupation. In writing (1.2) above, the simplification we have made relative to the equations of Scott et. al.[2-5] is to replace the dispersion of the lattice vibrations by a damping term. The essential physics of the removal of energy at a given site is described by our damped dispersionless (Einstein) oscillators. Intersite motion of the x-vibrations is not.

In the adiabatic (or infinite-relaxation) limit[16] wherein the vibrations described by the x's are assumed to reach their equilibrium positions rapidly relative to the evolution of the quasiparticle, the time derivatives of the x's in (1.2) are neglected and the resulting discrete nonlinear Schrödinger equation

$$i\,dc_m/dt = V(c_{m+1} + c_{m-1}) - \chi\,|c_m|^2\,c_m \qquad (1.3)$$

is taken as a point of departure for the analysis of the evolution of the quasiparticle. Although (1.3) in its stationary form had been obtained as early as 1959 by Holstein[16] and although much physics has been extracted from (1.3) or related equations through numerical analysis[4,5], exact solutions of (1.3) are not known in general. However, it has been found recently[7,8] that considerable insights can be gained into the essential physics of the system through exact analytical solutions of (1.3) when the system has only two sites (m = 1,2). Results available in the literature pertinent to this dimer version are: exact time evolution in terms of Jacobian elliptic functions for arbitrary initial conditions[7a,8a,8b], explicit demonstration of self-trapping transitions and other rich behaviour in the time evolution[7,8], evaluation of the stationary self-trapped states of the dimer[4,7b,8a], application of the analysis to specific experimentally realizable dimer systems[8a,8c], and evaluation of nonlinear memory functions[10,11]. Since a great deal of insight into the physics of the problem can thus be gained in the dimer (two-site) context, we present the analysis of finite-relaxation and finite temperature effects in the system of two sites.

In order to arrive as rapidly as possible at the essential physics of our problem, it is advisable, but not essential, to consider a simplification of the oscillator equation (1.2). Assume that the damping coefficient α is large enough to justify the neglect of the second derivative of the oscillator displacements. More formally, we take the limit $\omega\to\infty$, $\alpha\to\infty$, $\omega^2/\alpha = \Gamma$. The evolution of the oscillator displacements x towards their equilibrium positions now possesses a single characteristic "vibrational relaxation" time, viz. $\alpha/\omega^2 \equiv 1/\Gamma$, and (1.2) reduces to

$$dx_m/dt + \Gamma x_m = -(\chi\Gamma/E)\,|c_m|^2 \tag{1.4}$$

The constant factor in (1.2) has been expressed in (1.4) as $\chi\Gamma/E^2$ in terms of the vibrational relaxation rate Γ and of E and χ. The quantity E is the rate of change of the quasiparticle site energy with the oscillator displacement, and is of no importance here. The quantity χ, which we will call the nonlinearity parameter, occurs naturally in the study of the adiabatic dimer[7,8] and represents the lowering of the site energy of the quasiparticle that occurs as a result of the feedback phenomenon. Equation (1.4) describes the oscillator being driven by relaxation processes at a single rate Γ to its dynamic equilibrium position which is the product of $-(\chi/E)$ and the probability that the oscillator site is occupied by the quasiparticle.

The dimer (two-site) case of (1.1) and of (1.4) form the point of departure for our present analysis:

$$i\,dc_1/dt = Vc_2 + E\,x_1\,c_1 \tag{1.5}$$
$$i\,dc_2/dt = Vc_1 + E\,x_2\,c_2 \tag{1.6}$$
$$dx_1/dt + \Gamma x_1 = -(\chi\Gamma/E)\,|c_1|^2 \tag{1.7}$$
$$dx_2/dt + \Gamma x_2 = -(\chi\Gamma/E)\,|c_2|^2 \tag{1.8}$$

2. Basic Effects of Slow Relaxation in Nonlinear Systems

In this section, we focus attention on the simplest initial condition: the two oscillators are in their equilibrium positions in the absence of the quasiparticle, i.e. $x_1(0) = x_2(0) = 0$, and the quasiparticle is placed suddenly on one of the sites: $|c_1(0)|^2 = 1$, $|c_2(0)|^2 = 0$. A physical example of such sudden placement is the injection of a photoinduced carrier in a molecular crystal, or the creation of an electronic excitation through absorption of radiation[17]. It is found that the quasiparticle observable $p(t) \equiv$

$|c_1|^2 - |c_2|^2$ (the difference in the probabilities of occupation of the two sites), and the vibration observable $y(t) \equiv (-E/\chi)[x_1(t) - x_2(t)]$ (the difference between the displacements of the two oscillators normalized in the manner shown) obey the coupled evolution equations

$$d^2p(t)/dt^2 = Ap(t) - Bp^3(t) + C(t) \tag{2.1}$$
$$dy(t)/dt + \Gamma y(t) = \Gamma p(t) \tag{2.2}$$

The constants A and B in (2.1) are given, as in the adiabatic case[7] for initial occupation of one site, by

$$A = (\chi^2/2) - 4V^2; \qquad B = \chi^2/2 \tag{2.3}$$

and C(t), the correction term which marks the difference between the present analysis and the usual "adiabatic procedure" involving the replacement of y(t) by p(t), is given by

$$C(t) = -\chi^2 \{ y(t) \int ds\, y(s)[dp(s)/ds] + \tfrac{1}{2}[p(t)-p^3(t)]\} \tag{2.4}$$

On the basis of (2.1),(2.2) we have obtained striking results both through numerical analysis and approximate analytical methods. Numerical work shows a number of remarkable findings:

(1) The probabilities of the two site states settle at long times into constant values. Inspection establishes these values for this <u>non-adiabatic</u> dimer to be essentially the <u>stationary state</u> probabilities of the <u>adiabatic</u> dimer found first by Eilbeck et. al.[4] For $\chi \leqslant 2V$ these values equal 0.5. For $\chi \geqslant 2V$ they are given by $\tfrac{1}{2}\{1 \pm [1-(2V/\chi)^2]^{\frac{1}{2}}\}$. This is in contrast to earlier studies of damping[9] which were based on the procedure of appending stochastic Liouville equation terms to the dimer evolution and which resulted in an (unphysical) approach to equal probability distribution on the two sites.

(2) For values of the damping rate Γ which are large enough (with respect to other rates such as V in the system), the evolution of the probabilities <u>first follows the Jacobian elliptic function</u> dynamics of the adiabatic dimer deduced analytically by Kenkre and Campbell[7], <u>and then settles</u> into the stationary state values given by Eilbeck et. al.[4] The short time evolution is approximately described by $cn(2Vt \mid \chi/4V)$ for $\chi < 4V$ and by $dn(\tfrac{1}{2}\chi t \mid 4V/\chi)$ for $\chi > 4V$.

(2) A consequence of the conjunction of (a) and (b) is that, for appropriately large values of Γ, the variation of the nonlinearity parameter χ shows the coexistence of two transitions: the static transition at $\chi = 2V$ and the dynamic transition at $\chi = 4V$. The former is reflected in the long time behaviour in that the asymptotic value of the probabilities is 0.5 for $\chi < 2V$ but different from 0.5 for $\chi > 2V$. The latter is seen in the short time behaviour in that the probabilities are characteristic of the cn function and oscillate on both sides of 0.5 for $\chi < 4V$ but are characteristic of the dn function and oscillate only on one side of 0.5 for $\chi > 4V$.

(3) The probabilities exhibit a potentially misleading evolution for $2V < \chi < 4V$: at first their oscillations are damped to the value 0.5. After apparently settling into the value 0.5, they swing away into the stationary state values which are other than 0.5 and given in (a) above. This has considerable practical relevance. If numerical analysis were not carried out far enough, one could conclude erroneously that the dimer equilibrates to an equal-probability state on the two sites. Long time measurements with different characteristic times could lead to sharply different interpretations of the stationary behaviour of this system.

(4) For $\chi > 2V$, whether the long time limit of the probability of occupation of a given site is $\frac{1}{2}\{1+[1-(2V/\chi)^2]^{\frac{1}{2}}\}$ or $\frac{1}{2}\{1-[1-(2V/\chi)^2]^{\frac{1}{2}}\}$, is a function of the value of the damping rate Γ. Thus, a variation in Γ causes the equilibrium value to switch back and forth between the two options.

Analytical arguments which support the above numerical findings can be given for the highly physical case of fast vibrational relaxation. For the procedure used the reader is referred to ref. 12. The final result is a closed equation for p(t):

$$d^2p/dt^2+(\chi^2/2\Gamma)(1-p^2)(dp/dt) = A'p - Bp^3 \tag{2.5}$$

which is similar to but different from the adiabatic case:

$$d^2p/dt^2 = Ap - Bp^3 \tag{2.6}$$

(2.6) being simply (2.1) without the C(t) term. Comparison of (2.5) and (2.6) shows two differences. The non-adiabatic equation (2.5) has a "friction" term proportional to dp/dt which drives the p to its stationary state value, and the term proportional to p on its right hand side is not the constant A of the adiabatic case but is time-dependent:

$$A' = (\chi^2/2) - \{4V^2 - (\chi^2/T)\int ds[dp(s)/ds]^2\} \tag{2.7}$$

Equation (2.5) can be looked upon as describing a fictitious classical oscillator whose displacement is represented by the quantity p. The correction to A provided by the new term in A' clearly represents a dissociative antirestoring force acting on the fictitious oscillator since the factor $(\chi^2/T)\int ds[dp(s)/ds]^2$ multiplying p(t) in (2.7) is always positive. As $t \to \infty$, the value of dp/dt must either vanish, or oscillate, or tend to a nonzero constant, or grow without limit. The last three possibilities would cause (dp/dt) to have a positive nonvanishing value. The antirestoring force would then grow without bound and dissociate the fictitious oscillator. It is guaranteed, however, that this cannot happen since p, being the probability difference between the two dimer sites, is constrained to lie between -1 and 1. It follows that dp/dt vanishes as $t \to \infty$. It is thus possible to demonstrate analytically that the long time evolution of the probabilities governed by (2.5) is to stationary values. These values are given by putting the derivatives of p equal to zero in (2.5). Ignoring the trivial solution ($p(\infty) = 0$), we obtain, for the stationary probability difference,

$$p(\infty) = (1 - \{(8V^2/\chi^2) - (2/T)\int ds[dp(s)/ds]^2\})^{\frac{1}{2}} \tag{2.8}$$

The difference A' - A has a strong effect on the stationary states given by (2.8). Neglect of that difference would lead one to the erroneous conclusion that $p(\infty)$ equals $[1 - (8V^2/\chi^2)]^{\frac{1}{2}}$. The latter value corresponds not to the stationary states but to the average values around which the initially localized adiabatic dimer oscillates[7]. The correct adiabatic result for the stationary states is[4,7b,8a]

$$p(\infty) = [1 - [(4V^2/\chi^2)]^{\frac{1}{2}} \tag{2.9}$$

The integral $\int ds[dp(s)/ds]^2$ thus builds up from its initial zero value to $2TV^2/\chi^2$ as $t \to \infty$. Since dp/dt vanishes as $t \to \infty$, it is straightforward to return to the original equations of motion (2.2)-(2.5), put the x derivatives equal to zero, substitute the x values thus obtained in (2.2) and (2.3), and obtain the stationary state values of the probabilities by taking c_1 and c_2 to be proportional to $\exp(i\mathcal{E}t)$ where $\mathcal{E}$ is the energy of the stationary state. The result is indeed as given by Eilbeck et. al.[4]. These analytic arguments thus explain why the probabilities in the non-adiabatic dimer are driven to the stationary values of the adiabatic dimer.

3. Interplay of Nonlinearity and Quantum Phases in Nonadiabatic Systems

Consequences of the coupled equations (1.5)-(1.8) display a rich interplay of quantum phases in the initial conditions of the quasiparticle and the nonlinearity of the evolution equations. Consider, for simplicity, the initial oscillator displacements to be zero. Let the quasiparticle be placed almost equally on the two sites (for instance with 0.51 and 0.49 as the two site probabilities) but with the amplitudes in phase in one case and out of phase in the other. In other words, $c_1(0) = (0.51)^{\frac{1}{2}}$ in both cases but $c_2(0) = (0.49)^{\frac{1}{2}}$ in the first case but $-(0.49)^{\frac{1}{2}}$ in the second. It is found that while the probability of the initially occupied site tends to its stationary state value (0.958 or 0.042) for long times in both cases, the influence of the initial phase is pronounced: when the amplitudes are initially in phase, wild oscillations are seen in the probability whereas in the out-of-phase case the evolution is much quieter.

Detailed studies of the dependence of the evolution on initial phases uncover (see ref. 13) the considerable influence that initial phases have on the evolution. Among the features to be noticed are the approach to the stationary states for long times, the so-called "repulsion effect" derived by Tsironis and Kenkre[8b] for the adiabatic dimer, and a switching between stationary states as the initial phase is varied. The physical significance of the switching phenomenon is that, for extended systems, a quasiparticle placed in one location in a crystal could find itself self-trapped at quite different locations in the crystal, the precise spot of such self-trapping being dependent sensitively on the initial phase. This remarkable finding has relevance to the eventual localization of electronic excitations in biological systems such as photosynthetic units[18]. By the "repulsion effect" is meant an initial tendency of the probability to increase rather than to decrease in value[8b] which is observed for certain initial phases and which is a pure consequence of the interplay of nonlinearity and the quantum nature of the system. Another fascinating consequence of this interplay is a dramatic antidamping phenomenon which can be seen in the 0 phase difference case but not in the π phase difference case. It can be understood from a fast-relaxation analytic argument as follows. The generalization of (2.5) of section 2 abvoe to the case of arbitrary initial conditions treated in this seciton is

$$d^2p/dt^2 + D\,(dp/dt) = A'\,p - Bp^3 \tag{3.1}$$

the nonlinear damping coefficient D being given by

$$D = (\chi^2/2\Gamma)\,[(p_0{}^2 - p^2) - (4V/\chi)(\rho_{12}+\rho_{21})_0\,] \tag{3.2}$$

$$A' = A + (\chi^2/\Gamma)\int ds[dp(s)/ds]^2 \tag{3.3}$$

$$A = (\chi^2/2) - 4V^2 - 2V\chi\,(\rho_{12}+\rho_{21})_0 \tag{3.4}$$

The particular case of initial localization on one of the two sites is obtained by putting $p_0 = 1$ and $(\rho_{12}+\rho_{21})_0 = 0$, and has been described in section 2 above. One sees from (3.2) that, in the in-phase case, D can be negative as a result of the initial value $(\rho_{12}+\rho_{21})_0$ of the sum of the off-diagonal elements of the density matrix and thereby cause an initial increase in the probability oscillation amplitude (antidamping) eventually followed by an approach to the stationary state.

4. Concluding Remarks

The main physics to emerge from the analysis of nonadiabatic systems reported above is the recovery of the adiabatic evolution at short times and the approach towards the adiabatic stationary states at long times. The former occurs only if vibrational relaxation is fast on the scale of the other time constants of the system such as V, while the latter occurs for slow as well as fast relaxation. It is important to realize that adiabatic evolution does not show approach to the adiabatic stationary states because of the absence of damping agents. Nor does approach towards the stationary states occur if such damping is put in externally as in ref. 9. It is remarkable that it does occur naturally in the investigations reported above. Our manner of introducing damping in this study is in the vibrational relaxation process and therefore appropriate from the physics of the system. Our results have a bearing on some questions that have been often asked in the context of energy transfer in molecular crystals and photosynthetic systems[17-19]. One of these questions addresses the issue of whether it is ever possible for electronic excitations produced by the absorption of light in such systems to find themselves localized in various places in a manner that has been described as being like "raisins in a pudding"[18]. Our study shows explicitly that strong excitation-phonon interactions can indeed lead to such localization at "arbitrary" places in the system.

The replacement of the full relaxation of the vibrational system by a single-relaxation-time evolution that we have made in writing down (1.4) requires no apology since it is quite physical in many systems and its simplicity allows us to characterize the vibrational relaxation by a single rate. We have, however, also studied the dimer evolution in the absence of such replacement and find that it provides an excellent approximation to the actual evolution unless the damping in the relaxation is very low. Such low damping in the vibrational evolution is rare. It is, furthermore, straightforward to generalize our analysis to such systems.

The use of the damped dispersionless oscillator equation (1.2) in place of the undamped equation with dispersion used by Scott et.al.[2-5] requires comment. An equation with dispersion and no damping is unphysical for the dimer which we have undertaken to study, although it is certainly appropriate for long chains or crystals. In the absence of damping, the vibrational energy would not be depleted in the dimer. Physical dimers[11] are coupled to a reservoir (consisting, for instance, of other vibrational degrees of freedom) with which they exchange energy, and must therefore be treated either by stating explicitly the dynamics of the reservoir degrees of freedom or through damping terms as we have done in (1.2).

Finally, we mention that we have carried out (see refs. 15,20) studies of finite temperature on the systems described above by appending noise terms to the vibrational evolution equation and analyzing the resulting Langevin equations through extensions of standard stochastic methods. We have obtained[15,20] appropriate Fokker-Planck equations, calculated escape rates and found results[21] which appear to be in agreement with the numerical findings of Lomdahl and Kerr[22].

Acknowledgements

This work was carried out with the support of the DOE under contract no. DE-FG04-86ER45272.

References

1. A. S. Davydov, J. Theor. Biol. 38, 559 (1973); Usp. Fiz. Nauk. 138, 603 (1982) [Sov. Phys. Usp. 25, 898 (1982)] and references therein.
2. A. C. Scott, Phil. Trans. R. Soc. London A 315, 423 (1985).
3. A. C. Scott, Phys. Rev. A 26, 578 (1982).
4. J. C. Eilbeck, P. S. Lomdahl and A. C. Scott, Physica 16 D, 318 (1985).
5. J. M. Hyman, D. W. McLaughlan, and A. C. Scott, Physica 3 D, 23 (1981).
6. D. W. Brown, K. Lindenberg and B. J. West, Phys. Rev. A 35, 6169 (1987) and references therein.
7. (a) V. M. Kenkre and D. K. Campbell, Phys. Rev. B 34, 4959 (1986); (b) V. M. Kenkre, G. P. Tsironis and D. K. Campbell in Nonlinearity in Condensed Matter ed. A. R. Bishop, D. K. Campbell, P. Kumar and S. Trullinger (Springer, Berlin, 1987) p. 226.
8. (a) V. M. Kenkre and G. P. Tsironis, Phys. Rev. B 35, 1473 (1987); (b) G. P. Tsironis and V. M. Kenkre, Phys. Letters A 127, 209 (1988); (c) V. M. Kenkre and G. P. Tsironis, Chem. Phys.128,219 (1989).
9. G. P. Tsironis,V. M. Kenkre and D. Finley, Phys. Rev. B 37,4474 (1988).
10. H. -L. Wu and V. M. Kenkre, Phys. Rev. B 39,2664 (1989).
11. D. W. Brown, K. Lindenberg and B. J. West, Phys. Rev. B 37, 2946 (1988).
12. V. M. Kenkre and H. -L. Wu, Phys. Rev. B 39, 6907 (1989).
13. V. M. Kenkre and H. -L. Wu, Phys. Lett. A 135, 120 (1989).
14. H. -L. Wu, P. Grigolini and V. M. Kenkre,submitted for publication.
15. P. Grigolini, H. -L. Wu and V. M. Kenkre, Phys. Rev. B 40, 7045 (1989).
16. T. D. Holstein, Ann. Phys. 8, 325, 343 (1959).
17. See e.g. M. Pope and C. E. Swenberg, Electronic Processes in Organic Solids (Clarendon Press, Oxford, 1982).
18. R. S. Knox, in Bioenergetics of Photosynthesis (ed. by Govindjee; Academic Press,Inc., N.Y. 1975) p. 183.
19. See e.g. Energy Transfer Processes in Condensed Matter, (ed.Baldassare Di Bartolo, Plenum Press, N.Y. 1984).
20. V. M. Kenkre and P. Grigolini, Phys. Rev. B, submitted for publication.
21. V. M. Kenkre and P. Lomdahl, preprint.
22. P. Lomdahl and W. Kerr, Phys. Rev. Lett. 55, 1235 (1985).

Lecture Notes in Mathematics

Vol. 1236: Stochastic Partial Differential Equations and Applications. Proceedings, 1985. Edited by G. Da Prato and L. Tubaro. V, 257 pages. 1987.

Vol. 1237: Rational Approximation and its Applications in Mathematics and Physics. Proceedings, 1985. Edited by J. Gilewicz, M. Pindor and W. Siemaszko. XII, 350 pages. 1987.

Vol. 1250: Stochastic Processes – Mathematics and Physics II. Proceedings 1985. Edited by S. Albeverio, Ph. Blanchard and L. Streit. VI, 359 pages. 1987.

Vol. 1251: Differential Geometric Methods in Mathematical Physics. Proceedings, 1985. Edited by P.L. García and A. Pérez-Rendón. VII, 300 pages. 1987.

Vol. 1255: Differential Geometry and Differential Equations. Proceedings, 1985. Edited by C. Gu, M. Berger and R.L. Bryant. XII, 243 pages. 1987.

Vol. 1256: Pseudo-Differential Operators. Proceedings, 1986. Edited by H.O. Cordes, B. Gramsch and H. Widom. X, 479 pages. 1987.

Vol. 1258: J. Weidmann, Spectral Theory of Ordinary Differential Operators. VI, 303 pages. 1987.

Vol. 1260: N.H. Pavel, Nonlinear Evolution Operators and Semigroups. VI, 285 pages. 1987.

Vol. 1263: V.L. Hansen (Ed.), Differential Geometry. Proceedings, 1985. XI, 288 pages. 1987.

Vol. 1265: W. Van Assche, Asymptotics for Orthogonal Polynomials. VI, 201 pages. 1987.

Vol. 1267: J. Lindenstrauss, V.D. Milman (Eds.), Geometrical Aspects of Functional Analysis. Seminar. VII, 212 pages. 1987.

Vol. 1269: M. Shiota, Nash Manifolds. VI, 223 pages. 1987.

Vol. 1270: C. Carasso, P.-A. Raviart, D. Serre (Eds.), Nonlinear Hyperbolic Problems. Proceedings, 1986. XV, 341 pages. 1987.

Vol. 1272: M.S. Livšic, L.L. Waksman, Commuting Nonselfadjoint Operators in Hilbert Space. III, 115 pages. 1987.

Vol. 1273: G.-M. Greuel, G. Trautmann (Eds.), Singularities, Representation of Algebras, and Vector Bundles. Proceedings, 1985. XIV, 383 pages. 1987.

Vol. 1275: C.A. Berenstein (Ed.), Complex Analysis I. Proceedings, 1985–86. XV, 331 pages. 1987.

Vol. 1276: C.A. Berenstein (Ed.), Complex Analysis II. Proceedings, 1985–86. IX, 320 pages. 1987.

Vol. 1277: C.A. Berenstein (Ed.), Complex Analysis III. Proceedings, 1985–86. X, 350 pages. 1987.

Vol. 1283: S. Mardešić, J. Segal (Eds.), Geometric Topology and Shape Theory. Proceedings, 1986. V, 261 pages. 1987.

Vol. 1285: I.W. Knowles, Y. Saitō (Eds.), Differential Equations and Mathematical Physics. Proceedings, 1986. XVI, 499 pages. 1987.

Vol. 1287: E.B. Saff (Ed.), Approximation Theory, Tampa. Proceedings, 1985–1986. V, 228 pages. 1987.

Vol. 1288: Yu. L. Rodin, Generalized Analytic Functions on Riemann Surfaces. V, 128 pages, 1987.

Vol. 1294: M. Queffélec, Substitution Dynamical Systems – Spectral Analysis. XIII, 240 pages. 1987.

Vol. 1299: S. Watanabe, Yu.V. Prokhorov (Eds.), Probability Theory and Mathematical Statistics. Proceedings, 1986. VIII, 589 pages. 1988.

Vol. 1300: G.B. Seligman, Constructions of Lie Algebras and their Modules. VI, 190 pages. 1988.

Vol. 1302: M. Cwikel, J. Peetre, Y. Sagher, H. Wallin (Eds.), Function Spaces and Applications. Proceedings, 1986. VI, 445 pages. 1988.

Vol. 1303: L. Accardi, W. von Waldenfels (Eds.), Quantum Probability and Applications III. Proceedings, 1987. VI, 373 pages. 1988.

Lecture Notes in Physics

Vol. 330: J.M. Moran, J.N. Hewitt, K.Y. Lo (Eds.), Gravitational Lenses. Proceedings, 1988. XIV, 238 pages. 1989.

Vol. 331: G. Winnewisser, J.T. Armstrong (Eds.), The Physics and Chemistry of Interstellar Molecular Clouds mm and Sub-mm Observations in Astrophysics. Proceedings, 1988. XVIII, 463 pages. 1989.

Vol. 332: P. Flin, H.W. Duerbeck (Eds.), Morphological Cosmology. Proceedings, 1988. VII, 438 pages. 1989.

Vol. 333: I. Appenzeller, H.J. Habing, P. Léna (Eds.), Evolution of Galaxies – Astronomical Observations. Proceedings, 1988. X, 391 pages. 1989.

Vol. 334: L. Maraschi, T. Maccacaro, M.-H. Ulrich (Eds.), BL Lac Objects. XIII, 497 pages. 1989.

Vol. 335: A. Lakhtakia, V.K. Varadan, V.V. Varadan, Time-Harmonic Electromagnetic Fields in Chiral Media. VII, 121 pages. 1989.

Vol. 336: M. Müller, Consistent Classical Supergravity Theories. VI, 125 pages. 1989.

Vol. 337: A.P. Maclin, T.L. Gill, W.W. Zachary (Eds.), Magnetic Phenomena. Proceedings, 1988. VI, 142 pages. 1989.

Vol. 338: V. Privman, N.M. Švrakić, Directed Models of Polymers, Interfaces, and Clusters: Scaling and Finite-Size Properties. VI, 120 pages. 1989.

Vol. 339: F. Ehlotzky (Ed.), Fundamentals of Laser Interactions II. Proceedings, 1989. XI, 317 pages. 1989.

Vol. 340: M. Peshkin, A. Tonomura, The Aharonov-Bohm Effect. VI, 152 pages. 1989.

Vol. 341: E.F. Milone (Ed.), Infrared Extinction and Standardization. Proceedings, 1988. III, 79 pages. 1989.

Vol. 342: M. Balabane, P. Lochak, C. Sulem (Eds.), Integrable Systems and Applications. Proceedings, 1988. VII, 342 pages. 1989.

Vol. 343: M. Month, S. Turner (Eds.), Frontiers of Particle Beams; Observation, Diagnosis and Correction. Proceedings, 1988. IX, 509 pages. 1989.

Vol. 344: M. Rascle, D. Serre, M. Slemrod (Eds.), PDEs and Continuum Models of Phase Transitions. Proceedings, 1988. VII, 229 pages. 1989.

Vol. 345: H. Holden, A. Jensen (Eds.), Schrödinger Operators. Proceedings, 1988. V, 458 pages. 1989.

Vol. 346: D. Lüst, S. Theisen, Lectures on String Theory. VII, 346 pages. 1989.

Vol. 347: Ph. Blanchard, J.-P. Dias, J. Stubbe (Eds.), New Methods and Results in Non-linear Field Equations. Proceedings. 1987. VII, 133 pages. 1989.

Vol. 348: A. Bohm, M. Gadella, Dirac Kets, Gamow Vectors and Gel'fand Triplets. VII, 119 pages. 1989.

Vol. 349: Y.A. Kubyshin, J.M. Mourão, G. Rudolph, I.P. Volobujev, Dimensional Reduction of Gauge Theories, Spontaneous Compactification and Model Building. X, 80 pages. 1989.

Vol. 350: G. Tenorio-Tagle, M. Moles, J. Melnick (Eds.), Structure and Dynamics of the Interstellar Medium. Proceedings, 1989. XXI, 537 pages. 1989.

Vol. 351: A. Dervieux, B. Larrouturou (Eds.), Numerical Combustion. Proceedings, 1989. VII, 481 pages. 1989.

Vol. 352: K.B. Wolf (Ed.), Lie Methods in Optics II. Proceedings, 1988. XII, 197 pages. 1989.

Vol. 353: M. Barthes, J. Léon (Eds.), Nonlinear Coherent Structures. Proceedings, 1989. X, 280 pages. 1990.

GPSR Compliance

The European Union's (EU) General Product Safety Regulation (GPSR) is a set of rules that requires consumer products to be safe and our obligations to ensure this.

If you have any concerns about our products, you can contact us on ProductSafety@springernature.com

In case Publisher is established outside the EU, the EU authorized representative is:

Springer Nature Customer Service Center GmbH
Europaplatz 3
69115 Heidelberg, Germany

Batch number: 10620881

Printed by Printforce, the Netherlands